T0295466

Probing the Meaning of QUANTUM MECHANICS

Probability, Metaphysics, Explanation and Measurement

Probing the Meaning of QUANTUM MECHANICS

Probability, Metaphysics, Explanation and Measurement

Times of Entanglement

Shanghai, China, 21 – 22 September 2010

Editors

Diederik Aerts
Vrije Universiteit Brussel, Belgium

Jonas R. B. Arenhart
Federal University of Santa Catarina, Brazil

Christian de Ronde
Buenos Aires University, Argentina

Giuseppe Sergioli
Cagliari University, Italy

World Scientific

NEW JERSEY • LONDON • SINGAPORE • BEIJING • SHANGHAI • HONG KONG • TAIPEI • CHENNAI • TOKYO

Published by

World Scientific Publishing Co. Pte. Ltd.

5 Toh Tuck Link, Singapore 596224

USA office: 27 Warren Street, Suite 401-402, Hackensack, NJ 07601

UK office: 57 Shelton Street, Covent Garden, London WC2H 9HE

Library of Congress Cataloging-in-Publication Data

Names: Aerts, Diederik, 1953– editor. | Becker Arenhart, Jonas R., editor. | de Ronde, Christian, 1976– editor. | Sergioli, Giuseppe, editor.

Title: Probing the meaning of quantum mechanics : probability, metaphysics, explanation and measurement / editors Diederik Aerts, Vrije Universiteit Brussel, Belgium, Jonas Arenhart, Federal University of Santa Catarina, Brasil, Christian de Ronde, Buenos Aires University, Argentina, Giuseppe Sergioli, Cagliari University, Italy.

Description: New Jersey : World Scientific, [2024] | "Times of Entanglement Shanghai, China, 21–22 September 2010." | Includes bibliographical references.

Identifiers: LCCN 2023035605 | ISBN 9789811283581 (hardcover) | ISBN 9789811283598 (ebook) | ISBN 9789811283604 (ebook other)

Subjects: LCSH: Quantum theory.

Classification: LCC QC174.125 P76 2024 | DDC 530.12--dc23/eng/20230728

LC record available at https://lccn.loc.gov/2023035605

British Library Cataloguing-in-Publication Data

A catalogue record for this book is available from the British Library.

For any available supplementary material, please visit
https://www.worldscientific.com/worldscibooks/10.1142/13602#t=suppl

Printed in Singapore

CONTENTS

EXTENDING KOLMOGOROV'S AXIOMS FOR A GENERALIZED PROBABILITY THEORY ON COLLECTIONS OF CONTEXTS

KARL SVOZIL

Institute for Theoretical Physics, TU Wien,
Wiedner Hauptstrasse 8-10/136, 1040 Vienna, Austria
E-mail: svozil@tuwien.ac.at
http://tph.tuwien.ac.at/~svozil

Kolmogorov's axioms of probability theory are extended to conditional probabilities among distinct (and sometimes intertwining) contexts. Formally, this amounts to row stochastic matrices whose entries characterize the conditional probability to find some observable (postselection) in one context, given an observable (preselection) in another context. As the respective probabilities need not (but, depending on the physical/model realization, can) be of the Born rule type, this generalizes approaches to quantum probabilities by Auffêves and Grangier, which in turn are inspired by Gleason's theorem.

Keywords: Value indefiniteness; Kolmogorov Axioms of probability theory; Pitowsky's Logical Indeterminacy Principle; Quantum mechanics; Gleason theorem; Kochen-Specker theorem; Born rule.

1. Kolmogorov-type conditional probabilities among distinct contexts

A physical system or a mathematical entity may permit not only one "view" on it but may allow "many" such views. — Think of a single crystal luster whose light, depending on the viewpoint, may appear very different. Schrödinger [1, p. 15 & 95] quoted the Vedantic analogy of a *"many-faceted crystal which, while showing hundreds of little pictures of what is in reality a single existent object, does not really multiply that object. ... A comparison used in Hinduism is of the many almost identical images which a many-faceted diamond makes of some one object such as the sun."* Another example is the coordinatization or coding and encryption of a vector with respect to different bases — thereby in physical terms appearing as "coherent superpositions" (aka linear combinations) of the respective vectors of these bases. Still another example is the representation of an entity by isomorphic graphs.

This idea is grounded in epistemology and in issues related to the (empirical) cognition of ontology and might appear both trivial and sophistic at first glance. Nevertheless it may be difficult to find means or formal models exhibiting multiple contextual views of one and the same entity. Many conceptualizations of such situations are motivated by quantum complementarity [2–5].

A "view" or (used synonymously) "frame" [6] or "context" will be in full generality and thus informally (glancing at heuristics from quantum mechanics and partition logic) characterized as some domain or set of observables or properties which are

(i) *largest* or *maximal* in the sense that any extension yields redundancies,
(ii) yet at the same time in the *finest resolution* in the sense that the respective observables or properties are "no composite" of "more elementary" ones,
(iii) *mutually exclusive* in the sense that one property or observation excludes another, different property or observation, as well as
(iv) contains only *simultaneously measurable, compatible* observables or properties.

In what follows I shall develop a conceptual framework for very general probabilities on such collections of contexts. This amounts to an extension of Kolomogorov probabilities which are defined in a single context, to a multi-context situation. Scrutinized separately, every single context has "legit local" classical Kolomogorov probabilities. In addition to those local structures and measures, (intertwined) multi-context configurations and their probabilities have to be "joined", "woven", "meshed" or "stitched" together to result in consistent and coherent "global" multi-aspect views and probabilities.

In particular, one needs to cope with possible overlaps of contexts in common, intertwining, observables. Because two or more contexts need not (but may) be separated from one another; they may indeed intertwine in one or more common elements, and form complex propositional structures serving a variety of counterfactual [7,8] purposes [9–11].

This text is organized as follows: first, Kolmogorov's axioms or principles for probabilities are generalized to arbitrary event structures not necessarily dominated by the quantum formalism. Then these principles will be applied to quantum bistochasticity, as well as partition logics which offer an abundance of alternate configurations. Some "exotic" probabilities as well as possible generalizations by Cauchy's functional equations are briefly discussed. Throughout this article, only finite contexts will be considered.

2. Generalization of Kolmogorov's axioms to arbitrary event structures

Suppose that, as it is assumed for classical Kolmogorov probabilities, the elements $\mathbf{c}_1$ within any given single, individual finite context $\mathcal{C} = \{\mathbf{c}_1, \dots \mathbf{c}_n\}$ are mutually exclusive, compatible, and exhaustive; that is, the context contains a "maximal" set of mutually exclusive, compatible elements. Kolmogorov's axioms demand that (i) probabilities are non-negative; (ii) additivity of mutually exclusive events or outcomes $P(\mathbf{c}_i) + P(\mathbf{c}_j) = P(\mathbf{c}_i \cup \mathbf{c}_j)$; (iii) the probability of the tautology formed by the union of all elements in the context, adds up to one, that is, $\sum_{\mathbf{c}_i \in \mathcal{C}} P(\mathbf{c}_i) = P\left(\bigcup_{\mathbf{c}_i \in \mathcal{C}} \mathbf{c}_i\right) = 1$.

Inspired by the multi-context quantum case discussed later the following generalization to two- or, by induction, to a multi-context configuration is suggested: Suppose two arbitrary contexts $\mathcal{C}_1 = \{\mathbf{e}_1, \dots \mathbf{e}_n\}$ and $\mathcal{C}_2 = \{\mathbf{f}_1, \dots \mathbf{f}_m\}$. The conditional probabilities $P(\mathbf{f}_j|\mathbf{e}_i)$, with $1 \le j \le m$ and $1 \le i \le n$, which alternatively can be considered as either measuring the Bayesian degree of reasonable expectation representing a state of knowledge or as quantification of a personal belief [12] or the frequency of occurrence of "$\mathbf{f}_j$ given $\mathbf{e}_i$", can be arranged into a $(n \times m)$-matrix whose entries are $P(\mathbf{f}_j|\mathbf{e}_i)$, that is,

$$[P(\mathcal{C}_2|\mathcal{C}_1)] = [P(\{\mathbf{f}_1, \dots \mathbf{f}_m\}|\{\mathbf{e}_1, \dots \mathbf{e}_n\})] \equiv \begin{bmatrix} P(\mathbf{f}_1|\mathbf{e}_1) & \cdots & P(\mathbf{f}_m|\mathbf{e}_1) \\ \cdots & \cdots & \cdots \\ P(\mathbf{f}_1|\mathbf{e}_n) & \cdots & P(\mathbf{f}_m|\mathbf{e}_n) \end{bmatrix}. \quad (1)$$

Assume as axiom the following criterion: the conditional probabilities of the elements of the second context with respect to an arbitrary element $\mathbf{e}_k \in \mathcal{C}_1$ of the first context $\mathcal{C}_1$ are non-negative, additive, and, if this sum is extended over the entire second context $\mathcal{C}_2$, adds up to one:

$$\begin{aligned} P(\mathbf{f}_i|\mathbf{e}_k) + P(\mathbf{f}_j|\mathbf{e}_k) &= P[(\mathbf{f}_i \cup \mathbf{f}_j)|\mathbf{e}_k] \\ \sum_{\mathbf{f}_i \in \mathcal{C}_2} P(\mathbf{f}_i|\mathbf{e}_k) = P\left[\left(\bigcup_{\mathbf{f}_i \in \mathcal{C}_2} \mathbf{f}_i\right)|\mathbf{e}_k\right] &= 1. \end{aligned} \quad (2)$$

That is, the row sum taken within every single row of $[P(\mathcal{C}_2|\mathcal{C}_1)]$ adds up to one.

This presents a generalization of Kolmogorov's axioms, as it allows cases in which both contexts do not coincide. It just reduces to the classical axioms for single contexts if, instead of a single element $\mathbf{e}_k \in \mathcal{C}_1$ of the first context $\mathcal{C}_1$, the union of elements of this entire context $\mathcal{C}_1$ — and thus the tautology $\bigcup_{\mathbf{e}_i \in \mathcal{C}_1} \mathbf{e}_i$ — is inserted into (2).

We shall mostly be concerned with cases for which $n = m$; that is, the associated matrix is a row (aka right) stochastic (square) matrix. Formally, such a matrix A has nonnegative entries $a_{ij} \geq 0$ for $i, j = 1, \ldots, n$ whose row sums add up to one: $\sum_{j=1}^{n} a_{ij} = 1$ for $i = 1, \ldots, n$. If, in addition to the row sums, also the column sums add up to one — that is, if $\sum_{i=1}^{n} a_{ij} = 1$ for $j = 1, \ldots, n$ — then the matrix is called doubly stochastic. If J is a $(n \times n)$–matrix whose entries are 1, then a $(n \times n)$–matrix A is row stochastic if $\mathsf{AJ} = \mathsf{J}$.

It is instructive to ponder why intuitively those conditional probabilities should be arranged in right- but not in bistochastic matrices. Suppose a (physical or another model) system is in a state characterized by some element $\mathbf{e}_j \in \mathcal{C}_1$ of the first context $\mathcal{C}_1$. Then, if one takes the (union of elements of the) entire other context $\mathcal{C}_2$ — thereby exhausting all possible outcomes of the second "view" — the conditional probability for this system to be in *any* element of $\mathcal{C}_2$ given $\mathbf{e}_j \in \mathcal{C}_1$ should add up to one because this includes all that can be (aka happen or exist) with respect to the second "view". Indeed, if this conditional probability would not add up to one, say if it adds up to something strictly smaller or larger than one, then either some elements would be missing in, or be "external" to, the context $\mathcal{C}_2$, which cannot occur since by assumption contexts are "maximal".

On the other hand, if a particular element $\mathbf{f}_i \in \mathcal{C}_f$ of the second context $\mathcal{C}_2$ remains fixed and the column sum $\sum_{\mathbf{e}_j \in \mathcal{C}_2} P(\mathbf{f}_i | \mathbf{e}_j)$ extends over all $\mathbf{e}_j \in \mathcal{C}_2$ then there is no convincing reason why this column sum should add up to one. Indeed, as will be argued later, while quantum mechanics results in bistochastic matrices, generalized urn models resulting in partitions of (hidden) variables that will not induce bistochasticity.

3. Cauchy's functional equation encoding additivity

One way of looking at generalized global probabilities from "stitching" local classical Kolmogorov probabilities is to maintain the essence of the axioms — namely positivity, probability one (aka certainty) for tautologies, and, in particular, additivity. Additivity requires that, for mutually exclusive compatible events $\mathbf{c}_i$ and $\mathbf{c}_j$ within a given context, their probabilities can be expressed in terms of Cauchy-type functional equation

$P(\mathbf{c}_i) + P(\mathbf{c}_j) = P(\mathbf{c}_i \cup \mathbf{c}_j)$. With "reasonable" side assumptions, this amounts to the linearity of probabilities in the argument [13,14].

For operators in Hilbert spaces of dimensions higher than two — and, in particular, for linear operators A and B with an operator norm $|\mathsf{A}| = +\sqrt{\langle \mathsf{A} | \mathsf{A} \rangle}$ based on the Hilbert-Schmidt inner product $\langle \mathsf{A} | \mathsf{B} \rangle = \text{Trace}\,(\mathsf{A}^* \mathsf{B})$, where A^* stands for the adjoint of A — Cauchy's functional equation can be related to Gleason-type theorems [15–20].

The general case may involve other, hitherto unknown, arguments besides scalars and entities related to vector (or Hilbert) spaces. The discussion will not be extended to potential inputs and sources for generalized probabilities as the main interest is in developing a generalizing probability theory in the multi-context setting, but clearly these questions remain pertinent.

4. Examples of application of the generalized Kolmogorov axioms

4.1. *Quantum bistochasticity*

The multi-context quantum case has been studied in great detail with emphasis on motivating and deriving the Born rule [21,22] from elementary foundations. Recall that a context has been defined as the "largest" or "maximal" domain of both mutually exclusive as well as simultaneously measurable, compatible observables. In quantum mechanics "simultaneously measurability" transforms into *compatibility* and *commutativity*; that is, such observables are not complementary and can be jointly measured without restrictions. "Mutual exclusivity" is defind in terms of *orthogonality* of the respective observables. The spectral theorem asserts mutual orthogonality of unit eigenvectors $|\mathbf{e}_i\rangle$ and the associated orthogonal projection operators E_i formed by the dyadic product $\mathsf{E}_i = |\mathbf{e}_i\rangle\langle\mathbf{e}_i|$. A context can be equivalently represented by (i) an orthonormal basis, (ii) the respective one-dimensional orthogonal projection operators associated with the basis elements, or (iii) a single maximal operator (aka maximal observable) whose spectral sum is non-degenerate [9,23].

An essential assumption entering Gleason's derivation [6] of the Born rule for quantum probabilities is the validity of classical probability theory whenever the respective observables are compatible. Formally, this amounts to the validity of Kolmogorov probability theory for mutually commuting observables; and in particular, to the assumption of Kolmogorov's axioms within contexts.

Already Gleason pointed out [6] that it is quite straightforward to find an *ad hoc* probability satisfying this aforementioned assumption, which is based on the Pythagorean property: suppose (i) a quantized system is in a pure state $|\psi\rangle$ formalized by some unit vector, and (ii) some "measurement frame" formalized by an orthonormal basis $\mathcal{C} = \{|\mathbf{e}_1\rangle, \ldots, |\mathbf{e}_n\rangle\}$. Then the probabilities of outcomes of observable propositions associated with the orthogonal projection operators formed by the dyadic products $|\mathbf{e}_i\rangle\langle\mathbf{e}_i|$ of the vectors of the orthonormal basis can be obtained by taking the absolute square of the length of those projections of $|\psi\rangle$ onto $|\mathbf{e}_i\rangle$ along the remaining basis vectors, which amounts to taking the scalar products $|\langle\psi|\mathbf{e}_i\rangle|^2$. Since the vector associated with the pure state as well as all the vectors in the orthonormal system are of length one, and since these latter vectors (of the orthonormal system) are mutually orthogonal, the sum $\sum_{i=1}^{n} |\langle\psi|\mathbf{e}_i\rangle|^2$ of all these terms, taken over all the basis elements, needs to add up to one. The respective absolute squares are bounded between zero and one. In effect, the orthonormal basis "grants a view" of the pure quantum state. The absolute square can be rewritten in terms of a trace (over some arbitrary orthonormal basis) into the standard form known as the Born rule of quantum probabilities: $|\langle\psi|\mathbf{e}_i\rangle|^2 = \langle\psi|\mathbf{e}_i\rangle\langle\mathbf{e}_i|\psi\rangle = \langle\psi|\mathbf{e}_i\rangle\langle\mathbf{e}_i|\mathbb{I}_n\psi\rangle = \sum_{j=1}^{n}\langle\psi|\mathbf{e}_i\rangle\langle\mathbf{e}_i|\mathbf{g}_j\rangle\langle\mathbf{g}_j|\psi\rangle = \sum_{j=1}^{n}\langle\mathbf{g}_j|\underbrace{\psi\rangle\langle\psi|}_{=\mathsf{E}_\psi}\underbrace{\mathbf{e}_i\rangle\langle\mathbf{e}_i|}_{=\mathsf{E}_i}\mathbf{g}_j\rangle =$ $\text{Trace}(\mathsf{E}_\psi\mathsf{E}_i)$, where E_ψ and E_i are the orthogonal projection operators representing the state $|\psi\rangle$ and the (unit) vectors of the orthonormal basis $|\mathbf{e}_i\rangle$, respectively, and $\mathcal{C}' = \{|\mathbf{g}_1\rangle, \ldots, |\mathbf{g}_n\rangle\}$ is an arbitrary orthonormal basis, so that a resolution of the identity is $\mathbb{I}_n = \sum_{j=1}^{n} |\mathbf{g}_j\rangle\langle\mathbf{g}_j|$.

It is also well known that, at least from a formal perspective, unit vectors in quantum mechanics serve a dual role: On the one hand, they represent pure states. On the other hand, by the associated one-dimensional orthogonal projection operator, they represent an observable: the proposition that the system is in such a pure state [24,25]. Suppose now that we exploit this dual role by *expanding* the pure prepared state into a full orthonormal basis, of which its vector must be an element. (For dimensions greater than two such an expansion will not be unique as there is a continuous infinity of ways to achieve this.) Once the latter basis is fixed it can be used to obtain a "view" on the former (measurement) basis; and a completely symmetric situation/configuration is attained. We might even go so far as to say that which basis is associated with the "observed object" and with the "measurement apparatus," respectively, is purely a matter of convention and subjective perspective.

Therefore, as has been pointed out earlier, an orthogonal projection operator serves a dual role: on the one hand it is a formalization of a dichotomic observable — more precisely, an elementary yes-no proposition $\mathsf{E} = |\mathbf{x}\rangle\langle\mathbf{x}|$ associated with the claim that "the quantized system is in state $|\mathbf{x}\rangle$. And on the other hand it is the formal representation of a pure quantum state $|\mathbf{y}\rangle$, equivalent to the operator $\mathsf{F} = |\mathbf{y}\rangle\langle\mathbf{y}|$. By the Born rule the conditional probabilities are symmetric with respect to exchange of $|\mathbf{x}\rangle$ and $|\mathbf{y}\rangle$: let $\mathcal{C}' = \{|\mathbf{g}_1\rangle, \ldots, |\mathbf{g}_n\rangle\}$ be some arbitrary orthonormal basis of $\mathbb{C}^n$, then $P(\mathsf{E}|\mathsf{F}) = \text{Trace}(\mathsf{EF}) = \text{Trace}(\mathsf{FE}) = P(\mathsf{F}|\mathsf{E})$; or, more explicitly, $P(\mathsf{E}|\mathsf{F}) = \sum_{i=1}^{n} \langle\mathbf{g}_i|\mathbf{x}\rangle\langle\mathbf{x}|\mathbf{y}\rangle\langle\mathbf{y}|\mathbf{g}_i\rangle = \sum_{i=1}^{n} \langle\mathbf{x}|\mathbf{y}\rangle\langle\mathbf{y}|\underbrace{\mathbf{g}_i\rangle\langle\mathbf{g}_i}_{=\mathbb{I}_n}|\mathbf{x}\rangle = |\langle\mathbf{x}|\mathbf{y}\rangle|^2 = |\langle\mathbf{y}|\mathbf{x}\rangle|^2 = P(\mathsf{F}|\mathsf{E})$. Therefore, the respective conditional probabilities form a doubly stochastic (bistochastic) square matrix. This result is a special case of a more general result on quadratic forms on the set of eigenvectors of normal operators [26].

Consider two orthonormal bases aka two contexts. Their respective conditional probabilities can be arranged into a matrix form: The ith row jth column component corresponds to the conditional probability associated with the probability of occurrence of the jth element (observable) of the second context, given the ith element (observable) of the first context. By taking into account that cyclically interchanging factors inside a trace does not change its value this matrix needs to be not only row (right) stochastic but doubly stochastic (bistochastic) [21,22]; that is, the sum is taken within every single row and every single column adds up to one.

4.2. *Quasi-classical partition logics*

In what follows we shall study sets of partitions of a given set. They have models [27] based (i) on the finite automata initial state identification problem [28] as well as (ii) on generalized urns [29,30]. Partition logics are quasi-classical and value-definite in so far as they allow a separating set of "classical" two-valued states [9, Theorem 0]; and yet they feature complementarity. Many of these logics are *doubles* of quantum logics, such as for spin-state measurements; and thereby their graphs also allow faithful orthogonal representations [31]; and yet some of them have no quantum analog. Therefore, they neither form a proper subset of all quantum logics nor do they contain all logical structures encountered in quantum logics (they are neither continuous nor can they have a non-separating or nonexisting set of two-valued states). However, partition logics overlaps significantly with quantum logics, as they bear strong similarities with the structures arising in quantum theory.

If some (partition) logic which is a pasting [32–34] of contexts has a separating set of two-valued states [9, Theorem 0] then there is a constructive, algorithmic [35] way of finding a "canonical" partition logic [27], and, associated with it, all classical probabilities on it: first, find all the two-valued states on the logic, and assign consecutive number to these states. Then, for any atom (element of a context), find the index set of all two-valued states which are 1 on this atom. Associate with each one, say, the ith, of the two valued states a nonnegative weight $i \rightarrow \lambda_i$, and require that the (convex) sum of these weights $\sum_i \lambda_i = 1$ is 1. Since all two-valued states are included, the Kolmogorov axioms guarantee that the sum of measures/weights within each of the contexts in the logic exactly adds up to one.

It will be argued that in this case, and unlike for quantum conditional probabilities, the conditional probabilities, in general, do not form a bistochastic matrix.

4.2.1. *Two non-intertwining two-atomic contexts*

In the Babylonian spirit [36, p. 172] consider some anecdotal examples which have quantum doubles. The first one will be analogous to a spin-$\frac{1}{2}$ state measurement.

The logic in Fig. 1 enumerates the labels of the atoms (aka elementary propositions) according to the "inverse construction" — based on all four two-valued states on the logic — mentioned earlier, using all two-valued measures thereon [27]. With the identifications $\mathbf{e}_1 \equiv \{1,2\}$, $\mathbf{e}_2 \equiv \{3,4\}$, $\mathbf{f}_1 \equiv \{1,3\}$, and $\mathbf{f}_2 \equiv \{2,4\}$ we obtain all classical probabilities by identifying $i \rightarrow \lambda_i > 0$. The respective conditional probabilities are

$$
\begin{aligned}
[P(\mathcal{C}_2|\mathcal{C}_1)] &= [P(\{\mathbf{f}_1,\mathbf{f}_2\}|\{\mathbf{e}_1,\mathbf{e}_2\})] \equiv \begin{bmatrix} P(\mathbf{f}_1|\mathbf{e}_1) & P(\mathbf{f}_2|\mathbf{e}_1) \\ P(\mathbf{f}_1|\mathbf{e}_2) & P(\mathbf{f}_2|\mathbf{e}_2) \end{bmatrix} \\
&= \begin{bmatrix} \frac{P(\mathbf{f}_1\cap\mathbf{e}_1)}{P(\mathbf{e}_1)} & \frac{P(\mathbf{f}_2\cap\mathbf{e}_1)}{P(\mathbf{e}_1)} \\ \frac{P(\mathbf{f}_1\cap\mathbf{e}_2)}{P(\mathbf{e}_2)} & \frac{P(\mathbf{f}_2\cap\mathbf{e}_2)}{P(\mathbf{e}_2)} \end{bmatrix} = \begin{bmatrix} \frac{P(\{1,3\}\cap\{1,2\})}{P(\{1,2\})} & \frac{P(\{2,4\}\cap\{1,2\})}{P(\{1,2\})} \\ \frac{P(\{1,3\}\cap\{3,4\})}{P(\{3,4\})} & \frac{P(\{2,4\}\cap\{3,4\})}{P(\{3,4\})} \end{bmatrix} \\
&= \begin{bmatrix} \frac{P(\{1\})}{P(\{1,2\})} & \frac{P(\{2\})}{P(\{1,2\})} \\ \frac{P(\{3\})}{P(\{3,4\})} & \frac{P(\{4\})}{P(\{3,4\})} \end{bmatrix} = \begin{bmatrix} \frac{\lambda_1}{\lambda_1+\lambda_2} & \frac{\lambda_2}{\lambda_1+\lambda_2} \\ \frac{\lambda_3}{\lambda_3+\lambda_4} & \frac{\lambda_4}{\lambda_3+\lambda_4} \end{bmatrix},
\end{aligned}
\tag{3}
$$

as well as

$$
\begin{aligned}
[P(\mathcal{C}_1|\mathcal{C}_2)] &= [P(\{\mathbf{e}_1,\mathbf{e}_2\}|\{\mathbf{f}_1,\mathbf{f}_2\})] \\
&\equiv \begin{bmatrix} \frac{P(\{1\})}{P(\{1,3\})} & \frac{P(\{3\})}{P(\{1,3\})} \\ \frac{P(\{2\})}{P(\{2,4\})} & \frac{P(\{4\})}{P(\{2,4\})} \end{bmatrix} = \begin{bmatrix} \frac{\lambda_1}{\lambda_1+\lambda_3} & \frac{\lambda_3}{\lambda_1+\lambda_3} \\ \frac{\lambda_2}{\lambda_2+\lambda_4} & \frac{\lambda_4}{\lambda_2+\lambda_4} \end{bmatrix}.
\end{aligned}
\tag{4}
$$

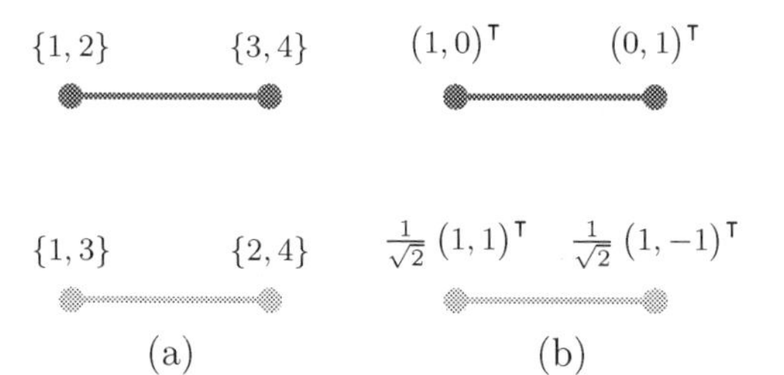

Fig. 1. Greechie orthogonality diagram of a logic consisting of two nonintertwining contexts. (a) The associated (quasi)classical partition logic representations obtained by an inverse construction using all two-valued measures thereon [27]; (b) a faithful orthogonal representation [37] rendering a quantum *double.*

4.2.2. *Two intertwining three-atomic contexts*

The L_{12} "firefly" logic depicted in Fig. 2 labels the atoms (aka elementary propositions) obtained by an "inverse construction" using all five two-valued measures thereon [27,38]. By design, it will be very similar to the earlier logic with four atoms. With the identifications $\mathbf{e}_1 \equiv \{1,2\}, \mathbf{e}_2 \equiv \{3,4\}, \mathbf{e}_3 = \mathbf{f}_3 \equiv \{5\}$, $\mathbf{f}_1 \equiv \{1,3\}$, and $\mathbf{f}_2 \equiv \{2,4\}$ we obtain all classical probabilities by identifying $i \rightarrow \lambda_i > 0$. The respective conditional probabilities are

$$
\begin{aligned}
[P(\mathcal{C}_2|\mathcal{C}_1)] &= [P(\{\mathbf{f}_1,\mathbf{f}_2,\mathbf{f}_3\}|\{\mathbf{e}_1,\mathbf{e}_2,\mathbf{e}_3\})] \\
&\equiv \begin{bmatrix} \frac{P(\{1\})}{P(\{1,2\})} & \frac{P(\{2\})}{P(\{1,2\})} & \frac{P(\emptyset)}{P(\{1,2\})} \\ \frac{P(\{3\})}{P(\{3,4\})} & \frac{P(\{4\})}{P(\{3,4\})} & \frac{P(\emptyset)}{P(\{3,4\})} \\ \frac{P(\emptyset)}{P(\{5\})} & \frac{P(\emptyset)}{P(\{5\})} & \frac{P(\{5\})}{P(\{5\})} \end{bmatrix} = \begin{bmatrix} \frac{\lambda_1}{\lambda_1+\lambda_2} & \frac{\lambda_2}{\lambda_1+\lambda_2} & 0 \\ \frac{\lambda_3}{\lambda_3+\lambda_4} & \frac{\lambda_4}{\lambda_3+\lambda_4} & 0 \\ 0 & 0 & 1 \end{bmatrix},
\end{aligned}
\tag{5}
$$

as well as

$$
\begin{aligned}
[P(\mathcal{C}_1|\mathcal{C}_2)] &= [P(\{\mathbf{e}_1,\mathbf{e}_2,\mathbf{e}_3\}|\{\mathbf{f}_1,\mathbf{f}_2,\mathbf{f}_3\})] \\
&\equiv \begin{bmatrix} \frac{P(\{1\})}{P(\{1,3\})} & \frac{P(\{3\})}{P(\{1,3\})} & \frac{P(\emptyset)}{P(\{1,3\})} \\ \frac{P(\{2\})}{P(\{2,4\})} & \frac{P(\{4\})}{P(\{2,4\})} & \frac{P(\emptyset)}{P(\{2,4\})} \\ \frac{P(\emptyset)}{P(\{5\})} & \frac{P(\emptyset)}{P(\{5\})} & \frac{P(\{5\})}{P(\{5\})} \end{bmatrix} = \begin{bmatrix} \frac{\lambda_1}{\lambda_1+\lambda_3} & \frac{\lambda_3}{\lambda_1+\lambda_3} & 0 \\ \frac{\lambda_2}{\lambda_2+\lambda_4} & \frac{\lambda_4}{\lambda_2+\lambda_4} & 0 \\ 0 & 0 & 1 \end{bmatrix}.
\end{aligned}
\tag{6}
$$

The conditional probabilities of the firefly logic, as depicted in Fig. 2(a), and enumerated in Eq. (6) form a right stochastic matrix. As mentioned earlier, given any particular outcome $\mathbf{f}_i$ of the second context corresponding to some respective row in the matrix (6), the row-sum of the conditional probabilities of all the conceivable mutually exclusive outcomes of the first context $\mathcal{C}_1 = \{\mathbf{e}_1,\mathbf{e}_2,\mathbf{e}_3\}$ must be one. However, the "transposed" statement is not true: the column-sum of the conditional probabilities of a particular element $\mathbf{e}_j$ with respect to all the mutually exclusive outcomes of the second context $C_2 = \{\mathbf{f}_1, \mathbf{f}_2, \mathbf{f}_3\}$, needs not be one.

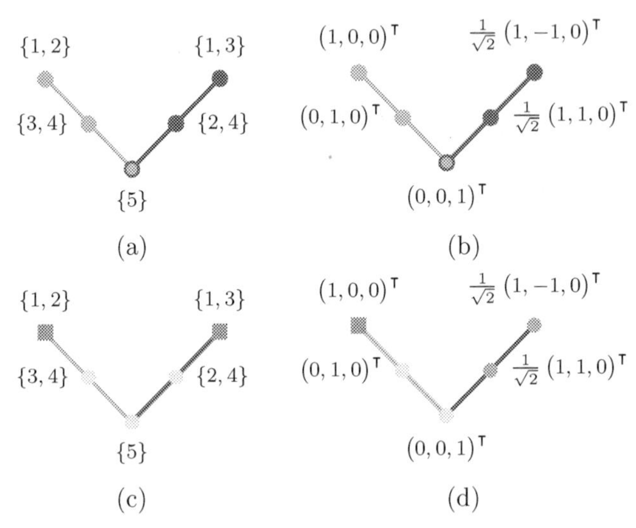

Fig. 2. Greechie orthogonality diagram of the L_{12} "firefly" logic. (a) The associated (quasi)classical partition logic representation obtained through in inverse construction using all two-valued measures thereon [27]; (b) a faithful orthogonal representation [37] rendering a quantum *double*; (c) "classical" two-valued measure number 1; (d) a pure quantum state prepared as $(1,0,0)^\intercal$. A red square and gray and green circles indicate value assignments 1, $\frac{1}{2}$ and 0, respectively.

Take, for example, the singular distribution case such that $\lambda_1 = 1$, and therefore, by positivity and convexity, $\lambda_{i\neq 1} = 0$, that is, $\lambda_2 = \lambda_3 = \lambda_4 = \lambda_5 = 0$. This configuration, depicted in Fig. 2(c), results in the following, partial (undefined components are indicated by the symbol "$\frac{0}{0}$") right stochastic matrix (7) derived from (6):

$$\begin{bmatrix} \frac{P(\{1\})}{P(\{1,3\})} & \frac{P(\{3\})}{P(\{1,3\})} & \frac{P(\emptyset)}{P(\{1,3\})} \\ \frac{P(\{2\})}{P(\{2,4\})} & \frac{P(\{4\})}{P(\{2,4\})} & \frac{P(\emptyset)}{P(\{2,4\})} \\ \frac{P(\emptyset)}{P(\{5\})} & \frac{P(\emptyset)}{P(\{5\})} & \frac{P(\{5\})}{P(\{5\})} \end{bmatrix} = \begin{bmatrix} \frac{\lambda_1}{\lambda_1+\lambda_3} & \frac{\lambda_3}{\lambda_1+\lambda_3} & 0 \\ \frac{\lambda_2}{\lambda_2+\lambda_4} & \frac{\lambda_4}{\lambda_2+\lambda_4} & 0 \\ 0 & 0 & 1 \end{bmatrix} = \begin{bmatrix} 1 & 0 & 0 \\ \frac{0}{0} & \frac{0}{0} & 0 \\ 0 & 0 & 1 \end{bmatrix}. \tag{7}$$

In such a case, in terms of, say, a generalized urn model, the observable proposition $\{2,4\}$ associated with the plaintext *"looked upon in the first color (in this case blue), the ball drawn from the urn shows the symbols 2 or 4"* will never occur; regardless of which ball type associated with the other context $\{1,2\}$, $\{3,4\}$, or $\{5\}$ one would have (counterfactually) drawn because the generalized urn is only loaded with balls of one type, namely the first type, with the symbol "$\{1,2\}$" painted on them in the first color, and the symbols "$\{1,3\}$" painted on them in the second color. (Instead of labels indicating the elements of the partition one may choose other symbols, such as $\{1,3\} \equiv a \equiv \{1,2\}$, $\{2,4\} \equiv b \equiv \{3,4\}$, and $c \equiv \{5\}$ in the respective colors [27,39].)

Ultimately one may say that it is the *discontinuity* of the two-valued measures which "prevents" the quasiclassical conditional probabilities to be arranged in a bistochastic matrix. A similar quantum realization could, for instance, be obtained by the three-dimensional faithful orthogonal representation [37 $\{1,2\} \equiv (1,0,0)^{\intercal}$, $\{3,4\} \equiv (0,1,0)^{\intercal}$, $\{5\} \equiv (0,0,1)^{\intercal}$, $\{1,3\} \equiv (1/\sqrt{2})\,(1,1,0)^{\intercal}$, and $\{2,4\} \equiv (1/\sqrt{2})\,(1,-1,0)^{\intercal}$. Preparition (aka "loading the quantum urn") with state $\{1,2\} \equiv (1,0,0)^{\intercal}$, as depicted in Fig. 2(d), yields the quantum bistochastic matrix

$$\left[P\left(\left\{\begin{pmatrix}1\\0\\0\end{pmatrix},\begin{pmatrix}0\\1\\0\end{pmatrix},\begin{pmatrix}0\\0\\1\end{pmatrix}\right\}\middle|\left\{\frac{1}{\sqrt{2}}\begin{pmatrix}1\\1\\0\end{pmatrix},\frac{1}{\sqrt{2}}\begin{pmatrix}1\\-1\\0\end{pmatrix},\begin{pmatrix}0\\0\\1\end{pmatrix}\right\}\right)\right] = \begin{bmatrix}\frac{1}{2}&\frac{1}{2}&0\\ \frac{1}{2}&\frac{1}{2}&0\\ 0&0&1\end{bmatrix}. \tag{8}$$

4.2.3. *Different intrinsically operational state preparation*

A different approach to partition logic would be to insist that only *intrinsical* — that is, for any embedded observer having access to means and methods available "from within" the system — operational state preparations should be allowed. In such a scenario it is operationally impossible for an observer with access to only one context — in the generalized urn model only one color — to single out the particular type of two-valued measure (aka ball). Thereby effectively any state preparation is reduced to the elements of the partition in the respective context (aka color).

Therefore, in the earlier firefly model depicted in Fig. 2, the intrinsic operational resolution is among the *subsets resulting from the unions of two-valued states* in $\{1,2\}$, $\{3,4\}$, and $\{5\}$ in the first context (aka color); and among $\{1,3\}$, $\{2,4\}$, and $\{5\}$ in the second context (aka color), as opposed to the single two-valued state discussed earlier in. Stated differently, an observer accessing a generalized urn in the first context (aka color) is not capable to differentiate between the first and the second two-valued measure (aka ball type), and would produce a mixture among them if asked to prepare the state $\{1,2\}$. Similarly, the observer would not be able to differentiate between the third and the fourth two-valued measure (aka ball type), and would thus produce a mixture between those when preparing the state $\{3,4\}$. However, the ball type $\{5\}$ is recognized and prepared without ambiguity. Indeed, if one assumes equidistribution (uniform mixtures [40, Assumption 1]) of measures (aka ball types), a very similar situation as in quantum mechanics [cf Fig. 2(d), Eq. (8)] would result as $\lambda_1 = \lambda_2 = \lambda_3 = \lambda_4 = \lambda_5 = \frac{1}{5}$ and one would thus "recover" the matrix in Eq. (8).

Pointedly stated there is an epistemic issue of state preparation: if one demands that the state has to be prepared by the distinctions accessible from a single context (aka color in the generalized urn model), then there is no way to prepare or access "ontologic states", say, selecting balls of type 1 (first two-valued measure) only. The difference is subtle: in the "ontic" state case one can resolve (and has access to) every single two-valued measure (aka ball type). In the "epistemic," intrinsic, operational state case one is limited to the operational procedures available — for example, one cannot "take off the colored glasses" in Wright's generalized urn model. That is, the resolution of balls is limited to whatever types can be differentiated in that color.

Whenever such a scenario is considered the respective matrices representing all conditional probabilities may be very different from the previous scenarios. Indeed, one may suspect that, with the assumption of preservation of equidistributed uniform mixtures across context changes, the respective matrices are bistochastic (at least for equidistributed urns) because of a certain type of "epistemic continuity:" the sum of the conditional probabilities for any particular outcome of the second context, relative to all other outcomes of the first context, should add up to unity.

4.2.4. *Pentagon/pentagram/house logic with five cyclically intertwining three-atomic contexts*

By now it should be clear how classical conditional probabilities work on partition logics. Consider one more example: the pentagon/pentagram/(orthomodular) house [33, p. 46 Fig. 4.4] logic in Fig. 3. Labels of the atoms (aka elementary propositions) are again obtained by an "inverse construction" using all 11 two-valued measures thereon [29]. take, for example, one of the two contexts $C_4 = \{\{2,7,8\},\{1,3,9,10,11\},\{4,5,6\}\}$ "opposite" to the context $C_1 = \{\{1,2,3\},\{4,5,7,9,11\},\{6,8,10\}\}$.

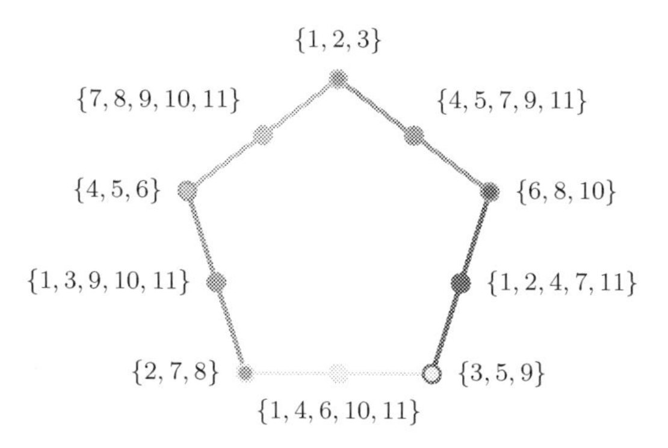

Fig. 3. Greechie orthogonality diagrams of the pentagon/pentagram/house logic.

With the identifications $\mathbf{e}_1 \equiv \{1,2,3\}$, $\mathbf{e}_2 \equiv \{4,5,7,9,11\}$, $\mathbf{e}_3 \equiv \{6,8,10\}$, $\mathbf{f}_1 \equiv \{2,7,8\}$, $\mathbf{f}_2 \equiv \{1,3,9,10,11\}$, and $\mathbf{f}_3 \equiv \{4,5,6\}$. The respective conditional probabilities are

$$
\begin{aligned}
&[P(\mathcal{C}_2|\mathcal{C}_1)] = [P(\{\mathbf{f}_1,\mathbf{f}_2,\mathbf{f}_3\}|\{\mathbf{e}_1,\mathbf{e}_2,\mathbf{e}_3\})] \\
&\equiv \begin{bmatrix}
\frac{P(\{2,7,8\}\cap\{1,2,3\})}{P(\{1,2,3\})} & \frac{P(\{1,3,9,10,11\}\cap\{1,2,3\})}{P(\{1,2,3\})} & \frac{P(\{4,5,6\}\cap\{1,2,3\})}{P(\{1,2,3\})} \\
\frac{P(\{2,7,8\}\cap\{4,5,7,9,11\})}{P(\{4,5,7,9,11\})} & \frac{P(\{1,3,9,10,11\}\cap\{4,5,7,9,11\})}{P(\{4,5,7,9,11\})} & \frac{P(\{4,5,6\}\cap\{4,5,7,9,11\})}{P(\{4,5,7,9,11\})} \\
\frac{P(\{2,7,8\}\cap\{6,8,10\})}{P(\{6,8,10\})} & \frac{P(\{1,3,9,10,11\}\cap\{6,8,10\})}{P(\{6,8,10\})} & \frac{P(\{4,5,6\}\cap\{6,8,10\})}{P(\{6,8,10\})}
\end{bmatrix} \\
&= \begin{bmatrix}
\frac{P(\{2\})}{P(\{1,2,3\})} & \frac{P(\{1,3\})}{P(\{1,2,3\})} & \frac{P(\emptyset)}{P(\{1,2,3\})} \\
\frac{P(\{7\})}{P(\{4,5,7,9,11\})} & \frac{P(\{11\})}{P(\{4,5,7,9,11\})} & \frac{P(\{4,5\})}{P(\{4,5,7,9,11\})} \\
\frac{P(\{8\})}{P(\{6,8,10\})} & \frac{P(\{10\})}{P(\{6,8,10\})} & \frac{P(\{6\})}{P(\{6,8,10\})}
\end{bmatrix} \\
&= \begin{bmatrix}
\frac{\lambda_2}{\lambda_1+\lambda_2+\lambda_3} & \frac{\lambda_1+\lambda_3}{\lambda_1+\lambda_2+\lambda_3} & 0 \\
\frac{\lambda_7}{\lambda_4+\lambda_5+\lambda_7+\lambda_9+\lambda_{11}} & \frac{\lambda_9+\lambda_{11}}{\lambda_4+\lambda_5+\lambda_7+\lambda_9+\lambda_{11}} & \frac{\lambda_4+\lambda_5}{\lambda_4+\lambda_5+\lambda_7+\lambda_9+\lambda_{11}} \\
\frac{\lambda_8}{\lambda_6+\lambda_8+\lambda_{10}} & \frac{\lambda_{10}}{\lambda_6+\lambda_8+\lambda_{10}} & \frac{\lambda_6}{\lambda_6+\lambda_8+\lambda_{10}}
\end{bmatrix}.
\end{aligned}
\tag{9}
$$

5. Greechie & Wright's twelfth dispersionless state on the pentagon/pentagram/house logic

Despite the aforementioned 11 two-valued states there exists another dispersionless state on cyclic pastings of an odd number of contexts; namely, a state being equal to $\frac{1}{2}$ on all intertwines/bi-connections [29,41]. This state and its associated probability distribution are neither realizable by quantum nor by classical probability distributions. In this case the conditional probabilities of any two distinct contexts $\mathcal{C}_i$ and $\mathcal{C}_j$, for $1 \leq i, j \leq 5$ are

$$[P(\mathcal{C}_i|\mathcal{C}_j)] \equiv \begin{bmatrix} \frac{1}{2} & 0 & \frac{1}{2} \\ 0 & 0 & 0 \\ \frac{1}{2} & 0 & \frac{1}{2} \end{bmatrix}. \tag{10}$$

6. Three-colorable dense points on the sphere

There exist dense subsets of the unit sphere in three dimensions which require just three colors for associating different colors within every mutually orthogonal triple of (unit) vectors [42–44] forming an orthonormal basis. By identifying two of these colors with the value "0", and the remaining color with the value "1" one obtains a two-valued measure on this "reduced" sphere. The resulting conditional probabilities are discontinuous.

7. Extrema of conditional probabilities in row and doubly stochastic matrices

The row stochastic matrices representing conditional probabilities form a polytope in $\mathbb{R}^{n^2}$ whose vertices are the n^n matrices $\mathbf{T}_i$, $i = 1, \ldots, n^n$, with exactly one entry 1 in each row [45, p. 49]. Therefore, a row stochastic matrix can be represented as the convex sum $\sum_{i=1}^{n^n} \lambda_i \mathbf{T}_i$, with nonnegative $\lambda_i \geq 0$ and $\sum_{i=1}^{n^n} \lambda_i = 1$.

For conditional probabilities yielding doubly stochastic matrices, such as, for instance, the quantum case, the Birkhoff theorem [26] yields more restricted linear bounds: it states that any doubly stochastic $(n \times n)$–matrix is the convex hull of $m \leq (n-1)^2 + 1 \leq n!$ permutation matrices. That is, if $\mathbf{A} \equiv a_{ij}$ is a doubly stochastic matrix such that $a_{ij} \geq 0$ and $\sum_{i=1}^{n} a_{ij} = \sum_{i=1}^{n} a_{ji} = 1$ for $1 \leq i, j \leq n$, then there exists a convex sum decomposition $\mathbf{A} = \sum_{k=1}^{m \leq (n-1)^2+1 \leq n!} \lambda_k \mathbf{P}_k$ in terms of $m \leq (n-1)^2 + 1$ linear independent permutation matrices $\mathbf{P}_k$ such that $\lambda_k \geq 0$ and $\sum_{k=1}^{m \leq (n-1)^2+1 \leq n!} \lambda_k = 1$.

8. Summary

I have attempted to sketch a generalized probability theory for multi-context configurations of observables which may or may not be embeddable into a single classical Boolean algebra. Complementarity and distinct contexts require an extension of the Kolmogorov axioms. This has been achieved by an additional axiom ascertaining that the conditional probabilities of observables in one context, given the occurrence of observables in another context, form a stochastic matrix. Various models have been discussed. In the case of doubly stochastic matrices, linear bounds have been derived from the convex hull of permutation matrices.

Acknowledgments

The author acknowledges the support by the Austrian Science Fund (FWF): project I 4579-N and the Czech Science Foundation (GAČR): project 20-09869L, as well as an invitation to the Santiago based IFICC, where enlightening discussions with Tomas Veloz and Philippe Grangier took place. All misconceptions and errors are mine.

References

1. E. Schrödinger, *My View of the World* (Cambridge University Press, Cambridge, UK, 1951), ISBN 9780521062244,9781107049710, https://doi.org/10.1017/CBO9781107049710.
2. D. Aerts, Lettere al Nuovo Cimento **34**(4), 107 (May 1982), https://doi.org/10.1007/bf02817207.
3. D. Aerts, Journal of Mathematical Physics **27**(1), 202 (1986), https://doi.org/10.1063/1.527362.
4. A. Y. Khrennikov, *Ubiquitous Quantum Structure* (Springer, Berlin Heidelberg, 2010), https://doi.org/10.1007/978-3-642-05101-2.
5. E. N. Dzhafarov, V. H. Cervantes, and J. V. Kujala, Philosophical Transactions of the Royal Society A: Mathematical, Physical and Engineering Sciences **375**(2106), 20160389 (2017), arXiv:1703.01252, https://doi.org/10.1098/rsta.2016.0389.
6. A. M. Gleason, Journal of Mathematics and Mechanics (now Indiana University Mathematics Journal) **6**(4), 885 (1957), https://doi.org/10.1512/iumj.1957.6.56050.
7. E. Specker, Dialectica **14**(2-3), 239 (1960), english traslation at https://arxiv.org/abs/1103.4537, arXiv:1103.4537, https://doi.org/10.1111/j.1746-8361.1960.tb00422.x.
8. K. Svozil, Quantum Reports **2**(2), 278 (2020), arXiv:1808.00813, https://doi.org/10.3390/quantum2020018.
9. S. Kochen and E. P. Specker, Journal of Mathematics and Mechanics (now Indiana University Mathematics Journal) **17**(1), 59 (1967), https://doi.org/10.1512/iumj.1968.17.17004.
10. A. A. Abbott, C. S. Calude, and K. Svozil, Journal of Mathematical Physics **56**(10), 102201(1, 102201 (2015), arXiv:1503.01985, https://doi.org/10.1063/1.4931658.
11. K. Svozil, Entropy **22**(6), 602 (May 2020), arXiv:1707.08915, https://doi.org/10.3390/e22060602.

12. J. Uffink, in C. Beisbart and S. Hartmann, eds., *Probabilities in Physics* (Oxford University Press, Oxford, UK, 2011), pp. 25–49, https://doi.org/10.1093/acprof:oso/9780199577439.003.0002.
13. J. Aczél, *Lectures on functional equations and their applications*, vol. 19 of *Mathematics in Science and Engineering* (Academic Press, New York-London, 1966), ISBN 9780080955254, translated by Scripta Technica, Inc. Supplemented by the author. Edited by Hansjorg Oser, https://www.elsevier.com/books/lectures-on-functional-equations-and-their-applications/aczel/978-0-12-043750-4.
14. D. Reem, Aequationes mathematicae **91**(2), 237 (Feb. 2017), https://doi.org/10.1007/s00010-016-0463-6.
15. P. Busch, Physical Review Letters **91**(12), 120403, 4 (2003), https://doi.org/10.1103/PhysRevLett.91.120403.
16. C. M. Caves, C. A. Fuchs, K. K. Manne, and J. M. Renes, Foundations of Physics. An International Journal Devoted to the Conceptual Bases and Fundamental Theories of Modern Physics **34**(2), 193 (2004), https://doi.org/10.1023/B:FOOP.0000019581.00318.a5.
17. H. Granström, *Gleason's theorem*, Master's thesis, Stockholm University (August 2006), http://3dhouse.se/ingemar/exjobb/helena-master.pdf.
18. V. J. Wright, *Gleason-type theorems and general probabilistic theories*, Ph.D. thesis, University of York (September 2019), http://etheses.whiterose.ac.uk/25354/.
19. V. J. Wright and S. Weigert, Journal of Physics A: Mathematical and Theoretical **52**(5), 055301 (Jan 2019), https://doi.org/10.1088%2F1751-8121%2Faaf93d.
20. V. J. Wright and S. Weigert, Foundations of Physics **49**(6), 594 (Jun 2019), https://doi.org/10.1007/s10701-019-00275-x.
21. A. Aufféves and P. Grangier, Scientific Reports **7**(2123), 43365 (1 (2017), arXiv:1610.06164, https://doi.org/10.1038/srep43365.
22. A. Aufféves and P. Grangier, Philosophical Transactions of the Royal Society A: Mathematical, Physical and Engineering Sciences **376**(2123), 20170311 (2018), arXiv:1801.01398, https://doi.org/10.1098/rsta.2017.0311.
23. P. R. Halmos, *Finite-Dimensional Vector Spaces*, Undergraduate Texts in Mathematics (Springer, New York, 1958), ISBN 978-1-4612-6387-6,978-0-387-90093-3, https://doi.org/10.1007/978-1-4612-6387-6.
24. J. von Neumann, *Mathematische Grundlagen der Quantenmechanik* (Springer, Berlin, Heidelberg, 1932, 1996), 2nd ed., ISBN 978-3-642-61409-5, 978-3-540-59207-5,978-3-642-64828-1, English translation in [46, https://doi.org/10.1007/978-3-642-61409-5.
25. G. Birkhoff and J. von Neumann, Annals of Mathematics **37**(4), 823 (1936), https://doi.org/10.2307/1968621.
26. M. Marcus, The American Mathematical Monthly **67**(3), 215 (1960), https://doi.org/10.1080/00029890.1960.11989480.
27. K. Svozil, International Journal of Theoretical Physics **44**, 745 (2005), arXiv:quant-ph/0209136, https://doi.org/10.1007/s10773-005-7052-0.

28. E. F. Moore, in C. E. Shannon and J. McCarthy, eds., *Automata Studies. (AM-34)* (Princeton University Press, Princeton, NJ, 1956), pp. 129–153, https://doi.org/10.1515/9781400882618-006.
29. R. Wright, in A. R. Marlow, ed., *Mathematical Foundations of Quantum Theory* (Academic Press, New York, 1978), pp. 255–274, ISBN 9780323141185, https://www.elsevier.com/books/mathematical-foundations-of-quantum-theory/marlow/978-0-12-473250-6.
30. R. Wright, Foundations of Physics **20**(7), 881 (1990), https://doi.org/10.1007/BF01889696.
31. K. Svozil, Soft Computing **24**, 10239 (Nov 2020), arXiv:1810.10423, https://doi.org/10.1007/s00500-019-04425-1.
32. R. J. Greechie, Journal of Combinatorial Theory. Series A **10**, 119 (1971), https://doi.org/10.1016/0097-3165(71)90015-X.
33. G. Kalmbach, *Orthomodular Lattices*, vol. 18 of *London Mathematical Society Monographs* (Academic Press, London and New York, 1983), ISBN 0123945801,9780123945808.
34. M. Navara and V. Rogalewicz, Mathematische Nachrichten **154**, 157 (1991), https://doi.org/10.1002/mana.19911540113.
35. J. Tkadlec, *2states. A Pascal program* (2009, 2017), private communication, electronic message from Aug 23nd, 2017.
36. O. Neugebauer, *Vorlesungen über die Geschichte der antiken mathematischen Wissenschaften. 1. Band: Vorgriechische Mathematik* (Springer, Berlin, Heidelberg, 1934), ISBN 978-3-642-95096-4,978-3-642-95095-7, https://doi.org/10.1007/978-3-642-95095-7.
37. L. Lovász, IEEE Transactions on Information Theory **25**(1), 1 (January 1979).
38. K. Svozil, in M. Burgin and C. S. Calude, eds., *Information and Complexity* (World Scientific, Singapore, 2016), vol. 6 of *World Scientific Series in Information Studies: Volume 6*, chap. Chapter 11, pp. 276–300, ISBN 978-981-3109-02-5,978-981-3109-04-9, arXiv:1509.03480, https://doi.org/10.1142/9789813109032_0011.
39. K. Svozil, in K. Engesser, D. M. Gabbay, and D. Lehmann, eds., *Handbook of Quantum Logic and Quantum Structures* (Elsevier, Amsterdam, 2009), pp. 551–586, ISBN 978-0-444-52869-8, arXiv:quant-ph/0609209, https://doi.org/10.1016/B978-0-444-52869-8.50015-3.
40. A. Aufféves and P. Grangier, *Stochastic, bistochastic, and unistochastic matrices for quantum probabilities* (2015), unpublished draft.
41. E. R. Gerelle, R. J. Greechie, and F. R. Miller, in C. P. Enz and J. Mehra, eds., *Physical Reality and Mathematical Description* (D. Reidel Publishing Company, Springer Netherlands, Dordrecht, Holland, 1974), pp. 167–192, ISBN 978-94-010-2274-3,978-90-277-0513-6,978-94-010-2276-7, https://doi.org/10.1007/978-94-010-2274-3.
42. C. D. Godsil and J. Zaks, *Coloring the sphere* (1988, 2012), University of Waterloo research report CORR 88-12, arXiv:1201.0486, https://arxiv.org/abs/1201.0486.

43. D. A. Meyer, Physical Review Letters **83**(19), 3751 (1999), arXiv:quant-ph/9905080, `https://doi.org/10.1103/PhysRevLett.83.3751`.
44. H. Havlicek, G. Krenn, J. Summhammer, and K. Svozil, Journal of Physics A: Mathematical and General **34**, 3071 (April 2001), arXiv:quant-ph/9911040, `https://doi.org/10.1088/0305-4470/34/14/312`.
45. A. Berman and R. J. Plemmons, *Nonnegative Matrices in the Mathematical Sciences* (Society for Industrial and Applied Mathematics (SIAM), Philadelphia, 1994), `https://doi.org/10.1137/1.9781611971262.fm`.
46. J. von Neumann, *Mathematical Foundations of Quantum Mechanics* (Princeton University Press, Princeton, NJ, 1955), ISBN 9780691028934, German original in [24, `http://press.princeton.edu/titles/2113.html`.

ARE NON-BOOLEAN EVENT STRUCTURES THE PRECEDENCE OR CONSEQUENCE OF QUANTUM PROBABILITY?

C. A. FUCHS* and B. C. STACEY

Physics Department, University of Massachusetts Boston,
Boston, MA 02125, USA
**E-mail: Christopher.Fuchs@umb.edu*
www.umb.edu

In the last five years of his life Itamar Pitowsky developed the idea that the formal structure of quantum theory should be thought of as a Bayesian probability theory adapted to the empirical situation that Nature's events just so happen to conform to a non-Boolean algebra. QBism too takes a Bayesian stance on the probabilities of quantum theory, but its probabilities are the personal degrees of belief a sufficiently-schooled agent holds for the consequences of her actions on the external world. Thus QBism has two levels of the personal where the Pitowskyan view has one. The differences go further. Most important for the technical side of both views is the quantum mechanical Born Rule, but in the Pitowskyan development it is a theorem, not a postulate, arising in the way of Gleason from the primary empirical assumption of a non-Boolean algebra. QBism on the other hand strives to develop a way to think of the Born Rule in a pre-algebraic setting, so that it itself may be taken as the primary empirical statement of the theory. In other words, the hope in QBism is that, suitably understood, the Born Rule is quantum theory's most fundamental postulate, with the Hilbert space formalism (along with its perceived connection to a non-Boolean event structure) arising only secondarily. This paper will avail of Pitowsky's program, along with its extensions in the work of Jeffrey Bub and William Demopoulos, to better explicate QBism's aims and goals.

Keywords: QBism; quantum foundations; Born rule; quantum logic; Pitowsky; Demopoulos; Bub.

1. Dedication by CAF

This paper was written too late to appear in the memorial volume for Itamar Pitowsky edited by Meir Hemmo and Orly Shenker [1], but it is never too late to tribute dear friends. Itamar Pitowsky and Bill Demopoulos came into my life as a package. I met Itamar in 1997 at the Sixth UK Conference on Conceptual and Mathematical Foundations of Modern Physics in Hull,

England. It was my first quantum foundations conference, and to tell the truth I was a little ashamed to be there: Indeed I had to work up the nerve to ask my postdoctoral advisor John Preskill if I might use some of my funding for such a frilly thing. And it was a frilly thing — only one talk in the whole meeting made any impression on me, Itamar's on simplifying Gleason's theorem. It blew me away. When I got home, I studied his paper [2] with meticulous care, not just for the math, but also for its programmatic character. I found something so alluring in how he presented the big picture of his efforts at the end, I pledged to one day write a paper myself ending with nearly the same words: "More broadly, Theorem 6 is part of the attempt to understand the mathematical foundations of quantum mechanics. In particular, it helps to make the distinction between its physical content and mathematical artifact clear."

Well, I must have made an impression on Itamar too, as he described me to Bill Demopoulos as "a Wunderkind of quantum information". (If we could only recapture our youths!) One thing led to another, from more than 1,300 email exchanges, to many long, thoughtful walks at Bill's farm in Ontario. Both men made an indelible impression on me. Professionally, what set them apart from other philosophers of physics is that, rather than try to dissuade me from developing QBism, they both were sources of constant encouragement. But there was so much more than our professional relationships between us. At least in the case of Bill's passing, it was not so sudden as Itamar's that I was able to tell him I love him.

This paper is dedicated to Itamar and Bill.

2. Introduction

QBism is a relative newcomer among interpretations of quantum mechanics. Though it has identifiable precursors, no mature statement of it existed in the literature before 2009–2010 [3,4]. Indeed, even the pre-QBist writings of the developers themselves contain much that QBism flatly contradicts [5]. One brief summary is that QBism is

> an interpretation of quantum mechanics in which the ideas of *agent* and *experience* are fundamental. A "quantum measurement" is an act that an agent performs on the external world. A "quantum state" is an agent's encoding of her own personal expectations for what she might experience as a consequence of her actions. Moreover, each measurement outcome is a personal event, an experience specific to the agent who incites it. Subjective judgments thus com-

> prise much of the quantum machinery, but the formalism of the theory establishes the standard to which agents should strive to hold their expectations, and that standard for the relations among beliefs is as objective as any other physical theory [6].

This is not the kind of move that one takes lightly. Only the stunning empirical success of quantum mechanics, coupled with the decades of confusion over its conceptual foundations, could have compelled the codification of thoughts like these into a detailed, systematized research program. The practical triumphs of quantum physics imply that the questions of conceptual foundations should be taken seriously, while the perennial debate, ceaseless and without clarification, implies that a radical step is necessary.

When compared against most interpretations of the quantum, QBism requires perspective shifts of an almost virtiginous character. For starters, it insists upon a stringently personalist Bayesianism. The problem with physicists and probability is that when one has vast heaps of data, pretty much any even half-baked interpretation of probability is good enough to scrape by. It is not until we push further into matters of principle that mediocrity begins to pose a real hazard. Fortunately, physicists are trained to shed intuitions that prove counterproductive. We could once get by with acting like the data points which help shape probability assignments actually *determine* them, but then, we also got along for quite a while thinking that heavy bodies must fall faster than light ones.

Likewise, foregrounding the concept of *agent* means taking a first-person perspective, and first-person singular at that [7]. This means rephrasing, and sometimes even rethinking, some discussions of quantum matters. For example, a first-person take on quantum communication protocols is that, fundamentally, Alice wants to be confident that Bob will react in the way she needs after she has intervened in the world by sending a message. For if Alice had no concern for the consequences returning to her or the potential suffering of those she regards as kin, then she would abandon the photonics and just throw a bottle into the sea.

Some changes in language end up being rather modest, though, in scale if not in eventual implication. QBism makes no category distinction between quantum states and probability distributions; the former are just examples of the latter in a coordinate system adapted for atomic and molecular physics [8]. Thus, the QBist avoids locutions like "the quantum state of the system" in favor of "*my* quantum state *for* the system" — not too many letters changed, but with a great deal of history behind the difference. A QBist exposition might say, "Suppose a physicist Alice ascribes a quantum state ρ," a turn of phrase that is philosophically justified while also sounding quite natural, since ascribing quantum states is a thing that working physicists actually do.

This example illustrates the peculiar flipside of QBist radicalism. After traversing the wilds of personalist Bayesian probability and the further realms of participatory realism [9], one arrives with a new appreciation of home. Certain habits of more innocent days turn out to be philosophically justified, or at least justifiable; one learns that one can do quantum foundations and quantum *physics* at the same time.

So, now that we've gotten a little bit of seeming almost-paradox on our fingers, it would be good to sort out what is and is not a shocking departure. How does QBism situate itself with respect to those other interpretational programs that share some of its morphology, either by convergence or common ancestry?

3. Two levels of personalism

Introducing a collection of correspondence, CAF identified three characteristics of QBism that distinguish it from prior interpretations [10].

> First is its crucial reliance on the mathematical tools of quantum information theory to reshape the look and feel of quantum theory's formal structure. Second is its stance that two levels of radical "personalism" are required to break the interpretational conundrums plaguing the theory. Third is its recognition that with the solution of the theory's conundrums, quantum theory does not reach an end, but is the start of a great journey.

We will return to the first point later (it is the motivation for the technical work that occupies most of our time not given over to teaching and administration). In this section, we will focus on the second point, the need for two levels of personalism. A pivotal moment in the history of QBism was the realization that one level did not suffice, even though that was already one more than many were willing to grant!

We have already met a significant part of the first layer in the introduction, where we noted that in QBism, quantum states are the possessions of the agent who asserts them. This is already enough to get booted from a philosophy-of-physics meeting, but we've hardly gotten started. Chasing the demands of self-consistency, we find that other elements of the quantum formalism must *also* be given the same status as quantum states, that is, the status of personalist Bayesian probability assignments. One important example is *the association of specific POVM elements with potential future experiences.* A positive-operator-valued measure, or POVM for short, is the basic notion of measurement in quantum information theory. Briefly put, a POVM is a collection of positive semidefinite operators on a Hilbert space that sum to the identity operator on that space, and each of the operators (or "effects") in the set corresponds to a possible outcome of the measurement that the POVM represents [11,12].

In contrast, in the tradition of Bub, Demopoulos, and Pitowsky one moves from quantum to classical by replacing the space of possible events with a non-Boolean algebra. Quantum probability is Bayesian probability on this algebraic structure. Events are tied to projectors on Hilbert space, and this binding is in essence an objective fact of nature [13]. This leads quickly to trouble [14]. For suppose that a physicist Alice starts her day with a subjective probability assignment, encoded as a density operator ρ. She then performs a measurement and obtains an outcome, which will be associated with some projector Π. She updates her state assignment accordingly, replacing ρ with Π. But now, *her probabilities have been locked to an objective value.* In order for the personalism of quantum-state assignments not to collapse into triviality, there has to be more flexibility. (For additional discussion on this point, see the 10 January 2007 entry drawn from correspondence with Bub, Demopoulos, Pitowsky, and Veiko Palge in Appendix B.)

The thorn that spurred the development of the second level of personalism was the old thought-experiment known as "Wigner's Friend" or the puzzle of the observer being observed [8]. This is a tale — *not* as old as time, though within the insular world of quantum foundations, it may seem so — a tale of two agents and one quantum system. The agents could be Wigner and his friend, or in the lingo of quantum information, Alice and Bob. The story begins with Alice and Bob having a conversation. They both agree, let us say, that a quantum state $|\psi\rangle$ expresses their beliefs about the system. (If they are inclined to the Bayesian school, they don't *have* to agree at this stage. $|\psi\rangle$ is a catalogue of expectations, or bluntly put, a compendium of beliefs. But let's say they do agree so far.) Moreover, they agree that at a given time, Alice will perform an experiment upon the system, a measurement which both Alice and Bob say is represented by some mathematical structure such that if Alice observes outcome i, she will update her expectations to the state vector $|i\rangle$. Then, Bob walks away and leaves the room. Alice is alone with the system. Time passes, and the agreed-upon moment of measurement goes by.

What, now, is the "correct" state that each physicist should have chosen for the system? Alice, everyone concurs, should have selected some vector $|i\rangle$. But what about Bob? Unless he holds Alice somehow exempt from quantum mechanics, he would express his beliefs about her, in principle, with a quantum state ρ, and a time-evolution operator family U_t with which he can adjust those expectations and calculate probabilities pertaining to different times. Before Bob interacts with Alice or the other system, he would write a joint state for the two, a quantity of the form

$$U_t\,(\rho \otimes |\psi\rangle\langle\psi|)\,U_t^\dagger \tag{1}$$

which will in general be an entangled state. Bob's state for the system will be a partial trace of this, typically a mixed state completely different from $|i\rangle$.

Either Alice or Bob could be deemed "incorrect"; for if either "had all the information" they would arguably be forced to adopt the state assignment of the other.

> And so the back and forth goes. Who has the *right* state of information? The conundrums simply get too heavy if one tries to hold to an agent-independent notion of correctness for otherwise personalistic quantum states. A QBist dispels these and similar difficulties by being conscientiously forthright. *Whose information?* "Mine!" *Information about what?* "The consequences (for *me*) of *my* actions upon the physical system!" It's all "I-I-me-me mine," as the Beatles sang [8].

Lately, there has been a resurgence of interest in Wigner's Friend — a growth industry, by the standards of quantum foundations. These recent elaborations of Wigner's Friend wrap it around a no-hidden-variables argument [15–18]. QBism, having answered the original conundrum on its own terms, finds nothing troubling about these elaborations. Pusey regards an extended Wigner's Friend puzzle as "actually good news" for QBism and for Rovelli's Relational Quantum Mechanics [17]. For a comparison and contrast between QBism and Rovellian RQM, see "FAQBism" [6]; for a critical evaluation of various statements made about QBism in this niche of the literature, see "QBism and Assumption (Q)" [19].

4. Gleason's theorem

Gleason's theorem is, in our experience, less well-known among physicists than some of its conceptual descendants and confrères. The original statement had the mildly intimidating language of finding "all measures on the closed subspaces of a Hilbert space" [20]. In more detail, Gleason defined a *frame function* of weight W for a separable Hilbert space $\mathcal{H}$ to be a function $f : \mathcal{H} \to \mathbb{R}$ such that if $\{x_i\}$ is an orthonormal basis of $\mathcal{H}$ then

$$\sum f(x_i) = W. \tag{2}$$

A frame function is *regular* if there exists a self-adjoint operator T acting on $\mathcal{H}$ such that $f(x)$ is given by an inner product

$$f(x) = (Tx, x) \tag{3}$$

for all $x \in \mathcal{H}$. Gleason proves, by ingenious and somewhat arduous means, that all nonnegative frame functions in three or more dimensions are regular.

If the weight of a frame function is 1, then that function begins to look like a probability assignment. In physics language, Gleason showed that if measurement outcomes are represented by vectors in orthonormal bases, then any consistent ascription of probabilities to measurement outcomes must take the form of the Born Rule. The definition of frame functions assumes that probability ascriptions are *noncontextual*: The function f sends a vector to a number in the unit interval, regardless of what basis that vector may be a part of [21]. Rather remarkably, Gleason's proof yields both the set of valid density operators and the Born Rule. Quite an economy of postulates! It is easy to see why an attempt to rebuild quantum theory on a better axiomatic footing might proceed by way of Gleason's theorem. As Wilce put it [22],

> The point to bear in mind is that, once the quantum-logical skeleton [the lattice of closed subspaces of $\mathcal{H}$] is in place, the remaining statistical and dynamical apparatus of quantum mechanics is essentially fixed. In this sense, then, quantum mechanics — or, at any rate, its mathematical framework — *reduces to* quantum logic and its attendant probability theory.

Why, then, have the QBist and QBist-adjacent efforts at reconstructing quantum theory skipped past Gleason's theorem? Part of the answer lies in that invocation of "the quantum-logical skeleton": As any good biologist knows, bones are not simple affairs [23], and presuming a working skeleton is asking for quite the sophisticated development already!

Gleason begins with a Hilbert space and orthonormal bases upon it. How, then, should we arrive at Hilbert space? Pitowsky presents an argument for getting to the point where Gleason can take over, a *representation theorem* for lattices that satisfy a certain set of conditions inspired by quantum mechanics [24]. He admits a gap in this argument, partly closed by another remarkable theorem, this one by Maria Pia Solèr [25,26]. But the QBist concern is more fundamental: Those lattice-theoretic conditions hail from *the quantum physics of the 1930s,* when "uncertainty" was the shocking feature, long before the discovery of more subtle mysteries like the violation of Bell inequalities. We can always take a fundamentally classical theory, a theory of local degrees of freedom, and *hide those variables* by imposing an "uncertainty principle" [27,28]. The resulting theories include features like the no-cloning theorem, incompatible observables, and even interference between paths in a Mach–Zehnder interferometer. This modern perspective, informed by studies into questions like what a quantum computer requires to outperform a classical emulation of one, takes the glamor away from the notions in which quantum logic was grounded.

The original version of Gleason's theorem fails if the Hilbert space is two-dimensional. Why? When $d = 2$, one cannot hold one basis vector in place and twirl the others around to generate multiple distinct bases. So, the constraints assumed by Gleason are too weak to carry real implicative weight. One might, for example, paint the northern hemisphere of the Bloch sphere all over with 1's, and the southern hemisphere with 0's. The resulting assignment of probabilities to von Neumann measurements — remember, orthonormal bases in $d = 2$ are pairs of antipodal points on the Bloch sphere — would be consistent with Gleason's rules, yet no choice of density operator ρ can yield those probabilities. This pesky difficulty is obviated if one follows the practice of quantum information theory and broadens the notion of *measurement* beyond the von Neumann definition to the wider class of POVMs. Doing so allows a proof of a Gleason-type result that is both simpler to prove and applicable in $d = 2$ where Gleason's was not [29–33].

The general good cheer created by shifting the fundamental notion of measurement to POVMs led QBism further from the quantum-logic tradition.

5. Non-Booleanity as a consequence

In this section, we will derive non-Booleanity as a consequence of a more fundamental manifestation of quantum strangeness, the nonexistence of intrinsic hidden variables. We will step through this argument for the case where it is simplest, a single qubit, which is already enough to derive a theory with noncommuting observables. Having gotten that far, we will briefly outline two approaches to generalizing the argument and thus deriving quantum theory in higher dimensions.

The centerpiece of the QBist program for reconstructing quantum theory is the concept of a *reference measurement.* This is an experiment with the property that if an agent possesses a probability distribution over its outcomes, she can calculate the probabilities that she should use for any other measurement. In classical particle mechanics, a reference measurement would be an experiment that reads off the position and momentum of all particles in the system, thereby locating it within phase space and allowing the agent to extrapolate the full dynamical trajectory. Any other experiment — say, observing the total angular momentum — is essentially a coarse-graining of the information that the reference measurement provides. If Alice expresses her uncertainty about the system's phase-space coordinates as a probability distribution over phase space, then she can calculate her probabilities for the possible results of a coarse-grained observation by convolving that "Liouville density" with an appropriate kernel.

Classical mechanics has the property that if two Liouville densities are distinguishable by a coarse-grained measurement, then they are nonoverlapping. Quantum physics violates this principle, in a way that we can make precise.

Consider the following scenario [34]. An agent Alice has a physical system of interest that she plans to perform either one of two experimental protocols upon. In one protocol, she will send the system directly into a measuring device and obtain an outcome. In the other, she will pass the system through a reference measurement first and then send it into the device from the first protocol. Let H_i denote the possible outcomes of the reference measurement and D_j the possible outcomes of the other. Then Alice can write her expectations for the results of these two protocols as $P(H_i, D_j)$, for sending the system into the D_j device via the reference H_i experiment first, and $Q(D_j)$, for going directly into the D_j device. Probability theory itself imposes no bond between these expectations: different conditions, different probabilities! But the additional assumption that the reference measurement simply reads off the pre-existing physical condition of the system, *an ontological assumption on top of probability theory,* does furnish a link:

$$Q(D_j) = \sum_i P(H_i, D_j) = \sum_i P(H_i)P(D_j|H_i). \tag{4}$$

The first equality here is a *physical* assumption, while the second is mandated by logic. That is, if Alice rejects the second equality — the "Law of Total Probability" — she is being Dutch-book incoherent, whereas if she rejects the first, she is just being skeptical about the applicability of classical intuition. Quantum physics provides a replacement for the first equality. Instead of marginalizing over H_i as in the previous expression, one performs a "quantum marginalization":

$$Q(D_j) = \mu\left(P(H_i, D_j)\right). \tag{5}$$

The exact form of the function μ will depend upon the choice of POVM used as the reference measurement.

To make this more concrete, let us take the case where the system of interest is a single qubit. Then a reference measurement can be any POVM whose effects span the space of Hermitian operators on the Hilbert space $\mathbb{C}^2$. For example, let the reference measurement be a POVM whose effects are proportional to rank-1 projectors that form a regular tetrahedron inscribed in the Bloch sphere. A convenient indexing is to let a and b take the values ± 1, and to write

$$H_{ab} = \frac{1}{4}\left(I + \frac{1}{\sqrt{3}}(a\sigma_x + b\sigma_y + ab\sigma_z)\right). \tag{6}$$

These four operators sum to the identity and span the space as desired. With

$$P(H_{ab}) = \mathrm{tr}(\rho H_{ab}) \tag{7}$$

by the Born Rule, we have

$$Q(D_j) = \mathrm{tr}(\rho D_j) = \sum_{ab}\left[3P(H_{ab}) - \frac{1}{2}\right] P(D_j|H_{ab}). \tag{8}$$

Here, we have defined

$$P(D_j|H_{ab}) = 2\mathrm{tr}(D_j H_{ab}) \tag{9}$$

as Alice's probability of obtaining the outcome D_j *given* that her preparation for the system is the state proportional to H_{ab}.

In this case, the function μ takes the form of the classical Law of Total Probability but with an elementwise affine map applied to the probability vector $P(H_{ab})$. Said another way, Eq. (8) *is* the Born Rule, written in explicitly probabilistic language throughout.

We see that picking a reference measurement establishes a mapping from density matrices into a probability simplex, thereby furnishing a wholly probabilistic way to write the quantum theory of a qubit. Importantly, in this representation not all probability vectors correspond to valid quantum states. The state space $\mathcal{P}$ is the ball in $\mathbb{R}^4$ within which

$$\frac{1}{6} \leq \sum_{a,b} P(H_{ab}) P'(H_{ab}) \leq \frac{1}{3}, \ \forall\ P, P' \in \mathcal{P}. \tag{10}$$

The pure states will be exactly those that saturate the upper bound on their Euclidean norm. Orthogonal states are *maximally distant,* saturating the lower bound, but the lower bound is not zero. As we prove in Appendix A, this is in fact a very general result, holding true for any reference measurement: Quantum states that are orthogonal in Hilbert space are overlapping probability distributions.

We can turn this idea around. Let's forget quantum theory for the moment and take the relation

$$Q(D_j) = \sum_{i=1}^{N} \left[\alpha P(H_i) - \beta\right] P(D_j|H_i) \tag{11}$$

as the defining axiom of a probabilistic theory. That is, let's say that this expression, which "breaks" classical intuition in the minimal way possible, is the empirical rule on top of probability theory that tells us how to "marginalize" over an experiment that is not merely ignored, but *unperformed.* Probability vectors $P(H)$ and conditional probability matrices $P(D|H)$ are consistent with one another if they always yield valid probability vectors $Q(D)$ when combined in this way. This provides a reciprocal consistency condition between the set of valid $P(H)$ and the set of valid $P(D|H)$, which we can turn into a condition on the set of valid $P(H)$ by making some fairly mild assumptions relating those two sets. For instance, we can posit that the state of maximal indifference, the flat probability vector, is a valid $P(H)$, and that posteriors from maximal indifference are valid priors. (For some conceptual background, see the 7 January 2011 entry drawn from correspondence with Demopoulos in Appendix B.) This implies that, if the $P(H)$ are column vectors, then any row in a matrix $P(D|H)$ is the transpose of some valid $P(H)$, up to overall scaling:

$$P(D_j|H_i) = \gamma P'(H_i). \tag{12}$$

The condition that $Q(D_j)$ be always nonnegative then yields

$$\alpha \sum_i P(H_i) P'(H_i) - \beta \sum_i P'(H_i) \geq 0, \tag{13}$$

which by normalization simplifies to

$$P \cdot P' \geq \frac{\beta}{\alpha}. \tag{14}$$

Pairs that saturate this bound are maximally distant from one another and imply a $Q(D_j)$ of 0. Moreover, because $Q(D)$, $P(H)$ and $P(D|H)$ must themselves be properly normalized, we have that

$$1 = \alpha - N\beta. \tag{15}$$

We follow the guiding principle that our theory should be as broadly applicable as possible, so the structures within it should have no "distinguishing marks or scars" that would make them peculiar to particular physical situations. Likewise, we want the sets we derive to be *maximal* with respect to the constraints we apply: To do otherwise would be to implicitly admit an additional constraint, and thus to be making more empirical assumptions on top of the basic "nonclassical marginalization" rule originally posited. Therefore, our set $\mathcal{P} \subset \mathbb{R}^N$ of valid probability vectors must have the property that any two points in it have a Euclidean inner product greater than β/α, and including *any additional point* would spoil that property.

This is already enough to make available some useful tools from higher-dimensional geometry. Given a point P in the probability simplex Δ, the *polar* of P is the set of all vectors in $\mathbb{R}^N$ whose elements sum to unity and which have an inner product with P that is greater than or equal to β/α. The polar of a set is the set of all vectors that are in the polars of all the points in the given set:

$$A^* = \left\{ u : \sum_i u(i) = 1,\ u \cdot v \geq \frac{\beta}{\alpha}\ \forall v \in A \right\}. \tag{16}$$

The requirement that a state space $\mathcal{P}$ be maximal implies that it is a *self-polar* subset of the probability simplex [34].

To move in the direction of quantum theory, we introduce a concept of dimension in terms of *Mutually Maximally Distant* sets. For any self-polar subset of the probability simplex, there must be a smallest sphere that encloses it, i.e., the circumscribing sphere of the state space. The states on this sphere are states of maximal confidence with respect to the reference measurement. We define a Mutually Maximally Distant set in $\mathcal{P}$ to be a set of points on the intersection of $\mathcal{P}$ and its circumscribing sphere such that any two points in the set saturate the lower bound on the inner product, β/α. This is the tool we need to establish an information-theoretic notion of "dimension": The dimension of a system is the largest possible size of a set of hypotheses, each of which are assertions of maximal confidence, and any two of which are maximally distinct from each other. (This is the natural counterpart of classical probability theory, where hypotheses of maximal confidence are Kronecker delta functions.) A brief derivation [34,35] shows that

$$d = 1 + \frac{r_{\text{out}}^2}{r_{\text{mid}}^2}, \tag{17}$$

where r_{out} is the radius of the circumscribing sphere and

$$r_{\text{mid}} = \frac{1}{\sqrt{N\alpha}} \tag{18}$$

is the radius of the *mid-sphere*, the sphere around the barycenter within which any two points always have an inner product that satisfies the lower bound.

The fundamental "bit" in our theory is a system whose state space $\mathcal{P}_{\text{bit}}$ meets the following condition. The maximal size of a set of points, any two of which are maximally distant from each other, is equal to exactly 2. In other words, given any hypothesis of maximal confidence about the system, there exists exactly one other hypothesis of maximal confidence fully distinct from it. (And if this applied only to *some* of those extremal hypotheses, then the state space would have "distinguishing marks and scars" indicating some other empirical constraint that we are not allowing at this level of generality.) For our state space $\mathcal{P}_{\text{bit}}$ to have this geometrical property and to satisfy the maximality condition given above, it must be a *ball* within the probability simplex in $\mathbb{R}^N$. Moreover, we can say which ball it is, because unless we invoke further empirical additions to probability theory, we have to use the distance scale that is already available to us, meaning that $\mathcal{P}_{\text{bit}}$ is the *inscribed ball of the probability simplex.* That is, to satisfy the condition $d = 2$, we must have $r_{\text{out}} = r_{\text{mid}}$, and both of them are in turn fixed to the value

$$r_{\text{out}} = r_{\text{mid}} = r_{\text{in}} = \frac{1}{\sqrt{N(N-1)}}. \tag{19}$$

We still have to fix the Euclidean dimension N for the fundamental "bit" system. There are multiple ways of doing so [34], each of which suggest different "foil theories" to quantum mechanics. However, even before fixing N, we know already that as long as the "nonclassical marginalization" rule (11) is indeed nonclassical (that is, $\beta > 0$), our theory will spurn explanation in terms of intrinsic hidden variables. With $N = 4$, we recover the Bloch ball and the theory of a single qubit, including all the POVMs that can be performed upon it. This is already enough to prove the impossibility of underlying hidden variables: While a qubit theory restricted to von Neumann measurements can be explained classically, fully general POVMs upon a qubit cannot [36].

In order to reach this conceptual point as quickly as possible, we have stepped only lightly upon the algebraic manipulations. More details are available in the literature [3,34,37,38].

Having derived the theory of a single qubit, one way forward would be to compose them. An argument based on Einstein locality, and using the postulate of a reference measurement to justify a Gleason-type noncontextuality condition, gives the tensor-product rule for composing state spaces [14,39]. This still leaves open the question of why the joint states for, say, a bipartite system of two qubits should be the *positive semidefinite* operators on the tensor-product space. A voyage into the literature yields at least one answer to this question [40,41], but the assumptions involved may feel unsatisfying — mathematically over-powered or physically under-motivated.

Another option, and the one we prefer, is to start with Eq. (11) and take it as the basic way to express the nonexistence of intrinsic hidden variables for systems of all dimensionalities d. In the same manner as before, this yields a theory defined by a *qplex,* a proper subset of the probability simplex in $\mathbb{R}^{d^2}$. Any two points P and P' in a qplex $\mathcal{P}$ satisfy the inequalities

$$\frac{1}{d(d+1)} \leq P \cdot P' \leq \frac{2}{d(d+1)}. \tag{20}$$

For details on how to go from the very general-looking Eq. (11) to these particular inequalities, see Appleby et al. [34] The remaining challenge is then to identify the qplexes that are isomorphic to quantum state space — the *Hilbert qplexes* — among all the possibilities, and to do so in the most economical possible way. Current research focuses on relaxing the conditions that are known to work, and understanding the geometrical structures that are necessary to establish the isomorphism, which turn out to be quite captivating entities [42–50].

Assuming that the Law of Total Probability is the way to "marginalize" over an unperformed experiment, and the desire for distinguishable states to be nonoverlapping distributions, are expressions of faith in hidden variables. QBism holds that the way to make progress is to let go of this faith and encourage something more interesting in its place. This analysis echoes Demopoulos' criticism of the early Putnam, in which Demopoulos points out how a worldview can reduce to being a hidden-variable one even if it was not formulated with a specific ontology — Bohmian or de Broglian, say — explicitly in mind [51]:

By claiming to avoid hidden variables, Putnam almost certainly meant that he could avoid a theory like Bohm's or de Broglie's. For Putnam, such theories are characterized by their invocation of special forces to account for the disturbance by measurement. Since he nowhere appeals to such forces, Putnam believes his account to be free of an appeal to hidden variables. The idea that his interpretation of the logical connectives assumes an underlying space of truth-value assignments — and that it is therefore a hidden variable interpretation of quantum mechanics after all — appeals to an abstract notion of hidden variable theory, one which was only beginning to emerge when Putnam wrote his paper, and one with which he was at the time almost certainly unfamiliar.

6. Interlude: The PBR theorem

In the previous section, we explored the consequences of abandoning the notion that *distinguishable* means *nonoverlapping.* This provides a convenient setting to discuss the Pusey–Barrett–Rudolph (PBR) theorem [52]. The PBR theorem is a "no hidden variables" argument, or to say it more precisely, an argument that hidden variables will end up being redundant. Following the tradition of what nowadays is called the "ontological models framework" [53], it tries to express quantum probabilities as being due to ignorance of underlying ontic states, and then shows that different quantum states must correspond to nonoverlapping distributions over the ontic state space. The crucial feature of the ontological models framework is that probabilities for experiment outcomes are calculated using the Law of Total Probability:

$$\mathrm{tr}\rho E = \int d\lambda\, P_\rho(\lambda) P(E|\lambda). \tag{21}$$

By adopting the Born Rule in the "quantum marginalization" form of Eq. (11) as a fundamental postulate, we spurn this framework from first principles onward. And because QBism lies outside of the ontological models framework, preferring to find its realism on a more subtle and stimulating level, the PBR theorem has little to say about it. Indeed, the PBR theorem has no *direct* impact on QBism at all, though it may have the indirect effect of indicating that half-hearted attempts at interpreting quantum theory in an informational way do not go far enough.

The PBR theorem relies upon a compatibility criterion between quantum states that was first codified by Caves, CAF and Schack [54], based on an earlier proposal by Peierls [55]. Bub has an admirably concise exposition of the PBR theorem, in an essay where he also raises the intriguing question of proving analogues of it in nonclassical theories that are not quantum mechanics. He shows that an analogue of the PBR theorem can be proved in a nonclassical and nonquantum setting that he calls "Bananaworld" [56].

7. Demopoulos

The interpretational work of Demopoulos owes much to Pitowsky. He credits Pitowsky for advancing and defending "the view that the ψ–function represents a state of belief about a system rather than its physical state" [57]. Demopoulos regards the probabilities in quantum physics as being probabilities over "effects", which are explicitly *not* the effects that comprise POVMs [51]. Rather, they "are to be thought of as the traces of particle interactions on systems for which we have 'admissible' theoretical descriptions in terms of their dynamical properties" [57]. It is ultimately difficult to draw a line between this and, for instance, Pauli's imagery of data recorded "by objective registering apparatus, the results of which are objectively available for anyone's inspection" [58]. Therefore, we find Demopoulos' interpretation vulnerable to critiques in the Wigner's Friend tradition.

That said, there is much in Demopoulos' writing that aligns with QBism, and in the interests of solidarity and emotional positivity, we will spend much of this section talking about that.

First, let us get out of the way one of the more significant divisions between QBism and the position that Demopoulos articulated. From the standpoint of the former, the latter tends to slide back to a rather Bohrian view that quantum phenomena must be treated through classical intermediary apparatus, with a definition of *classical* that may have more to do with stable record-keeping than with the details of Newtonian or Maxwellian dynamics.

Quoting a letter that CAF wrote to Demopoulos on 9 January 2011 [10]:

> I didn't see that you strengthened your defense against my charge of a kind of dualism. Maybe I'm still missing something. When you write, "Such systems are epistemically accessible to an extent that systems which are characterized only in terms of their eternal properties and their effects are not," I would think that you're admitting that there are (at least) two distinct kinds of systems in

the world: Those that have a certain type of epistemic accessibility and those that do not, and that that epistemic accessibility is not characterized by things to do with the observer's situation, but rather the with system's itself. [...] And you don't weaken my accusation any either by writing on page 18: "This kind of conceptual dependence, does not preclude the application of quantum mechanics to systems that record effects. Although it is largely a matter of convenience which systems are, and which are not, taken to record effects, it is not wholly a matter of convenience." The phrase 'not wholly a matter of convenience' again points me to an ontic distinction between two types of system. Again, a reason for my saying, "dualism". Similarly with respect to your sentence, "This leaves entirely open the empirical question of why it is that some systems appear to be amenable to descriptions that are expressible in the framework of classical mechanics." I.e., you don't explain why, but there is an empirical distinction between two kinds of system.

I'm not saying there's anything necessarily philosophically wrong with a dualism; mostly it is that it just doesn't match my taste, and doesn't feel like the right direction for moving physics forward. It adds a bigger burden on the physicist than I'd rather him have: For now, for each lump of matter, he'll have to — in some mysterious way — come to a conclusion of whether it supports dynamical properties or not. How does he do that? Where you leave us is where Bohr left us — as far as I can tell, in just telling us "it must be so, but I'm not going to tell you where/how to make that distinction" ... i.e., that there must be two kinds of system for us to build our evidentiary base for the quantumly treatable ones at all.

Imagine my going up to Rainer Blatt and saying, "Rainer, I just found this nice rock on the beach. You think you might be able to use it as a component in that quantum computer you want to build?" Asher Peres would say, "Of course he can use it as a component; it is just a matter of money. With enough money, any old rock can be polished into a quantum computer." But if it ain't so, then it must be a burden on physics to say when it can and when it cannot be done. My guess is that Rainer will never be able to codify and make explicit such a criterion of distinction; he'll never be able to tell me which tests he must perform to certify my rock ineligible for quantum computation.

When discussing Gleason's work, Demopoulos considers *two-valued measures,* maps from the closed subspaces of a Hilbert space to the pair $\{0, 1\}$. He notes [57],

> It is evident that generalized two-valued measures and generalized truth-value assignments are formally interchangeable with one another, whatever the conceptual differences between probability measures and truth-value assignments.

QBism stresses the "conceptual differences": Its preferred school of Bayesianism forcefully rejects the idea that a probability-1 assignment indicates pre-existing physical truth. This leads to dismissing the EPR criterion of reality.[a] As CAF wrote to Demopoulos on 29 November 2007 [10], regarding a paper eventually published in 2010 [51]:

> Where I do take issue with what you write is the very difficult issue of probability 1. If I understand you correctly, I disagree with the first sentence of the first full paragraph from page 22: "The representation of an elementary particle as a function which, when presented with an experimental configuration, yields an effect, is interchangeable with its representation as a class of propositions only when the effects are predictable with 0-1 probability."
>
> For, I would say, not even then.

Additional discussion on this point can be found in the 24 June 2006 entry drawn from correspondence with Bub in Appendix B.

Having established this difference in content and in taste, we can now proceed to matters of greater agreement. Demopoulos writes [51],

> Instead of providing a solution to the long-standing issues involving measurement and the paradoxes, the discovery of "the logic of quantum mechanics" revealed a new and different problem, namely, the impossibility of interpreting the probabilities of the theory so

[a]Likewise, it implies that "incompatible" quantum-state assignments do not have to indicate different intrinsic physical conditions of a system. Two quantum states ρ and ρ' are incompatible in the Peierlsian sense if there exists a measurement for which ascribing ρ implies the outcome probabilities $(0, 1)$, whereas ascribing ρ' implies the outcome probabilities $(1, 0)$. Shifting one's state assignment from ρ to ρ' thus changes a probability-0 assignment into probability-1, but the agent has no obligation to regard this as a shift of a pre-existing physical property, for the same reason that they do not take EPR as gospel.

> that every proposition belonging to a particle is non-contextually determinately true or false. Rather than refuting classical logic, this discovery should be seen as refuting the idea that the probabilities are interpretable as the probabilities of such propositions. Ironically, the consequences of quantum mechanics for logic are almost the precise opposite of what Quine and Putnam imagined they might be. If anything, the problem of hidden variables upholds the centrality of classical logic in our theorizing about the physical world, while allowing that the Boolean algebraic structures, which are so closely associated with classical logic, are not appropriate for every use of probability in physics.

QBism finds much to like here. Rather than stressing the "classical logic" language, QBism speaks in terms of *internal self-consistency*: Within a given scenario fixed by an agent's choice of action, the agent's mesh of beliefs should hold together, i.e., they should be Dutch-book coherent. Moreover, the notion of Dutch-book coherence is, after a fashion, built upon "classical logic". That is, one way of expressing the basic precept of it is that an agent should not assign different probabilities to events whose descriptions can be translated to one another by Boolean manipulations. The Law of Total Probability itself follows from Dutch-book coherence, but it is "not appropriate for every use of probability in physics".

QBism has been accused of being "instrumentalism" with tiresome regularity. In this regard, it is helpful to recall Demopoulos' response to that charge being leveled at his own interpretation. Replying to the accusation that his view would make quantum physics "a mere heuristic for prediction", Demopoulos noted [51],

> What is at issue is the interpretation of the quantum algorithm for assigning probabilities and for reasoning from the position of uncertainty which such probabilities necessarily signify. This leaves a considerable degree of realism intact. First, there is nothing in the view that denies the existence of elementary particles [...]. Secondly, the view allows that there is a plethora of properties which elementary particles are unproblematically represented as having; these include all of their eternal or non-dynamical properties. Both they and the particles themselves are real objects of theoretical investigation.

QBism reserves the right to quarrel with the letter of this creed, but it is in accord with the spirit of it.[b] The technical work outlined in §5 above is a fundamentally realist investigation of exactly the type that Demopoulos indicates.

8. Conclusions

In this paper, we have availed ourselves of Pitowsky's program and its extensions, particularly the work of William Demopoulos, to better explicate QBism's aims and goals.

The Pitowskyan interpretation of quantum mechanics is, chiefly, the idea that the formal structure of quantum theory should be seen as a Bayesian probability theory, adapted to the empirical situation that Nature's events conspire to conform to a non-Boolean algebra. QBism also gives a Bayesian reading of quantum probabilities, but it instead regards them as the personal degrees of belief a savvy agent might hold for personal experiences arising from her actions on the external world. Thus QBism has two levels of personalism, whereas the Pitowskyan view has only one.

As we have seen, the differences go further and get into the technical meat of quantum information. The Born Rule is crucially important for the technical developments of both views, but in the Pitowskyan tradition it is a theorem, derived per Gleason from the underlying assumption of a non-Boolean algebra. In contrast, QBism strives to place the Born Rule in a pre-algebraic setting, so that it itself may shine forth as the primary empirical statement of the theory. The QBist approach to reconstructing quantum theory hopes that, in the right language, the Born Rule is quantum theory's most fundamental postulate, with the Hilbert space formalism arising as a consequence. There remains a certain *à la carte* aspect to the reconstruction of quantum theory from the Born Rule; the aspiring reconstructor has at some junctures her choice of secondary postulates to invoke, though they mostly have the character of asking that the resulting structures be as mathematically unremarkable as possible so that all the *physics* can be loaded into the Born Rule. We have stepped through this in the special case of a single qubit and the POVMs thereupon.

The distinctions between QBism and the Bub–Demopoulos–Pitowsky school manfiest in shifts of technical emphasis. We have seen, for example,

[b]For example, a working field theorist might devalue the word *particle,* or at least drop many connotations of it. But her motivation to do so would be based on the Unruh effect or on spacetime curvature invalidating a choice of positive-frequency subspace [59]. These reasons would have little to do with the "interpretation of quantum mechanics" as that topic has socially been defined. For more on QBism and QFT, see "FAQBism" [6].

how the QBist reconstruction program plays down Gleason's theorem and the quantum-logic tradition. On a deeper and more conceptual level, the two diverge in their theories of truth. There is, at least implicitly, a sentiment for a correspondence theory of truth in how Bub, Pitowsky and Demopoulos approach the quantum. By contrast, QBism invokes a Jamesian, pragmatist attitude. This is a move that is unfamiliar to many in the present-day philosophy-of-physics circles, as evidenced by Timpson's needing to define it for the professionals who read *Stud. Hist. Phil. Mod. Phys.* [60] with a footnote:

> Pragmatism is the position traditionally associated with the nineteenth and early twentieth century American philosophers P[ei]rce, James and Dewey; its defining characteristic being the rejection of *correspondence* notions of truth in which truths are supposed to mirror an independently existing reality after which we happen to seek, in favour of the thought that truth may not be separated from the process of enquiry itself. The caricature slogan for the pragmatist's replacement notion (definition) of truth is that 'Truth is what works!' in the business of the sincere and open investigation of nature.

(The authors admit a fondness for James, not just on account of his ideas but also because he writes to be understood — indeed, to make the life of the mind a thing that is *felt*.)

This turn to pragmatism has concrete expression in the QBist rejection of the EPR criterion and sidelining of the ontological models framework. To a QBist, trying to recast all quantum probabilities as ordinary, neoclassical ignorance is to miss the point entirely. One way or another, such efforts end as baroque restatements of their starting point. Much more intriguing are nontrivial reproductions of *portions* of quantum theory within such a framework, since these illustrate how some "quantum" behaviors are only *weakly nonclassical.* Among these are the no-cloning theorem and the fact that there is generally no good Boolean "and" operation for experiments — for spin-$\hat{x}$ and spin-$\hat{z}$ measurements upon a bit system, to pick an easy example. In turn, this demonstration indicates that quantum logic put its emphasis in the wrong place. This is no slight to its pioneers, only a retrospective judgment based on discoveries after their time. (As scientists, we belong to the last profession of romantics [61], the last to believe in geniuses — and John von Neumann was certainly one of those.)

Like we are with many subjects, we are quite willing to raid the quantum-logic bookshelf to find inspiration for technical conjectures [62], but non-Booleanity is not where we are seeking the fundamental lesson of quantum mechanics. For us it is a consequence, not a presumption. More broadly, writing the Born Rule as Eq. (11) is part of the attempt to understand the mathematical foundations of quantum mechanics. In particular, it helps to make the distinction between its physical content and mathematical artifact clear.

Acknowledgments

This research was supported in part by the John Templeton Foundation. The opinions expressed in this publication are those of the authors and do not necessarily reflect the views of the John Templeton Foundation. CAF was further supported in part by the John E. Fetzer Memorial Trust, and grants FQXi-RFP-1612 and FQXi-RFP-1811B of the Foundational Questions Institute and Fetzer Franklin Fund, a donor advised fund of Silicon Valley Community Foundation.

Appendix A. Even orthogonal quantum states are overlapping probability distributions

We begin by generalizing Eq. (6) to an arbitrary four-outcome reference measurement $\{H_i\}$ for a qubit. This is the minimum number of outcomes necessary for a qubit reference measurement; any fewer, and the POVM elements could not span the state space [63]. Given any qubit state ρ, at most one element of the set $\{H_i\}$ can be orthogonal to it, and so at most one entry in the vector $P(H)$ can equal zero. Consequently, any two qubit states ρ and ρ' will have probabilistic representations with overlapping support.

To generalize further and encompass systems of arbitrary dimension, as well as reference measurements that are not necessarily minimal, first we show that we can focus our attention on pure states. We can write any density matrix as a convex combination of rank-1 projectors, say in the matrix's eigenbasis, so if we have two quantum states

$$\rho := \sum_i c_i |\psi_i\rangle\langle\psi_i|,\ \rho' := \sum_j c_j' |\psi_j'\rangle\langle\psi_j'|, \tag{A.1}$$

then their Hilbert–Schmidt inner product is just a weighted average:

$$\mathrm{tr}\rho\rho' = \sum_{ij} c_i c_j' \left|\langle\psi_i|\psi_j'\rangle\right|^2. \tag{A.2}$$

Let us say that the probabilistic representations of these basis vectors are

$$s_i(k) := \mathrm{tr} H_k|\psi_i\rangle\langle\psi_i|,\ s'_j(k) := \mathrm{tr} H_k|\psi'_j\rangle\langle\psi'_j|. \tag{A.3}$$

Then, the inner product of ρ and ρ' is a weighted average of the "B-inner products" of these probability vectors:

$$\mathrm{tr}\rho\rho' = \sum_{ij} c_i c'_j\, s_i^{\mathrm{T}} B s'_j, \tag{A.4}$$

where the matrix B is the inverse of the Gramian of the reference measurement:

$$[B^{-1}]_{mn} = \mathrm{tr} H_m H_n. \tag{A.5}$$

The B-inner product of two valid probability vectors is bounded below by zero. What does this imply for the ordinary Euclidean inner product, i.e., the dot product? In particular, can the dot product ever be zero itself? To understand this, it suffices to consider two pure states, since per the above discussion, mixing cannot lower their inner products. Suppose that $|\psi\rangle$ and $|\psi'\rangle$ are two different pure states in dimension d. They define a subspace of the full space of Hermitian operators on $\mathbb{C}^d$. Let H_k^{P} denote the images of the reference-measurement effects projected into this subspace. Any quantum state living within this subspace will have the same inner products with the H_k^{P} as it did with the original H_k. Thus,

$$s(k) := \langle\psi|H_k|\psi\rangle = \langle\psi|H_k^{\mathrm{P}}|\psi\rangle, \tag{A.6}$$

and likewise for s', the probabilistic representation of $|\psi'\rangle$.

Imagine that $s \cdot s'$ were zero. Because each entry in either vector is nonnegative, this can only happen when s and s' have completely disjoint supports. In other words, some of the H_k^{P} will be orthogonal to $|\psi\rangle\langle\psi|$, and some will be orthogonal to $|\psi'\rangle\langle\psi'|$. And for each value of k, at least one of these options will obtain.

In the qubit-sized subspace defined by $|\psi\rangle$ and $|\psi'\rangle$, there is a unique state orthogonal to $|\psi\rangle$ and a unique state orthogonal to $|\psi'\rangle$. Call these $|\psi_\perp\rangle$ and $|\psi'_\perp\rangle$; they are antipodal to $|\psi\rangle$ and $|\psi'\rangle$ respectively on the Bloch sphere for this subspace. Each of the H_k^{P} must be proportional either to $|\psi_\perp\rangle$ or to $|\psi'_\perp\rangle$. But this means that the H_k^{P} cannot span the subset of state space given by $|\psi\rangle$ and $|\psi'\rangle$ — the reference measurement cannot be "informationally complete" on this subspace, for any such measurement must have at least *four* distinct outcomes to cover a qubit. Consequently, the proposal that $s \cdot s' = 0$ contradicts the assumption that we had a working reference measurement in the first place. And so, for any reference measurement H_k whatsoever, the dot product of valid probability vectors will always be greater than zero, even when the corresponding quantum states are orthogonal.

Appendix B. Additional excerpts of correspondence

Itamar Pitowsky to CAF, 5 May 2001:

> Of course, you can print [our correspondence]. The whole thing is a wonderful idea. Historians of science often complain that published articles never tell even half the story of science, because they don't let you see the false starts, misleading intuitions, errors, or even how a sound idea is baked. Correspondence makes the story come alive, but it usually takes almost a century to publish, and it's done only in cases of the likes of Einstein. ... Your collection arrives on the scene almost in real time, very nice.

CAF to Itamar Pitowsky, 26 June 2002:

> I too used to think that the [partial Boolean algebra] approach was the way to go if one wanted to build up a theory along quantum logical lines. But now, I'm not so convinced of it. That is because I am starting to think that quantum mechanics is more analogous to the epistemological theory Richard Jeffrey calls "radical probabilism" than anything else. From that view, there are "probabilities all the way down" with one never getting hold of the truth values of *any* propositions. Rüdiger Schack and I just discovered a wealth of material on Jeffrey's webpage `http://www.princeton.edu/~bayesway/`.
>
> In any case, I think what this leads to is that we ought to be focusing much more on characterizing quantum mechanics solely in terms of the "logic" of POVMs than anything else — these being the structures analogous to what crops up in Jeffrey's "probability kinematics." Thus, if one is looking to characterize PBAs, the best task might be to focus on the kinds of PBAs that POVMs generate, rather than the ones of Kochen and Specker based solely on standard measurements. [...] Beyond that, I am now of the mind that all one really ever needs for understanding quantum mechanics is a *single* Boolean algebra that is kept safely in the background (solely) for reference. The rest of the theory (and indeed all real-world measurement) is about probability kinematics with respect to that algebra.

CAF to Bill Demopoulos, 21 April 2004:

As you argued in your earlier letter to me (one from last year sometime), our views — or maybe just our languages — may not be so incompatible as one might think. However, I am left with the feeling that this is only a contingent feature of the particular stages of the game we happen to be at, at the moment. In particular, from my own view, I think it is quite important that we strive to stop thinking of quantum states as states of knowledge about the *truth value* of this or that proposition (even if truth value is not invariant with respect to 'experimental arrangement' — the idea you are toying with). My feeling is that the imagery of measurement outcomes mapping to truth values (in this context anyway) will only cloud our vision for how to take the next big step.

What is the next big step? I think it is a deeper understanding of how — very literally — the world "is in the making" (to use a Jamesian phrase). To try to make that idea at least graspable (if not either clear or consistent yet), and to try to show you quantum theory's role in all this, let me attach four letters I've written recently. They're contained within the attached files.[c] I think they are my best statements to date of what I am shooting for; and I think that goal fundamentally conflicts with the idea of "measurement" propositions having truth values in the conventional sense.

That is not to say, however, that I am yet ready to give up on the idea of physical systems having autonomous properties. The question is, what can still be pinned down as a property in the conventional sense?

CAF to Bill Demopoulos, 17 February 2006:

Let me try to consider a situation and 1) try to imagine what you would say of it (but probably in my idiosyncratic language), followed by 2) what I think I would say of it ... and then see if there is a substantial distinction.

[c]In the large samizdat [10], see the 24 June 2002 note "The World is Under Construction" to Wiseman; the 27 June 2002 note "Probabilism All the Way Up" to Wiseman; the 12 August 2003 note "Me, Me, Me" to Mermin and Schack; and the 18 August 2003 note "The Big IF" to Sudbery and Barnum.

Start with a finite dimensional Hilbert space, say of dimension 3, and imagine it indicative of some real physical system within an observer's concern. From that Hilbert space, let us form all possible sets of three mutually orthogonal one-dimensional projection operators. That is, let us consider all possible sets of the form $\{P_1, P_2, P_3\}$.

What is it that you would say of those sets? If I understand you correctly, it is this. Each such set $\{P_1, P_2, P_3\}$ corresponds to a set of mutually exclusive properties that the system can possess. At any given time, one of those projectors will have a truth value 1 and the other two will have values 0. Now consider a potentially different such set $\{Q_1, Q_2, Q_3\}$; again, at any given time, one of those projectors will have a truth value 1 and the other two will values 0. What is interesting in your conception, if I understand it, is that even if two elements happen to be identified between those two sets — for instance, if $P_1 = Q_3$ — there is *no requirement* that P_1 and Q_3 need have the same truth value; P_1 might have the truth value 0, whereas Q_3 might have the truth value 1. Another way to say this is that the truth-value assignments depend upon the whole set and not simply the individual projection operators. For you, all the identification $P_1 = Q_3$ amounts to is that the *probability* for the truth value of P_1 within the set $\{P_1, P_2, P_3\}$ is the *same* as the *probability* for the truth value of Q_3 within the set $\{Q_1, Q_2, Q_3\}$. (If you were a Bayesian about probabilities — though I don't think you are — you would say, "Well P_1 has whatever truth value it does, and Q_3 has whatever truth value it does (each within their appropriate set of mutually exclusive triples), but my degree of belief about the truth value of P_1 is the same as my degree of belief about the truth value of Q_3. That is the rule I am going to live by.") Then it follows from Gleason's theorem that there exist no probability assignments for the complete (i.e., continuously infinite) set of triples that are not of the quantum mechanical form. In particular, one can never sharpen one's knowledge to a delta function assignment for *each* triple. This is how you cash out the idea of an 'incompletely knowable domain.'

That is a novel idea, and if I understand it correctly, I like it.

However, now let me contrast my characterization of you with what I think has been my working conception. I prefer not to think of the triples $\{P_1, P_2, P_3\}$ as sets of mutually exclusive *properties* inherent within the system all by itself, but rather *actions* that can be taken upon the system by an external *agent*. Each *set* of such projectors corresponds to a distinct action; what the individual elements within each set represent are the (generally unpredictable) *consequences* of that action. What are the consequences in operational terms? Distinct sensations within the agent. The reason I insist on calling them consequences, rather than "sensations" full stop, is because I want to make it clear that the domain of what we are talking about is sensations that come about through the action of an agent *upon* the external world.

The essential idea of [what was to become QBism] is that no element of a set $\{P_1, P_2, P_3\}$ has a truth value before the action of the agent. Rather the truth value — if you want to call it that (maybe it is not the best terminology) — is generated (or given birth to) in the process. At that point, one of the P_i stands in autonomous existence (within the agent), whereas the other two fall.

I hope I have characterized both of us accurately!

Here is the question that has been troubling me. Is there any real distinction (one that makes an pragmatic difference) between our views? You say the truth value is there and revealed by the measurement, and I say it's made by the measurement and wasn't there beforehand. So what?

If there is a pragmatic distinction, Steven van Enk and I through discussions this week have come to believe that it may show up most clearly in how you and I would treat counterfactuals with regard to measurement. Let us take a situation where an agent ascribes a quantum state ρ to the system; contemplating the measurement $\{P_1, P_2, P_3\}$, we know that he will ascribe probabilities according to the Born rule $\mathrm{tr}(\rho P_i)$ for the various outcomes. Suppose he now performs that measurement and actually gets value P_2.

What does getting that outcome teach him about the quantum system? I think you would say it reveals which of the three mutually exclusive properties the system actually had. On the other hand, I would say it teaches him nothing about the system per se; the outcome P_2 is just the consequence of his action. What is the implication of this on counterfactuals? Here's at least one.

Suppose after you get your outcome, you contemplate magically having performed a distinct measurement $\{Q_1, Q_2, Q_3\}$ instead. I think you're careful to point out in your paper that the knowledge of P_2 carries no implication for what you would have found with this other imaginary measurement. But what happens if you conceptually transform this measurement $\{Q_1, Q_2, Q_3\}$ to one closer and closer to the original, i.e., to $\{P_1, P_2, P_3\}$? In the *limit* when the two are identical again, I think you would say that knowledge of the outcome P_2 in the original case implies that P_2 will also be the outcome in the limiting counterfactual case. But what would I say? From my conception, there is no reason at all to believe that the limiting counterfactual case will give rise to the same outcome P_2. The best one can do, either in the original case or the counterfactual case, is to say that an outcome i will arise with probability $\mathrm{tr}(\rho P_i)$. In fact, a counterfactual analysis with this kind of result may be the very meaning of the idea that quantum measurements are generative of their outcomes.

CAF to Jeffrey Bub, 24 June 2006:

Bubism 1. *The transition from classical to quantum mechanics involves replacing the representation of properties as a Boolean lattice, i.e., as the subsets of a set, with the representation of properties as a certain sort of non-Boolean lattice.*

The? I would rather say *one possible way of looking at* the transition from classical to quantum mechanics involves blah, blah, blah. And, you partially recover from this a few paragraphs later where you write:

Bubism 2. *Of course, other ways of associating propositions with features of a Hilbert space are possible, and other ways of assigning truth values, including multi-valued truth value assignments and contextual truth value assignments. Ultimately, the issue here concerns what we take as the salient structural change involved in the transition from classical to quantum mechanics, and this depends on identifying quantum propositions that take the same probabilities for all quantum states.*

But let me hang on this point for a moment despite your partial recovery. For when you say things like, "Fuchs misses the essential point," you should realize that that judgement (at most) comes from within a context very different from the one I am working in.

I would, for instance, never say "the representation of properties in quantum mechanics involves as a certain sort of non-Boolean lattice." That is just not the context I'm working in. Similarly, I would not say, as you say in the next section, "Somehow, a measurement process enables an indeterminate property, that is neither instantiated nor not instantiated by a system in a given quantum state, to either instantiate itself or not with a certain probability." — i.e., I would not say that a measurement process instantiates any *property* at all for a quantum system.

Instead, the setting for our quantum Bayesian program (i.e., the particular one of Caves, Schack, and me), is one where all the *properties* intrinsic to a quantum system are timeless and have no dynamical character whatsoever — moreover, those properties have nothing to do with particular quantum state assignments or particular quantum measurement outcomes. In that way, the idea of a non-Boolean lattice simply doesn't apply to them.

John Sipe recently made a nice write-up of our view for his book that, I think, brings this one difference between you and me into pretty stark relief. Maybe it's worthwhile to quote it at length, as it may lay the groundwork for a good bit of our later discussion:

> This interpretation shares some features with operationalism. *Measurements*, for example, are understood in a manner close to that adopted by an operationalist. They are characterized by POVMs, and those abstract elements are associated with tasks in the laboratory undertaken with gadgets that are part of the primitives of the theory. The result of any such measurement is simply one of a possible number of outcomes, and there is no talk of these measurements "revealing" the value of any variable, in the sense that an arbitrarily precise position measurement in classical mechanics is often described as revealing the position of a particle. Yet, compared to the operationalist's quiet, unassuming terminology of "tasks" and "outcomes," advocates of this interpretation adopt a more active manner of speaking, referring to "actions" (or even "interventions")

undertaken by an agent, and the "consequences" that those actions elicit.

This indicates a role for the observer (or agent) in this interpretation that is more significant than the role played by such a person in operational quantum mechanics. The significance of that role becomes clear when we consider the reference of density operators in this interpretation. Density operators do not refer to sets of tasks that define *preparations*, as they do in operational quantum mechanics. Rather, a density operator is taken to encode the beliefs of an agent concerning the probabilities of different consequences of possible future actions. While these beliefs may be *informed* by knowledge of the tasks involved in setting up the particular gadgetry associated with a preparation, they are not *determined* by it. Hence there is not a unique, "correct" density operator necessarily associated with each preparation procedure, as there is in operational quantum mechanics. In the present view two different researchers, one more skilled in quantum mechanics than the other, could adopt different density operators after being identically informed of the details of a particular preparation procedure. One density operator might be more successful than the other in predicting the possible consequences of future actions, but each would be the correct density operator *for that agent* insofar as it correctly encoded that agent's beliefs.

Thus, while the abstract elements in the theory associated with measurements are identified with tasks in the laboratory, as in operationalism, the abstract elements in the theory associated with preparations are identified with beliefs of the agent, signaling a kind of empiricist perspective.

So in contrast to operational quantum mechanics, where density operators are necessarily updated following a measurement — since the combination of the previous preparation and the measurement constitutes a new preparation, and an operationalist associates the new density operator with *that* — in this view there is no necessary updating of a density operator in the light of measurement outcomes, since there is no *necessary* connection between the consequences of an agent's action (more prosaically, "measurement outcomes") and his or her beliefs. After all, foolish researchers, like foolish men and women more generally, could choose not to modify their beliefs concerning the consequences of future actions despite their knowledge of the consequences of recent ones. And note that even wise researchers will not update their beliefs concerning future actions until they *know* the consequences of recent ones; hence a wise researcher's "personal density operator" (the only kind of density operator there is in this view!) will not change until that researcher is actually aware of a measurement outcome.

Other abstract elements in the theory, such as the dimension of the Hilbert space, and the dimensions of various factor spaces, are actually associated with instances of attributes of physical objects. Hence with respect to the reference of *these* abstract elements this interpretation is realist. The manner in which this works can best be seen by first reviewing the role measurement outcomes play in revealing aspects of the universe in realist classical mechanics, and then comparing that with the role such outcomes play in this interpretation of quantum mechanics.

An arbitrarily precise position measurement of a bead moving along a wire, in realist classical mechanics, reveals the position of the particle, the instance (say, $x = 10\,\mathrm{cm}$) of a particular attribute (bead position) of a physical object (bead) that actually exists in Nature. In contrast, a usual Stern–Gerlach device oriented along the z direction *does not*, in this interpretation of quantum mechanics, reveal the z-component of angular momentum, or for that matter anything else. The particular outcome of one experimental run is simply a consequence of performing the experiment. Nonetheless, repeated experimentation *does* reveal that the electron associated with the atom passing through the device should be taken as a spin-1/2 particle. Here the attribute under consideration is taken to be *internal angular momentum*, and the instance — the irreducible representation appropriate to the particle of interest — *spin-1/2*. The role of an "instance of an attribute" in this interpretation is not to specify one of a number of possible expressions of existence, as it is in realist classical mechanics, but rather to specify one class of possible beliefs — the one that the theory recommends — about the consequences of future interventions of a particular type.

Note that, at least within nonrelativistic physics, the instances of the attributes in this interpretation are fixed. A spin-1/2 particle remains a spin-1/2 particle. Thus there are no dynamical variables in this theory, only nondynamical variables analogous to the mass of a particle in nonrelativistic classical mechanics. The point of physics is to identify these nondynamical variables. Repeated interventions by experimentalists, and the careful noting of the range of consequences that those interventions elicit, is how these fixed instances are discovered.

In this interpretation of quantum mechanics, with its mix of operationalist, empiricist, and realist identification of abstract elements in the theory, these fixed instances specify the [[agent independent features]] of the "quantum world," and it is the business of physics to figure them out. This is done by experimentation, and the theoretical linking of basis vectors in the appropriate Hilbert space with various measurements, providing an "anchor" for those basis kets to our experience, the consequences of our actions. Particularly significant is the Hamiltonian operator and its basis kets [[in Caves' particular version of all this]]. As time evolves during what is colloquially described as "unitary evolution," we have the option to modify our beliefs or to modify the anchors of those beliefs; the first strategy corresponds to the usual Schrödinger picture, the second to the Heisenberg picture.

Regardless of the strategy, the [[properties intrinsic to the]] quantum world of this interpretation [[are]] a fixed, static thing. [[This aspect of the quantum world]] is a frozen, changeless place. Dynamics refers not to the quantum world, but only to our actions, our experiences, and our beliefs as agents. Or, more poetically (à la Chris), life does not arise from our interventions; it is our interventions.

John doesn't represent us correctly in every detail of this presentation — for the purpose at hand, it only seemed essential to modify him in a few instances, which I have have marked with double brackets [[·]] — but I would say he is roughly on track, and he certainly gets it that we are not concerned with the usual way of ascribing properties to quantum systems via the values of measurement outcomes or probability-1 predictions (i.e., the eigenvector-eigenvalue link).

Which brings me back again to your paper:

Bubism 3. *For a quantum state, the properties represented by Hilbert space subspaces are not partitioned into two such mutually exclusive and collectively exhaustive sets: some propositions are assigned no truth value. Only propositions represented by subspaces that contain the state are assigned the value 'true,' and only propositions represented by subspaces orthogonal to the state are assigned the value 'false.' This means that propositions represented by subspaces that are at some non-zero or non-orthogonal angle to the ray representing the quantum state are not assigned any truth value in the state, and the corresponding properties must be regarded as indeterminate or indefinite: according to the theory, there can be no fact of the matter about whether these properties are instantiated or not.*

You see, my way of looking at things wouldn't even allow me to say what you say here. It is just a very different world that I am working in.

To try to make this point, let me quote a couple of emails I wrote to Bas van Fraassen a few months ago. It started with my saying this:

> The way I view quantum measurement now is this. When one performs a "measurement" on a system, all one is really doing is taking an *action* on that system. From this view, time evolutions or unitary operations etc., are not actions that one can take on a system; only "measurements" are. Thus the word measurement is really a misnomer — it is only an action. In contradistinction to the old idea that a measurement is a query of nature, or a way of gathering

information or knowledge about nature, from this view it is just an action on something external — it is a kick of sorts. The "measurement device" should be thought of as being like a prosthetic hand for the agent — it is merely an extension of him; in this context, it should not be thought of as an independent entity beyond the agent. What quantum theory tells us is that the formal structure of all our possible actions (perhaps via the help of these prosthetic hands) is captured by the idea of a Positive-Operator-Valued Measure (or POVM, or so-called "generalized measurement"). We take our actions upon a system, and in return, the system gives rise to a reaction — in older terms, that is the "measurement outcome" — but the reaction is in the agent himself. The role of the quantum system is thus more like that of the philosopher's stone; it is the catalyst that brings about a transformation (or transmutation) of the agent.

Reciprocally, there [[may]] be a transmutation of the system external to the agent. But the great trouble in quantum interpretation — I now think — is that we have been too inclined to jump the gun all these years: We have been misidentifying where the transmutation indicated by quantum mechanics (i.e., the one which quantum theory actually talks about, the "measurement outcome") takes place. It [[may]] be the case that there are also transmutations in the external world (transmutations in the system) in each quantum "measurement", BUT that is not what quantum theory is about. [[Quantum mechanics]] is only a hint of that more interesting transmutation. [[Instead, the main part of quantum mechanics is about how]] the agent and the system [[together bring about]] a little act of creation that ultimately has an autonomy of its own.

which led to the following dialogue:

Fraassenation 1. *Writers on the subject have emphasized that the main form of measurement in quantum mechanics has as result the value of the observable at the end of the measurement — and that this observable may not even have had a definite value, let alone the same one, before.*

Your phrase "*may not* even have a definite value" floated to my attention. I guess this floated to my attention because I had recently read the following in one of the Brukner/Zeilinger papers,

> Only in the exceptional case of the qubit in an eigenstate of the measurement apparatus the bit value observed reveals a property already carried by the qubit. Yet in general the value obtained by the measurement has an element of irreducible randomness and therefore cannot be assumed to reveal the bit value or even a hidden property of the system existing before the measurement is performed.

I wondered if your "may not" referred to effectively the same thing as their disclaimer at the beginning of this quote. Maybe it doesn't. Anyway, the Brukner/Zeilinger disclaimer is a point that Caves, Schack, and I now definitely reject: From our view all measurements are generative of a *non*-preexisting property regardless of the quantum state. I.e., measurements never reveal "a property already carried by the qubit." For this, of course, we have to adopt a Richard Jeffrey-like analysis of the notion of "certainty" — i.e., that it too, like any probability assignment, is a state of mind — or one along (my reading of) Wittgenstein's — i.e., that "certainty is a tone of voice" — to make it all make sense, but so be it.

and

> **Fraassenation 2.** *Suppose that an observer assigns eigenstate* $|a\rangle$ *of* A *to a system on the basis of a measurement, then predicts with certainty that an immediate further measurement of* A *will yield value* a*, and then makes that second measurement and finds* a*. Don't you even want to say that the second measurement just showed to this observer, as was expected, the value that* A *already had? He does not need to change his subjective probabilities at all in response to the 2nd measurement outcome, does he?*

It is not going to be easy, because this in fact is what Schack and I are actually writing a whole paper about at the moment — this point has been the most controversial thing (with the Mermin, Unruh, Wootters, Spekkens, etc., crowd) that we've said in a while, and it seems that it's going to require a whole paper to do the point justice. But I'll still try to give you the skinny of it:

- Q: He does not need to change his subjective probabilities at all in response to the 2nd measurement outcome, does he?
- A: No he doesn't.
- Q: Don't you even want to say that the second measurement just showed to this observer, as was expected, the value that A already had?
- A: No I don't.

The problem is one of the very consistency of the subjective point of view of quantum states. The task we set before ourselves is to completely sever any supposed connections between quantum states and the actual, existent physical properties of the quantum system. It is only from this — if it can be done, and of course we try to argue it can be done — that we get any "interpretive traction" (as Chris Timpson likes to say) for the various problems that plague QM. [[...]]

This may boil down to a difference between the Rovellian and the Bayesian/Paulian approach; I'm not clear on that yet. [[...]] Rovelli relativizes the states to the observer, even the pure states, and with that — through the eigenstate-eigenvalue link — the values of the observables. I'm not completely sure what that means in Rovelli-world yet, however.

I, on the other hand, do know that I would say that a measurement intervention is always generative of a new fact in the world, whatever the measurer's quantum state for the system. If the measurer's state for the system *happens* to be an eigenstate of the Hermitian operator describing the measurement intervention, then the measurer will

> be confident, *certain* even, of the consequence of the measurement intervention he is about to perform. But that *certainty* is in the sense of Jeffrey and Wittgenstein above — it is a "tone of voice" of utter confidence. The world could still, as a point of principle, smite the measurer down by giving him a consequence that he predicted to be impossible. In a traditional development — with ties to a correspondence theory of truth — we would then say, "Well, that proves the measurer was wrong with his quantum state assignment. He was wrong before he ever went through the motions of the measurement." But as you've gathered, I'm not about traditional developments. Instead I would say, "Even from my view there is a sense in which the measurer's quantum state is *wrong*. But it is *made wrong* by the *actual* consequence of the intervention — it is made wrong on the fly; its wrongness was not determined beforehand." And that seems to be the main point of contention.

Particularly this is going to be a key point when I finally come to the analysis in Section 7 of your paper.

CAF to Veiko Palge and Bill Demopoulos, 14 November 2006:

The best answer I can give you, I think, is that neither of these kinds of structures [sigma-algebras or noncommutative generalizations] map onto what I'm thinking. The main reason for this is that I think it is incorrect to think of the process of "quantum measurement" either 1) as the *revelation* of a property inherent in the system under observation, or 2) as the *production* of such a property in the system. And without that, I don't think there is enough glue to bind the events occurring in quantum measurements together into an algebraic structure (say, a lattice or a Boolean algebra, or even a partial Boolean algebra, where there are Boolean algebras tied together at the edges) — at least not in any useful sense that intrigues me as a physicist. [...] I think there is a fruitful similarity between what [Bill] and I are seeking, even if we ultimately diverge. It is this: In both our views — and they are the only places I've ever seen this style of idea — even when two measurements share a common element (say a given projector P_i), there is no implication of a common truth value being imposed on P_i across the measurements. The reason for this for Bill is that the system's properties are bound up with the whole orthogonal family of projectors the individual P_i happens to be embedded within. In contrast, the reason for this for me is that I don't think of quantum measurement outcomes as signifying properties intrinsic to the system — they are simply consequences of actions for me. What makes the element P_i identified across measurements — for both of us — is not truth value, but that the *probabilities* for P_i are identical in both cases. What this means particularly from my Bayesian way of thinking is that a judgment is being made: I, the agent, am identifying *this* potential outcome of this measurement with *that* potential outcome of that measurement because I judge their probabilities equal under all imaginable circumstances. (See Section 4.1 of my `quant-ph/0205039`.) It is not that they are identified in Nature itself. Thus, for me, I think, there is no good sense in which they lie in the same event space at all.

CAF to Veiko Palge, 10 January 2007:

Let me start with your first sentence: "It seems that one of the main reasons you reject a well-structured event space is that it assumes a realistic interpretation: its elements would correspond to intrinsic properties of quantum systems." That is the wrong direction of reasoning — though I am probably the cause of this misimpression through the restrictive choices of readings I recommended to you. It is not that I reject a well-structured event space because it *assumes* its elements would correspond to intrinsic properties of quantum systems, but rather this is the *result* of a thoroughgoing subjective interpretation of probabilities within the quantum context. What cannot be forgotten is that quantum-measurement outcomes, by the usual rules, *determine* posterior quantum states. And those posterior quantum states in turn determine further probabilities.

Thus, if one takes the timid, partial move that Itamar and Jeff Bub, say, advocate — i.e., simply substituting one or another non-Boolean algebra for the space of events, and leaving the rest of Bayesian probability theory seemingly intact — then one ultimately ends up re-objectifying what had been initially supposed to be subjective probabilities. That is: When I look at the click, and note that it is value i, and value i is rigidly — or I should say, *factually*

— associated with the projector Π_i in some nonBoolean algebra, then I have no choice (through Lüders rule) but to assign the posterior quantum state Π_i to the system. This means the new quantum state Π_i will be as factual as the click. And any new probabilities (for the outcomes of further measurements) determined from this new quantum state Π_i will also be factual.

So, the starting point of the reasoning is to *assume* that there is a category distinction between probabilities and facts (this is the subjectivist move of de Finetti and Ramsey). Adding the ingredient of the usual rules of quantum mechanics, one derives a dilemma: If there is a rigid, factual connection between the clicks i and elements Π_i of an algebraic structure, then probabilities are factual after all. Holding tight to my assumption of a category distinction between facts and probabilities, I end up rejecting the idea that there is a unique, factual mapping between i and Π_i. [...]

[Now let] me return to your paragraph before signing off. If I read this question of yours in isolation: "Can't one just take the elements as corresponding to clicks and blips in the measurement apparatus?" Then at one level my answer is, "Of course; I've never said otherwise." What is at issue here is whether the events — the clicks and blips themselves — fall within the kind of algebraic structure you speak of, or whether it is something else (something a conceptual layer above the events) that falls within it. From the [proto-QBist] point of view, for a single device with clicks i, one agent might associate the clicks with a set of orthogonal projectors Π_i, and another agent might associate them with a set of noncommuting effects (i.e., POVM elements) E_i. This was the sense in which I meant there is no stand-alone event space at all: For us, the algebraic structure of the events (the clicks and blips), their level of commutivity or noncommutivity and whatnot, is just as subjective as the quantum state. The clicks i themselves are objective (in the sense of not being functions of the subject's beliefs or degrees of belief), but their association with a particular set of operators E_i is a subjective judgment.

I hope this completely answers your question now.

But let me extend the discussion a little to try to give you a more positive vision of what we're up to. The starting point is the category distinction between facts and probabilities applied to the quantum measurement context. From this we glean that quantum operations and quantum states are of the same level of subjectivity. But that is not our ending point. Because, implicit in everything we have said there are these autonomous, realistically-interpreted quantum systems: The agent has to interact with the quantum system to receive his quantum measurement outcome. No quantum system, no measurement outcome. Thus the [proto-QBist] position is more than a kind of positivism or operationalism. The objects of the external world with which we interact have a certain kind of active power, and we become aware of the presence of that active power particularly in the course of quantum measurement. When we kick on a quantum system it surprises us with a kick back.

Can we say anything more explicit about the active power? Can we give it some mathematical shape? Yes, I think we can, but that is a research project. Still, I think the hints for it are already in place ... and indeed they are in the quantum formalism, as we would expect them to be. One of the hints is this: The Born transformation rule. When we make probability assignments in quantum mechanics, we are assuming more than de Finettian / Ramseyian coherence. We assume that if we set the probabilities for the outcomes of this measurement this way, then we should set the probabilities for the outcomes of that measurement that way and the rule of transformation is a linear one. This, from the [proto-QBist] view, is the content of the Born rule (see the last section in the attached paper). And it is empirical, contingent: A different world than the one we live in might have had a different transformation rule. So, if we're looking for something beyond personal probabilities in quantum mechanics, that is a point to take seriously. It hints of some deep property of our world, and I'd like to know what that property is. [...]

Though [the proto-QBists] banish the algebraic structure of Hilbert space from having anything to do with a fundamental event space (and in this way their quantum Bayesianism differs from the cluster of ideas Pitowsky and Bub are playing with), they do not banish the algebraic structure from playing any role whatsoever in quantum mechanics. It is just that the algebraic structure rears its head at the conceptual level of coherence rather than in a fundamental event space. It is not that potential events are objectively tied to together in an algebraic way, but that our gambling commitments (normatively) *should be.* This is another point of contact between my view and Bill's.

CAF to Bill Demopoulos, 7 January 2011:

Take a POVM consisting of operators E_k. I.e., these are positive semi-definite operators that sum to the identity operator. Thus they might seem like mysterious abstract entities (perhaps properties of the system, or properties of the 'measuring device', or maybe something still weirder, though surely agent-independent). But, via a fiducial SIC POVM in the sky (as I talk about in my papers), one can map these operators to a set of conditional probabilities $p(i|k)$ in a one-to-one fashion. That is, these operators contain nothing over and above the prescription of a probability assignment. There is nothing to them mathematically but that — nothing is lost by thinking of them in these terms, as probability assignments. Similarly, the philosophical point should be this: That there is nothing conceptually more than this either.

So, what should one call the "outcome" of a quantum measurement? Well, for the agent it is an experience "k" ... but what meaning does that have for him? How could he articulate it? What does it mean to him? How will he change his behavior with regard to it? The answer is: It is $p(i|k)$. It is a probability distribution conditioned on k. The only operational handle the agent has on k is through the assignment he makes, $p(i|k)$.

But all probabilities (for the personalist Bayesian) are subjectively given (i.e., functions of the agent only, his history and experiences, not the object). And thus, I guess, my resistance to part of your way of thinking about effects. (Of course, as you should know by now, I am in serious agreement with you in other ways ... in fact, probably with you more than with any other living philosopher.)

References

1. M. Hemmo and O. Shenker, editors, *Quantum, Probability, Logic: The Work and Influence of Itamar Pitowsky* (Springer Nature, London, 2020).
2. I. Pitowsky, "Infinite and Finite Gleason's Theorems and the Logic of Indeterminacy," *Journal of Mathematical Physics* **39** (1997), 218–228.

3. C. A. Fuchs and R. Schack, "Quantum-Bayesian Coherence," arXiv: 0906.2187 (2009). A condensed version was later printed as *Reviews of Modern Physics* **85** (2013), 1693.
4. C. A. Fuchs, "QBism, the Perimeter of Quantum Bayesianism," arXiv: 1003.5209 (2010).
5. B. C. Stacey, "Ideas abandoned en route to QBism," arXiv:1911.07886 (2019).
6. J. B. DeBrota and B. C. Stacey, "FAQBism," arXiv:1810.13401 (2018).
7. N. D. Mermin, "Why QBism is not the Copenhagen interpretation and what John Bell might have thought of it," in *Quantum [Un]Speakables II: Half a Century of Bell's Theorem*, edited by R. Bertlmann and A. Zeilinger (Springer, Berlin, 2017), pp. 83–94; arXiv:1409.2454.
8. C. A. Fuchs and B. C. Stacey, "QBism: Quantum Theory as a Hero's Handbook." In *Proceedings of the International School of Physics "Enrico Fermi," Course 197 – Foundations of Quantum Physics*, edited by E. M. Rasel, W. P. Schleich and S. Wölk (Italian Physical Society, 2019), pp. 133–202. arXiv: 1612.07308.
9. C. A. Fuchs, "On Participatory Realism," in *Information and Interaction: Eddington, Wheeler, and the Limits of Knowledge*, edited by I. T. Durham and D. Rickles (Springer, Berlin, 2016), pp. 113–134. arXiv:1601.04360.
10. C. A. Fuchs, *My Struggles with the Block Universe* (2014). Edited by B. C. Stacey, with a foreword by M. Schlosshauer. arXiv:1405.2390.
11. A. Peres, *Quantum Theory: Concepts and Methods* (Kluwer, 1993).
12. M. A. Nielsen and I. Chuang, *Quantum Computation and Quantum Information,* tenth anniversary edition (Cambridge University Press, 2010).
13. J. Bub and I. Pitowsky, "Two dogmas about quantum mechanics," arXiv: 0712.4258 (2007).
14. C. A. Fuchs, "Quantum mechanics as quantum information (and only a little more)," arXiv:quant-ph/0205039 (2002).
15. D. Frauchiger and R. Renner, "Quantum theory cannot consistenty describe the use of itself," *Nature Communications* **9** (2018), 3711, arXiv:1604.07422.
16. Č. Brukner, "On the quantum measurement problem," arXiv:1507.05255. In *Quantum [Un]Speakables II* (Springer-Verlag, 2017).
17. M. Pusey, "An inconsistent friend," *Nature Physics* **14,** 10 (2018), 977–978.
18. J. Bub, "'Two Dogmas' Redux," arXiv:1907.06240 (2019).
19. B. C. Stacey, "On QBism and Assumption (Q)," arXiv:1907.03805 (2019).
20. A. M. Gleason, "Measures on the closed subspaces of a Hilbert space," *Indiana University Mathematics Journal* **6,** 4 (1957), 885–93.
21. H. Barnum, C. M. Caves, J. Finkelstein, C. A. Fuchs and R. Schack, "Quantum probability from decision theory?," *Proceedings of the Royal Society of London A* **456** (2000), 1175–82, arXiv:quant-ph/9907024.
22. A. Wilce, "Quantum Logic and Probability Theory." In *The Stanford Encyclopedia of Philosophy* (Spring 2017 Edition), edited by E. N. Zalta.
23. R. Black (writing as B. Switek), *Skeleton Keys: The Secret Life of Bone* (Riverhead Books, 2019).
24. I. Pitowsky, "Quantum mechanics as a theory of probability." In *Physical*

Theory and its Interpretation: Essays in Honor of Jeffrey Bub, edited by W. Demopoulos and I. Pitowsky (Springer, 2006). arXiv:quant-ph/0510095.

25. M. P. Solèr, "Characterization of hilbert spaces by orthomodular spaces," *Communications in Algebra* **23** (1995), 219–43.
26. A. Prestel, "On Solèr's characterization of Hilbert spaces," *Manuscripta Mathematica* **86** (1995), 225–38.
27. R. W. Spekkens, "Evidence for the epistemic view of quantum states: A toy theory," *Physical Review A* **75** (2007), 032110, arXiv:quant-ph/0401052.
28. R. W. Spekkens, "Reassessing claims of nonclassicality for quantum interference phenomena," PIRSA:16060102 (2016).
29. P. Busch, "Quantum states and generalized observables: A simple proof of Gleason's theorem," *Physical Review Letters* **91** (2003), 120403, arXiv:quant-ph/9909073.
30. C. M. Caves, C. A. Fuchs, K. K. Manne and J. M. Renes, "Gleason-type derivations of the quantum probability rule for generalized measurements," *Foundations of Physics* **34** (2004), 193–209, arXiv:quant-ph/0306179.
31. H. Granström, *Gleason's Theorem*. Master's thesis, Stockholm University, 2006.
32. V. J. Wright and S. Weigert, "A Gleason-type theorem for qubits based on mixtures of projective measurements," *Journal of Physics A* **52** (2019), 055301, arXiv:1808.08091.
33. V. J. Wright and S. Weigert, "Gleason-type theorems from Cauchy's functional equation," *Foundations of Physics* **49** (2019), 594–606.
34. M. Appleby, C. A. Fuchs, B. C. Stacey and H. Zhu, "Introducing the Qplex: A novel arena for quantum theory," *European Physical Journal D* **71** (2017), 197, arXiv:1612.03234.
35. M. Appleby, Å. Ericsson and C. A. Fuchs, "Properties of QBist state spaces," *Foundations of Physics* **41** (2011), 564–79, arXiv:0910.2750.
36. R. W. Spekkens, "Contextuality for preparations, transformations, and unsharp measurements," *Physical Review A* **71** (2005), 052108, arXiv:quant-ph/0406166.
37. C. A. Fuchs and B. C. Stacey, "Some negative remarks on operational approaches to quantum theory." In *Quantum Theory: Informational Foundations and Foils,* edited by G. Chiribella and R. W. Spekkens (Springer, 2016), arXiv:1401.7254.
38. B. C. Stacey, "Quantum theory as symmetry broken by vitality," arXiv:1907.02432 (2019).
39. H. Barnum, C. A. Fuchs, J. M. Renes and A. Wilce, "Influence-free states on compound quantum systems," arXiv:quant-ph/0507108 (2005).
40. H. Barnum, S. Beigi, S. Boixo, M. B. Elliott and S. Wehner, "Local quantum measurement and no-signaling imply quantum correlations," *Physical Review Letters* **104** (2010), 140401, arXiv:0910.3952.
41. G. de la Torre, L. Masanes, A. J. Short and M. P. Müller, "Deriving quantum theory from its local structure and reversibility," *Physical Review Letters* **109** (2012), 90403, arXiv:1110.5482.

42. G. Zauner, *Quantendesigns: Grundzüge einer nichtkommutativen Designtheorie*. PhD thesis, University of Vienna, 1999. `http://www.gerhardzauner.at/qdmye.html`.
43. J. M. Renes, R. Blume-Kohout, A. J. Scott and C. M. Caves, "Symmetric informationally complete quantum measurements," *Journal of Mathematical Physics* **45** (2004), 2171–2180, `arXiv:quant-ph/0310075`.
44. A. J. Scott and M. Grassl, "Symmetric informationally complete positive-operator-valued measures: A new computer study," *Journal of Mathematical Physics* **51** (2010), `arXiv:0910.5784`.
45. M. Appleby, S. T. Flammia, and C. A. Fuchs, "The Lie Algebraic Significance of Symmetric Informationally Complete Measurements," *Journal of Mathematical Physics* **52** (2011), 022202.
46. M. Appleby, C. A. Fuchs, and H. Zhu, "Group Theoretic, Lie Algebraic and Jordan Algebraic Formulations of the SIC Existence Problem," *Quantum Information and Computation* **15** (2015), 61–94, `arXiv:1312.0555v2`.
47. M. Appleby, S. Flammia, G. McConnell and J. Yard, "SICs and algebraic number theory," *Foundations of Physics* **47** (2017), 1042–59, `arXiv:1701.05200`.
48. C. A. Fuchs, M. C. Hoang and B. C. Stacey, "The SIC question: History and state of play," *Axioms* **6** (2017), 21, `arXiv:1703.07901`.
49. J. B. DeBrota and B. C. Stacey, "Lüders Channels and the Existence of Symmetric-Informationally-Complete Measurements," *Physical Review A* **100** (2019), 062327, `arXiv:1907.10999`.
50. J. B. DeBrota, C. A. Fuchs and B. C. Stacey, "Symmetric Informationally Complete Measurements Identify the Irreducible Difference between Classical and Quantum," forthcoming in *Physical Review Research*, `arXiv:1805.08721`.
51. W. Demopoulos, "Effects and propositions," *Foundations of Physics* **40** (2010), 368–89, `arXiv:0809.0659`.
52. M. F. Pusey, J. Barrett and T. Rudolph, "On the reality of the quantum state," *Nature Physics* **8** (2012), 475, `arXiv:1111.3328`.
53. N. Harrigan and R. W. Spekkens, "Einstein, incompleteness, and the epistemic view of quantum states," *Foundations of Physics* **40** (2010), 125, `arXiv:0706.2661`.
54. C. M. Caves, C. A. Fuchs and R. Schack, "Conditions for compatibility of quantum-state assignments," *Physical Review A* **66** (2002), 062111, `arXiv:quant-ph/0206110`.
55. R. Peierls, *More Surprises in Theoretical Physics* (Princeton University Press, 1991).
56. J. Bub, "Bananaworld: Quantum mechanics for primates," `arXiv:1211.3062` (2012).
57. W. Demopoulos, "Generalized probability measures and the framework of effects." In *Probability in Physics*, edited by Y. Ben-Menahem and M. Hemmo (Springer, 2012).
58. W. Pauli, *Writings on Physics and Philosophy*, edited by C. P. Enz and K. von Meyenn, (Springer-Verlag, 1994).

59. R. M. Wald, *Quantum Field Theory in Curved Spacetime and Black Hole Thermodynamics* (The University of Chicago Press, 1994).
60. C. Timpson, "Quantum Bayesianism: A study," *Studies in History and Philosophy of Modern Physics* **39** (2008), 579–609, `arXiv:0804.2047`.
61. J. Gleick, *Genius: The Life and Science of Richard Feynman* (Pantheon, 1992).
62. B. C. Stacey, "Sporadic SICs and exceptional Lie algebras," `arXiv:1911.05809` (2019).
63. J. B. DeBrota, C. A. Fuchs and B. C. Stacey, "The Varieties of Minimal Tomographically Complete Measurements," forthcoming in *International Journal of Quantum Information*, `arXiv:1812.08762` (2020).

"B" IS FOR BOHR

ULRICH MOHRHOFF

Sri Aurobindo International Centre of Education, Pondicherry 605002, India
E-mail: ujm@auromail.net

It is suggested that the "B" in QBism rightfully stands for Bohr. The paper begins by explaining why Bohr seems obscure to most physicists. Having identified the contextuality of physical quantities as Bohr's essential contribution to Kant's theory of science, it outlines the latter, its own contextuality (human experience), and its decontextualization. In order to preserve the decontextualization achieved by Kant's theory, Bohr seized on quantum phenomena as the principal referents of atomic physics, all the while keeping the universal context of human experience at the center of his philosophy. QBism, through its emphasis on the individual experiencing subject, brings home the intersubjective constitution of objectivity more forcefully than Bohr ever did. If measurements are irreversible and outcomes definite, it is because the experiences of each subject are irreversible and definite. Bohr's insights, on the other hand, are exceedingly useful in clarifying the QBist position, attenuating its excesses, and enhancing its internal consistency.

Keywords: Bohr; contextuality; decontextualization; experience; intersubjectivity; objectivity; Kant; QBism.

1. Introduction

I have this "madly optimistic" (Mermin called it) feeling that Bohrian–Paulian ideas will lead us to the next stage of physics. That is, that thinking about quantum foundations from their point of view will be the beginning of a new path, not the end of an old one.

— *Christopher A. Fuchs* [1]

The beginning of the 21st century saw the launch of a new interpretation of quantum mechanics, by Carlton Caves, Chris Fuchs, and Ruediger Schack [2]. Initially conceived as an extended personalist Bayesian theory of probability called "Quantum Bayesianism," it has since been re-branded as "QBism," the term David Mermin [3] prefers, considering it "as big a break with 20th century ways of thinking about science as Cubism was with 19th century ways of thinking about art." The big break lies not in the emphasis that the mathematical apparatus of quantum mechanics is a probability calculus — that ought to surprise no one — but in this *plus* a radically subjective Bayesian interpretation of probability *plus* a radically subjective interpretation of the events to which (and on the basis of which) probabilities are assigned.

Recently the referent of the "B" in QBism became moot. While Mermin [4] at one time suggested that it should stand for Bruno de Finetti ("Quantum Brunoism"), he now endorses the term Bettabilitarianism [5] suggested by Chris Fuchs, which was coined by Oliver Wendell Holmes, Jr [6]:

> I must not say necessary about the universe.... We don't know whether anything is necessary or not. I believe that we can *bet* on the behavior of the universe in its contact with us. So I describe myself as a *bet*tabilitarian.

In the present paper, I shall argue that the "B" in QBism rightfully stands for Niels Bohr.

The paper is organized as follows. Section 2 explains why Bohr nowadays seems obscure to most physicists. Section 3 identifies the contextuality of physical variables as Bohr's essential contribution to Kant's theory of science.[a] Section 4 outlines Kant's theory of science, its own contextuality — human experience — and its decontextualization. Section 5 distinguishes three kinds of realism: the "good" (internal) realism of Kant and Bohr, the "bad" (direct or naive) realism recently defended by John Searle [20], and the "ugly" realism associated with the representative theory of perception.

For Kant, objectivity meant the possibility of thinking of experiences as experiences of objects. Realizing that in the new field of experience opened up by the quantum theory this possibility no longer exists, Bohr supplemented the object-oriented language of everyday discourse with the language of quantum phenomena. This is discussed in Sec. 6. Section 7 explains why Bohr insisted on the need to use (i) "ordinary language" or "plain language" or "the common human language" and (ii) "classical concepts" or "the terminology of classical physics" yet never mentioned "classical language" nor "the language of classical physics" (i.e., not once in his *Collected Works*). Section 8 addresses the apparent conflict in Bohr's writings

[a] Affinities between Bohr's theory of science and Kant's have been noted by a number of scholars [8–10,12–19].

between invocations of "irreversible processes" and "irreversible amplification effects" on the one hand and "the essential irreversibility inherent in the very concept of observation" on the other. Section 9 briefly surveys the philosophically rather barren period between the passing of Niels Bohr and the advent of QBism.

The discussion of QBism begins in Sec. 10, in which it is argued that by admitting incoherent superpositions of alternatives involving distinct cognitive states, the QBist solutions to "Wigner's friend" [21] and similar conundrums overshoot their marks. Section 11 concerns the placement of Bell's shifty split [22]. QBists see only one alternative to placing the Heisenberg cut inside the objective world, to wit, between the objective world and the private experiences in which it originates. There is, however, another alternative, which is to place it between the objective world and the unspeakable domain beyond the reach of our concepts, which only becomes speakable by saying in ordinary language "what we have done and what we have learned" [BCW 7:349].[b] This, I content, is where it should be placed, and where to all intents and purposes it was placed by Bohr.

The concluding section deals with how QBism relates to Bohr, beginning with certain misreadings of Bohr that QBists share with the majority of current interpreters of quantum mechanics. My final verdict is that QBism, through its emphasis on the individual experiencing subject, brings home the intersubjective constitution of our common external world more forcefully than Bohr ever did. (The time wasn't ripe for this then. Perhaps it is now.) Bohr's insights, on the other hand, are exceedingly useful in clarifying the QBist position, attenuating its excesses, and enhancing its internal consistency.

2. Why Bohr seems obscure

As a philosopher Niels Bohr was either one of the great visionary figures of all time, or merely the only person courageous enough to confront head on, whether or not successfully, the most imponderable mystery we have yet unearthed. — N. David Mermin [26]

Today, Bohr seems obscure to most physicists. Catherine Chevalley [27] has identified three reasons for this deplorable state of affairs. The first is that Bohr's mature views, which "remained more or less stable at least over the

[b]In what follows, BCW (followed by volume and page number) refers to the *Collected Works* of Niels Bohr [23–25].

latter thirty years of Bohr's life" [28] (i.e., since at least 1932), have come to be equated with one variant or another of the Copenhagen interpretation. The latter only emerged in the mid-1950's, in response to David Bohm's hidden-variables theory [29] and the Marxist critique of Bohr's alleged idealism, which had inspired Bohm. The second reason is that Bohr's readers will usually not find in his writings what they expected to find, while they will find a number of things that they did not expect. What they expect is a take on problems arising in the context of Ψ-ontology — the spurious reification of a probability calculus — such as the problem of objectification or the quantum-to-classical transition. What they find instead is discussions of philosophical issues such as the meanings of "objectivity," "truth," and "reality" and the roles of language and communication. The third reason is that the task of making sense of quantum mechanics is seen today as one of grafting a metaphysical narrative onto a mathematical formalism, in a language that is sufficiently vague philosophically to be understood by all and sundry. For Bohr, as also for Werner Heisenberg and Wolfgang Pauli, the real issues lay deeper. They judged that the conceptual difficulties posed by quantum mechanics called in question the general framework of contemporary thought, its concepts, and its criteria of consistency.

3. Contextuality

It was Immanuel Kant, the most important philosopher of the modern era, who first demonstrated that it was possible to provide a scientific theory with much stronger justification than mere empirical adequacy. The kind of argument inaugurated by him to this end begins by assuming that a certain proposition **p** is true, and then shows that another proposition **q**, stating a precondition for the truth of **p**, must also be true: if **q** were not true, **p** could not be true. For his immediate purpose the relevant proposition **p** was that empirical knowledge is possible, and the corresponding proposition **q** was that certain universal laws must hold.

"Reason," Kant [30, p. 109] wrote, "must approach nature with its principles in one hand ... and, in the other hand, the experiments thought out in accordance with these principles." The concepts in terms of which reason's principles are formulated owe their meanings to our cognitive faculties of intuition[c] and thought. They allow us to ask meaningful questions, and to make sense of the answers we obtain by experiment and observation:

[c]The German original, *Anschauung*, covers both visual perception and visual imagination.

"what reason would not be able to know of itself and has to learn from nature, it has to seek in the latter" but it has to do this "in accordance with what reason itself puts into nature." What Kant did not anticipate was that experiments would come to play the same constitutive role as our cognitive faculties do in defining the terms of our discourse with nature. The insight that certain questions are *contextual* — that they have no answers unless their answers are elicited by actual experiments — is due to Bohr. To hitch the definition of physical quantities to the experimental conditions under which they are observed, is Bohr's ground-breaking contribution to Kant's theory of science [31]. It has, moreover, been spectacularly borne out by the no-go theorems of John Bell [32], Simon Kochen and Ernst Specker [33], and Alexander Klyachko and coworkers [34].

4. Decontextualization

Bohr's contextuality, however, was not the first to play a role in natural philosophy. From the end of the 17th century onwards, it was widely accepted by philosophers that objects existed relative to a context, to wit, human experience. By placing the subject of empirical science squarely into the context of human experience, Kant dispelled many qualms that had been shared by thinkers at the end of the 18th century — qualms about the objective nature of geometry, about the purely mathematical nature of Newton's theory, about the unintelligibility of action at a distance, about Galileo's principle of relativity, to name a few.

Concerning the laws of geometry, which apply to objects constructed by us in the space of our imagination, the question was why they should also apply to the physical world. Kant's answer was that they apply to objects perceived as well as to objects imagined because visual perception and visual imagination share the same space.[d] As to the mathematical nature of Newtonian mechanics, it was justified, not by the Neo-Platonic belief that the book of nature was written in mathematical language, but by its being a precondition of scientific knowledge. What made it possible to conceive of appearances as aspects of an objective world was the mathematical regularities that obtain between them. Newton's refusal to explain action at a

---

[d]It is noteworthy that Kant's argument applies, not to Euclidean geometry specifically, which was the only geometry known in Kant's time, but to geometry in general, and thus to whichever geometry is best suited to formulating the laws of physics. It has even been said [35] that Kant's theory of science set in motion a series of re-conceptualizations of the relationship between geometry and physics that eventuated in Einstein's theories of relativity.

distance was similarly justified, inasmuch as the only intelligible causality available to us consists in lawful mathematical relations between phenomena: for the Moon to be causally related to the Earth was for the Moon to stand in a regular mathematical relation to the Earth. As to the principle of relativity, ditto: lawful mathematical relations only exist between phenomena, and thus only between objects or objective events, but never between a particular phenomenon and space or time itself.[e]

Kant's premise was that "space and time are only forms of sensible intuition, and therefore only conditions of the existence of the things as appearances." It follows

> that we have no concepts of the understanding and hence no elements for the cognition of things except insofar as an intuition can be given corresponding to these concepts, consequently that we can have cognition of no object as a thing in itself, but only insofar as it is an object of sensible intuition, i.e. as an appearance; from which follows the limitation of all even possible speculative cognition of reason to mere objects of *experience*. Yet ... even if we cannot *cognize* these same objects as [i.e., *know* them to be] things in themselves, we at least must be able to *think* them as things in themselves. For otherwise there would follow the absurd proposition that there is an appearance without anything that appears. [30, p. 115]

Kant was the first to show that the predictive success of a scientific theory does not have to be attributed to some empirically inaccessible correspondence between the structure of the theory and the structure of the real world. Needless to say, this had to be done without calling into question the objectivity of the theory, i.e., in a way that allowed people to think of phenomena as appearances of things "out there." We must be able to *decontextualize* the objective world, to forget that it depends on us. And if there is only the single universal context of human experience, this is easily done. We are free to think of perceived objects as faithful representations of real objects (things in themselves), free to forget that the apparently mind-independent system of objects "out there" was a mental construct, and that the concepts that were used in its construction are meaningless outside the context of human experience.

---

[e]Here, too, it would be an anachronism to argue that Kant singled out Galilean relativity, which was the only relativity known in his time. His argument holds for every possible principle of relativity, including Einstein's.

5. Realism good, bad, and ugly

In an essay written during the last year of his life [36], Erwin Schrödinger expressed his astonishment at the fact that despite "the absolute hermetic separation of my sphere of consciousness" from everyone else's, there was "a far-reaching structural similarity between certain parts of our experiences, the parts which we call external; it can be expressed in the brief statement that we all live in the same world." This similarity, Schrödinger avowed, was "not rationally comprehensible. In order to grasp it we are reduced to two irrational, mystical hypotheses," one of which[f] is "the so-called hypothesis of the real external world." Schrödinger left no room for uncertainty about what he thought of this hypothesis. To invoke "the existence of a real world of bodies which are the causes of sense impressions and produce roughly the same impression on everybody ... is not to give an explanation at all; it is simply to state the matter in different words. In fact, it means laying a completely useless burden on the understanding." It means uselessly translating the statement "everybody agrees about something" into the statement "there exists a real world which causes everybody's agreement." Instead of explaining the fact expressed by the first statement, the second merely reinforces its incomprehensibility, for the relation between this real world and those aspects of our experiences about which there is agreement, is something we cannot know. The causal relations we know are internal to those of our experiences about which we agree.

In ancient and medieval philosophy, to *be* was either to be a substance or to be a property of a substance. Substance was self-existent; everything else depended for its existence on a substance. With Descartes, the human conscious subject assumed the role of substance: to *be* became either to be a subject or to exist as a representation for a subject. Thus was born the representative theory of perception. In the eyes of philosopher John Searle [20, p. 23], the move from the older view that "we really perceive real ob-

[f]The alternative hypothesis, which he endorsed, was "that we are all really only various aspects of the One" [36]: the multiplicity of minds "is only apparent, in truth there is only one mind. This is the doctrine of the Upanishads. And not only of the Upanishads" [38]. The One in question is the ultimate subject, from which we are separated by a veil of self-oblivion. The same veil (according to the Upanishads) also prevents us from perceiving the ultimate object, as well as its identity with the ultimate subject. If "to Western thought this doctrine has little appeal," Schrödinger remarked [39], it is because our science "is based on objectivation, whereby it has cut itself off from an adequate understanding of the Subject of Cognizance, of the mind." To him, this was "precisely the point where our present way of thinking does need to be amended, perhaps by a bit of blood-transfusion from Eastern thought."

jects" to the view that we only perceive sense impressions was "the greatest single disaster in the history of philosophy over the past four centuries." A disaster it was indeed, not least because it continues to muddy the scientific waters when it comes to sensory perception.[g]

The representative theory of perception poses this dilemma: either the gap between representations and the objects they are supposed to represent can never be bridged, or the world is reduced to representations. Either science deals with objects in the real world, in which case we have no justifiable idea of how we come to have representations, or it deals with representations, in which case we have no justifiable knowledge of the real world. Transcendental philosophy, inaugurated by Kant and continued in the 20th century by Edmund Husserl [41], emerged as a critique of the representative theory. In an attempt to defend the older, direct realism, Searle has invoked the fact that we are able communicate with other human beings, using publicly available meanings in a public language. For this to work, he argued [20, p. 276], we have to assume common, publicly available objects of reference:

> So, for example, when I use the expression "this table" I have to assume that you understand the expression in the same way that I intend it. I have to assume we are both referring to the same table, and when you understand me in my utterance of "this table" you take it as referring to the same object you refer to in this context in your utterance of "this table."

The implication then is that

> you and I share a perceptual access to one and the same object. And that is just another way of saying that I have to presuppose that you and I are both seeing or otherwise perceiving the same public object. But that public availability of that public world is precisely the direct realism that I am here attempting to defend.

[g]The standard scientific account of perception begins by positing the mind-independent existence of a real world. Objects in this world are said to emit light or sound waves, which are said to stimulate peripheral nerve endings (retinas or ear drums). The stimulated nerves are said to send signals to the brain, where neural processes are said to give rise to perceptual experience. The trouble with this account is that, beginning by invoking events in an independently existing real world, it leads to the conclusion that we have access only to our experiences, and hence no access to this independently existing real world. If brain processes take place in this world, then no one has any idea how to cross the "explanatory gap" [40] from brain processes to subjective experiences, and if brain processes form part of the experienced world — if they themselves are objects of perceptual experience — they cannot be what gives rise to perceptual experience.

Searle points out that his argument is transcendental in Kant's sense. Here **p** is the assumption that we are able to communicate with each other in a public language, and **q** is the conclusion that there must be publicly available objects in a public world about which we can communicate in a public language. The actual implication of his argument, however, is not what Searle claims it to be. The key role that language plays in establishing the rationally incomprehensible correspondence between the "external parts" of our internal experiences, has already been emphasized by Schrödinger:

> What does establish it is *language*, including everything in the way of expression, gesture, taking hold of another person, pointing with one's finger and so forth, though none of this breaks through that inexorable, absolute division between spheres of consciousness. [36]

But direct realism does not merely hold that you and I are both perceiving the same public object; it also attributes the agreement between our respective "spheres of consciousness," which so astonished Schrödinger, to our perceiving an *independently* existing real world. Searle's transcendental argument merely allows us to communicate with each other *as if* direct realism were true. What he succeeds in defending against the "ugly" representative realism, therefore, is not the "bad" direct realism but the "good" *internal realism* inaugurated by Immanuel Kant and defended (among others) by Hilary Putnam and Bernard d'Espagnat.[h]

6. Bohr: From objects to quantum phenomena

The hallmark of empirical knowledge is objectivity. To Kant, objectivity meant the possibility of thinking of appearances as experiences of *objects*. His inquiry into the preconditions of empirical science was therefore an inquiry into the conditions that make it possible to organize sense impressions into (identifiable) objects. Yet in the new field of experience opened up by the quantum theory, this possibility no longer seemed to exist. As Schrödinger [45] wrote,

> Atoms — our modern atoms, the ultimate particles — must no longer be regarded as identifiable individuals. This is a stronger deviation from

[h]Putnam assumed the existence of a mind-independent real world but insisted that it does not dictate its own descriptions to us: "talk of ordinary empirical objects is not talk of things-in-themselves but only talk of things-for-us" [42] — "we don't know what we are talking about when we talk about 'things in themselves'" [43]. D'Espagnat [44], for his part, stressed the necessity of distinguishing between an empirically inaccessible veiled reality and an intersubjectively constructed objective reality.

> the original idea of an atom than anybody had ever contemplated. We must be prepared for anything.

For the present-day physicist, it is not easy to understand the bewilderment that the founders and their contemporaries experienced in the early days of the quantum theory:

> All the verities of the preceding two centuries, held by physicists and ordinary people alike, simply fell apart — collapsed. We had to start all over again, and we came up with something that worked just beautifully but was so strange that nobody had any idea what it meant except Bohr, and practically nobody could understand him. So naturally we kept probing further, getting to smaller and smaller length scales, waiting for the next revolution to shed some light on the meaning of the old one. [46]

That revolution never came. Quantum mechanics works as beautifully in the nucleus as it does in the atom; and it works as beautifully in the nucleon as it does in the nucleus, seven or eight orders of magnitude below the level for which it was designed. (It also works beautifully many orders of magnitude above that level, as for example in a superconductor.) It is therefore past time to try more seriously to understand what Bohr intended to drive home.

"Without sensibility no object would be given to us," Kant [30, p. 193] wrote, "and without understanding none would be thought." And again: "we have no concepts of the understanding ... except insofar as an intuition can be given corresponding to these concepts" [30, p. 115]. Bohr could not have agreed more, insisting as he did that meaningful physical concepts have not only mathematical but also visualizable content. Such concepts are associated with pictures, like the picture of a particle following a trajectory or the picture of a wave propagating in space. In the classical theory, a single picture could accommodate all of the properties a system can have. When quantum theory came along, that all-encompassing picture fell apart. Unless certain experimental conditions obtained, it was impossible to picture the electron as following a trajectory (which was nevertheless a routine presupposition in setting up Stern–Gerlach experiments and in interpreting cloud-chamber photographs), and there was no way in which to apply the concept of position. And unless certain other, incompatible, experimental conditions obtained, it was impossible to picture the electron as a traveling wave (which was nevertheless a routine presupposition in interpreting the scattering of electrons by crystals), and there was no way in which to apply the concept of momentum.

Bohr settled on the nexus between pictures, physical concepts, and experimental arrangements as key to "the task of bringing order into an entirely new field of experience" [47]. If the visualizable content of physical concepts cannot be described in terms of compatible pictures, it has to be described in terms of something that *can* be so described, and what can be so described are the experimental conditions under which the incompatible physical concepts can be employed. What distinguishes these experimental conditions from the quantum systems under investigation is their accessibility to direct sensory experience.

What Bohr added to Kant's theory of science was his insight that empirical knowledge was not necessarily limited to what is *directly* accessible to our senses, and that, therefore, it does not have to be *solely* a knowledge of sense impressions organized into objects. It can also be a knowledge of objects that are *not* objects of sensible intuition but instead are constituted by experimental conditions, which *are* sense impressions organized into objects. This is why "the objective character of the description in atomic physics depends on the detailed specification of the experimental conditions under which evidence is gained" [BCW 10:215]. Quantum mechanics does not do away with objects of sensible intuition but supplements them with *quantum phenomena*:

> all unambiguous interpretation of the quantum mechanical formalism involves the fixation of the external conditions, defining the initial state of the atomic system concerned and the character of the possible predictions as regards subsequent observable properties of that system. Any measurement in quantum theory can in fact only refer either to a fixation of the initial state or to the test of such predictions, and it is first the combination of measurements of both kinds which constitutes a well-defined phenomenon. [BCW 7:312]

7. Bohr: Concepts and language

The transition in Bohr's thinking from the familiar epistemology of objects to an epistemology supplemented with quantum phenomena resulted from his insight that "the facts which are revealed to us by the quantum theory ... lie outside the domain of our ordinary forms of perception" [BCW 6:217]. As early as 1922, Bohr opined that the difficulties physicists were facing at the time were "of such a kind that they hardly allow us to hope, within the world of atoms, to implement a description in space and time of the kind corresponding to our usual sensory images" [BCW 10:513–514]. By 1926, the mature (non-relativistic) theory was in place, and by 1929 Bohr's thoughts had gelled into what to my mind remains the most astute understanding of quantum mechanics to date.

In his writings of that year, abundant reference is made to "our (ordinary) forms of perception," time and space. As in: quantum theory has "justified the old doubt as to the range of our ordinary forms of perception when applied to atomic phenomena" [BCW 6:209]; "at the same time as every doubt regarding the reality of atoms has been removed, ... we have been reminded in an instructive manner of the natural limitation of our forms of perception" [BCW 6:237]. This limitation was "brought to light by a closer analysis of the applicability of the basic physical concepts in describing atomic phenomena" [BCW 6:242]. That is to say, the natural limitation of our forms of perception both implies and is implied by a natural limitation of the applicability of our basic physical concepts, which is a consequence of the uncertainty relations. Yet "in spite of their limitation, we can by no means dispense with those forms of perception which colour our whole language and in terms of which all experience must ultimately be expressed" [BCW 6:283]. In other words, the conceptual framework of quantum physics is the same as that of classical physics, the difference being that in quantum physics its applicability is limited.

"When speaking of a conceptual framework," Bohr wrote, "we merely refer to an unambiguous logical representation of relations between experiences" [BCW 10:84]. In Bohr's time and the cultural environment in which he lived, Kant's theory of science still exercised considerable influence. There can be little doubt that the unambiguous logical representation of relations between experiences that Bohr had in mind, was in all important respects the conceptual framework staked out by Kant, providing the general structure of an object-oriented language.

In Kant's theory of science, the relevant relations between experiences fall under the logical categories of substance, causality, and interaction. The logical relation between a (logical) subject and a predicate makes it possible for us to think of a particular nexus of perceptions as the properties of a *substance*, connected to it as predicates are connected to a subject. It makes it possible for me to think of perceptions as connected not by me, in my experience, but in an object "out there" in the public world. The logical relation between antecedent and consequent (if ... then...) makes it possible for us to think of what we perceive at different times as properties of substances connected by *causality*. It makes it possible for me to think of asynchronous perceptions as connected not merely in my experience but also objectively, by a causal nexus "out there." And the category of community or reciprocity, which Kant associated with the disjunctive relation (either... or...), makes it possible for us to think of what we perceive in different locations as properties of substances connected by a *reciprocal action.* It makes it possible for me to think of simultaneous perceptions as connected not only in my experience but objectively. (Kant thought that by establishing a reciprocal relation, we establish not only an objective spatial relation but also an objective relation of simultaneity.)

But if we are to be able to think of perceptions as properties of substances, or as causally connected, or as affecting each other, the connections must be regular. For perceptions to be perceptions of a particular kind of thing (say, an elephant), they must be connected in an orderly way, according to a concept denoting a lawful concurrence of perceptions. For perceptions to be causally connected, like (say) lightning and thunder, they must fall under a causal law, according to which one perception necessitates the subsequent occurrence of another. (By establishing a causal relation falling under a causal law, we also establish an objective temporal relation.) And for perceptions to be reciprocally connected, like (say) the Earth and the Moon, they must affect each other according to a reciprocal law, such as Newton's law of gravity. It is through lawful connections in the "manifold of appearances" that we are able to think of appearances as perceptions of a self-existent system of causally evolving (and thus re-identifiable) objects, from which we, the experiencing subjects, can remove ourselves.

Even in a field of experience in which the concepts required to bundle sense impressions into objects cannot be applied, one has to rely on the common object-oriented language. Where one cannot speak of objects, one has to speak of quantum phenomena, i.e., of experimental arrangements and results indicated by measuring instruments:

> The argument is simply that by the word "experiment" we refer to a situation where we can tell others what we have done and what we have learned and that, therefore, the account of the experimental arrangement and of the results of the observations must be expressed in unambiguous language with suitable application of the terminology of classical physics. [BCW 7:349]

Two expressions are significant here: "unambiguous language" and "the terminology of classical physics." Presently (November 2020) a combined Google search for "Bohr" and "classical language" (the latter term including the quotes) yields over 10,000 results, while a Google search for "Bohr" and "language of classical physics" yields more than 76,000 results. By contrast, searching the 13 volumes of the *Complete Works* of Niels Bohr does not yield a *single* occurrence of either "classical language" or "language of classical physics." It is the ubiquity in the secondary literature of these latter expressions that is chiefly responsible for the widespread misconceptions about Bohr's thinking.

While Bohr insisted on the use of classical *concepts* (or, the terminology of classical physics, or simply "elementary physical concepts" [BCW 7:394]), the *language* on the use of which he insisted was "ordinary language" [BCW 7:355], "plain language" [BCW 10:159], the "common human language" [BCW 10:157–158], or the "language common to all" [10:xxxvii]. A distinction must therefore be drawn between the role that classical concepts played in Bohr's thinking and the role that was played by the common human language. The common human language is the object-oriented language of everyday discourse, while classical concepts are not proprietary to classical physics but denote attributes that owe their meanings to our forms of perception, such as position, orientation, and the ones that are defined in terms of invariances under spacetime transformations.

One day during tea at his institute, Bohr was sitting next to Edward Teller and Carl Friedrich von Weizsäcker. Von Weizsäcker [48] recalls that when Teller suggested that "after a longer period of getting accustomed to quantum theory we might be able after all to replace the classical concepts by quantum theoretical ones," Bohr listened, apparently absent-mindedly, and said at last: "Oh, I understand. We also might as well say that we are not sitting here and drinking tea but that all this is merely a dream." If we are dreaming, we are unable to tell others what we have done and what we have learned. Therefore

> it would be a misconception to believe that the difficulties of the atomic theory may be evaded by eventually replacing the concepts of classical physics by new conceptual forms. ... the recognition of the limitation of our forms of perception by no means implies that we can dispense with our customary ideas or their direct verbal expressions when reducing our sense impressions to order. [BCW 6:294]

Or, as Heisenberg [49, p. 56] put it, "[t]here is no use in discussing what could be done if we were other beings than we are."[i] Bohr's claim that

[i]Heisenberg thought it possible that the forms of perception of other beings, and hence their concepts, could be different from ours: our concepts "may belong to the species 'man,' but not to the world as independent of men" [49, p. 91].

the "classical language" (i.e., plain language supplemented with the terminology of classical physics) was indispensable, has also been vindicated by subsequent developments in particle physics:

> This [claim] has remained valid up to the present day. At the *individual* level of clicks in particle detectors and particle tracks on photographs, all measurement results have to be expressed in classical terms. Indeed, the use of the familiar physical quantities of length, time, mass, and momentum-energy at a subatomic scale is due to an extrapolation of the language of classical physics to the non-classical domain. [50, p. 162]

It is therefore an irony that Bohr, seeing Kant as arguing for the necessary validity and unlimited reach of classical concepts, regarded complementarity as an alternative to Kant's theory of science, thus drawing the battle lines in a way which put Kant and himself on opposing sides. Just as Kant did not argue for the universal validity of Euclidean geometry *in particular* (see Note d), nor for Galilean relativity *in particular* (see Note e), so his arguments did not, in effect, establish the *unlimited reach* of classical concepts. As his arguments merely established the validity of whichever geometry, and whichever principle of relativity, was most convenient, so they established the necessary validity of classical concepts but not their unlimited reach. What Kant did not anticipate was the possibility of an empirical knowledge that, while being obtained *by means of* sense impressions organized into objects, was not a knowledge *of* sense impressions organized into objects.

8. Bohr and the irreversibility of measurements

If the terminology of quantum phenomena is used consistently, then nothing — at any rate, nothing we know how to think about — happens between "the fixation of the external conditions, defining the initial state of the atomic system concerned" and "the subsequent observable properties of that system" [BCW 7:312]. Any story purporting to detail a course of events in the interval between a system preparation and a subsequent observation is inconsistent with "the essential wholeness of a quantum phenomenon," which "finds its logical expression in the circumstance that any attempt at its subdivision would demand a change in the experimental arrangement incompatible with its appearance" [BCW 10:278]. What, then, are we to make of the following passages [emphases added]?

> [E]very well-defined atomic phenomenon is closed in itself, since its observation implies a permanent mark on a photographic plate *left by the impact of an electron* or similar recordings *obtained by suitable amplification devices of essentially irreversible functioning.* [BCW 10:89]

> Information concerning atomic objects consists solely in the marks they make on these measuring instruments, as, for instance, a spot *produced by the impact of an electron on a photographic plate* placed in the experimental arrangement. The circumstance that such marks are *due to irreversible amplification effects* endows the phenomena with a peculiarly closed character pointing directly to the irreversibility in principle of the very notion of observation. [BCW 10:120]

> In this connection, it is also essential to remember that all unambiguous information concerning atomic objects is derived from the permanent marks — such as a spot on a photographic plate, *caused by the impact of an electron* — left on the bodies which define the experimental conditions. Far from involving any special intricacy, the *irreversible amplification effects on which the recording of the presence of atomic objects rests* rather remind us of the essential irreversibility inherent in the very concept of observation. [BCW 7:390; BCW 10:128]

If a well-defined atomic phenomenon is closed, how can something happen between the fixation of the external conditions and a permanent mark on a photographic plate? Does not the interposition of the impact of an electron and/or of subsequent amplification effects amount to a subdivision of the phenomenon in question?

Ole Ulfbeck and Aage Bohr [51] have shed light on this issue. For them, clicks in counters are "events in spacetime, belonging to the world of experience." While clicks can be classified as electron clicks, neutron clicks, etc., "there are no electrons and neutrons on the spacetime scene" and "there is no wave function for an electron or a neutron but [only] a wave function for electron clicks and neutron clicks." "[T]here is no longer a particle passing through the apparatus and producing the click. Instead, the connection between source and counter is inherently non-local." The key to resolving the issue at hand is that each click has an "onset" — "a beginning from which the click evolves as a signal in the counter." This onset

> has no precursor in spacetime and, hence, does not belong to a chain of causal events. In other words, the onset of the click is not the effect of something, and it has no meaning to ask how the onset occurred.... [T]he occurrence of genuinely fortuitous clicks, coming by themselves, is recognized as the basic material that quantum mechanics deals with....

> [T]he wave function enters the theory not as an independent element, but in the role of encoding the probability distributions for the clicks.... [T]he steps in the development of the click, envisaged in the usual picture, are not events that have taken place on the spacetime scene.... [T]he downward path from macroscopic events in spacetime, which in standard quantum mechanics continues into the regime of the particles, does not extend beyond the onsets of the clicks.

If irreversible amplification effects — the steps in the development of the click — only occur "in the usual picture," then they neither modify nor subdivide the quantum phenomenon in which — through an illegitimate extension of the object-oriented language of classical physics — they are said to occur.

For Niels Bohr, "the physical content of quantum mechanics is exhausted by its power to formulate statistical laws governing observations obtained under conditions specified in plain language" [BCW 10:159]. A quantum phenomenon thus has a statistical component, which correlates events in the world of experience. The so-called irreversible amplification effects belong to this statistical component. The unmediated step from the source to the onset of the click, and the subsequent unmediated steps in the development of the click, are steps in a gazillion of alternative sequences of possible outcomes of *unperformed* measurements, and unperformed measurements do not affect the essential wholeness of a quantum phenomenon.

Niels Bohr, moreover, strongly cautioned against the terminology of "disturbing a phenomenon by observation" and of "creating physical attributes to objects by measuring processes" [BCW 7:316]. If there is nothing to be disturbed by observation, if even the dichotomy of objects and attributes created for them by measuring processes is unwarranted, then it is not just the measured property that is constituted by the experimental conditions under which it is observed; it is the quantum system itself that is so constituted. Recently this point was forcefully made by Brigitte Falkenburg in her monograph *Particle Metaphysics* [50, pp. 205–206]:

> [O]nly the experimental context (and our ways of conceiving of it in classical terms) makes it possible to talk in a sloppy way of *quantum objects*.... Bare quantum "objects" are just bundles of properties which underlie superselection rules and which exhibit non-local, acausal correlations.... They seem to be Lockean empirical substances, that is, collections of empirical properties which constantly go together. However, they are only individuated by the experimental apparatus in which they are measured or the concrete quantum phenomenon to which they

> belong.... They can only be individuated as context-dependent quantum *phenomena*. Without a given experimental context, the reference of quantum concepts goes astray. In this point, Bohr is absolutely right up to the present day.

In the following passages [emphases added], Bohr goes beyond invoking irreversible amplification effects, apparently arguing that the quantum features involved in the atomic constitution of a measurement apparatus (or the statistical element in its description) can be neglected because the relevant parts of a measurement apparatus are sufficiently large and heavy.

> In actual experimentation this demand [that the experimental arrangement as well as the recording of observations be expressed in the common language] is met by the specification of the experimental conditions by means of bodies like diaphragms and photographic plates *so large and heavy that the statistical element in their description can be neglected.* The observations consist in the recording of permanent marks on these instruments, and the fact that the amplification devices used in the production of such marks involves essentially irreversible processes presents no new observational problem, but merely stresses the element of irreversibility inherent in the definition of the very concept of observation. [BCW 10:212]

> In actual physical experimentation this requirement [that we must employ common language to communicate what we have done and what we have learned by putting questions to nature in the form of experiments] is fulfilled by using as measuring instruments rigid bodies like diaphragms, lenses, and photographic plates *sufficiently large and heavy to allow an account of their shape and relative positions and displacements without regard to any quantum features inherently involved in their atomic constitution....* The circumstance that [recordings of observations like the spot produced on a photographic plate by the impact of an electron] involve essentially irreversible processes presents no special difficulty for the interpretation of the experiments, but rather stresses the irreversibility which is implied in principle in the very concept of observation. [BCW 10:165]

How can the size and weight of a measuring device justify

- the irreversibility in principle of the very notion of observation [BCW 10:120],
- the essential irreversibility inherent in the very concept of observation [BCW 7:390; BCW 10:128],

— the irreversibility which is implied in principle in the very concept of observation [BCW 10:165], or
— the element of irreversibility inherent in the definition of the very concept of observation [BCW 10:212]?

The only irreversibility that can justify the irreversibility of observations is the incontestable irreversibility of human sensory experience. For Bohr, "the emphasis on the subjective character of the idea of observation [was] essential" [BCW 10:496]. If, as he insisted, the description of atomic phenomena nevertheless has "a perfectly objective character," it was "in the sense that no explicit reference is made to any *individual* observer and that therefore ... no ambiguity is involved in the communication of information" [BCW 10:128, emphasis added]. It was never in the sense that no reference was made to the community of communicating observers or to the incontestable irreversibility of their experiences.

Like Kant, Bohr was concerned with the possibility of an objective knowledge of phenomena (or appearances, or experiences). Kant primarily addressed the possibility of organizing appearances into a world of objects. This involved concepts that allowed the individual not only to think of appearances as a world of external objects but also to communicate with others about a world of external objects. Bohr primarily addressed the requisite possibility of communication. This involved the use of concepts that allowed the individual not only to communicate with others about a world of external objects but also to think of appearances as a world of external objects. Bohr and Kant were on the same page.

What distinguishes the objectivity that could be achieved in the age of Kant from the objectivity that can be achieved in the age of quantum mechanics is that a complete elision of the subject is no longer feasible. Having asserted that "we can have cognition of no object as a thing in itself, but only insofar as it is an object of sensible intuition, i.e. as an appearance," Kant (as we have seen) went on to affirm that

> even if we cannot *cognize* these same objects as things in themselves, we at least must be able to *think* them as things in themselves. For otherwise there would follow the absurd proposition that there is an appearance without anything that appears.[j] [30, p. 115]

[j]Kant's note: "To *cognize* an object, it is required that I be able to prove its possibility (whether by the testimony of experience from its actuality or *a priori* through reason). But I can *think* whatever I like, as long as I do not contradict myself. . . ."

If the only relevant context is human experience, or if the reach of human sensory experience is unlimited (as classical physics takes it to be), the elision of the subject can be achieved. It is possible to transmogrify calculational tools into objective physical processes with some measure of consistency [52]. But if the experimental context is relevant as well, or if the reach of human sensory experience is limited, the elision of the subject is a lost cause, and so is the transmogrification of calculational tools into physical mechanism or natural processes. If one nevertheless wants to establish the irreversibility of measurements and the definiteness of outcomes without invoking the definiteness and irreversibility of human sensory experience, one has no choice but to invoke such "macroscopic" features as the size and weight of a measurement apparatus. This (to appropriate a well-known passage by John Bell),

> is like a snake trying to swallow itself by the tail. It can be done — up to a point. But it becomes embarrassing for the spectators even before it becomes uncomfortable for the snake. [22]

In the words of Bernard d'Espagnat [10], quantum mechanics practically compels us

> to adopt the idea that was, in fact, at the very core of Kantism and constitutes its truly original contribution to philosophical thinking, to wit, the view that things and events, far from being elements of a "reality per se," are just phenomena, that is, elements of our experience.

So what were Bohr's intentions in invoking, in lieu of the irreversibility of human experience, the size and weight of the measurement apparatus? Was it in order to appease the naïve realistic inclinations of lesser minds? Or was it in order to indicate the price one had to pay (but which he personally was not willing to pay) for exorcising every possible reference to human sensory experience? My suggested answer is "yes" to both questions.

9. The intervening years

In 1932 John von Neumann [53] developed the mathematical part of the quantum theory into an autonomous formal language. In doing so he turned the theory into a mathematical formalism that was in need of a physical interpretation. In the 1950s, interpreting quantum mechanics began to turn into a growth industry. First David Bohm presented his hidden-variables interpretation [29], then Hugh Everett put forward his relative-state interpretation [54], whereupon Heisenberg entered the fray, arguing that the

Copenhagen interpretation was the only viable interpretation [49, Chaps. 3 & 8]. He thereby transformed Bohr's views into just another interpretation of a mathematically formulated theory. Historically, Bohr's reply [55] to the argument by Einstein, Podolsky, and Rosen [56] was widely accepted as a definitive refutation by the physics community. During the "shut up and calculate" period of the post-war years, Bohr's perspective was lost. His paper, which only treated the mathematical formalism in a footnote, is now widely seen as missing the point.

By transmogrifying a probability algorithm — the wave function or the so-called state vector — into a bona fide physical state, adopting the eigenvalue–eigenstate link,[k] and modeling the "process" of measurement as a two-stage affair ("pre-measurement" followed by "objectification"), von Neumann created what is commonly known as the measurement problem but is more appropriately called "the disaster of objectification" [58]. This is how quantum mechanics came to be labeled as "the great scandal of physics" [59], as a theory that "makes absolutely no sense" [60], and as "the silliest" of all the theories proposed in the 20th century [61].

What is responsible for these mischaracterizations should not be hard to detect: think of a quantum state's dependence on time as the time-dependence of an evolving physical state, rather than as the dependence of probabilities on the time of the measurement to the possible outcomes of which they are assigned, and you have two modes of evolution whereas in reality there is not even one.

Nevertheless, today the reasons for these mischaracterizations *are* hard to detect. One of these reasons is that whereas a junior-level classical mechanics course devotes a considerable amount of time to different formulations of classical mechanics, even graduate-level courses often emphasize one particular formulation of quantum mechanics almost to the exclusion of all variants, of which there are (at least) nine [62]. It would seem reasonable to expect that an interpretation of quantum mechanics be based on features that are common to all formulations of the theory, not on the mathematical idiosyncrasies of a particular formulation, such as the wave-function formulation. What is common to all formulations is that they afford tools for calculating correlations between measurement outcomes.

[k]Thus formulated by Dirac [57]: "The expression that an observable 'has a particular value' for a particular state is permissible in quantum mechanics in the special case when a measurement of the observable is certain to lead to the particular value, so that the state is an eigenstate of the observable."

Another reason is the axiomatic method by which quantum mechanics is now typically taught. First students are told that the state of a quantum-physical system is (or is represented by) a normalized element of a Hilbert space. Then they are told that observables are (or are represented by) self-adjoint operators, and that the possible outcomes of a measurement are the eigenvalues of such an operator. Then comes a couple of axioms concerning the time evolution of states — unitary *between* measurements and as stipulated by the projection postulate *at the time of* a measurement. A further axiom stipulates that the states of composite systems are (or are represented by) vectors in the tensor product of the Hilbert spaces of the component systems. And finally, almost as an afterthought, comes an axiom about probabilities, the Born rule. This is how the *The Ashgate Companion to Contemporary Philosophy of Physics* [59] comes to distinguish between a "bare quantum formalism," which it describes as "an elegant piece of mathematics" that is "prior to any notion of probability, measurement etc.," and a "quantum algorithm," which it describes as "an ill-defined and unattractive mess," whose business is to extract "empirical results" from the former. In actual fact, there is no such thing as a bare quantum formalism. Every single axiom of the theory only makes sense as a feature of a probability calculus [63].

It is beyond doubt that significant progress was made during the roughly four decades between the passing of Niels Bohr and the advent of QBism. We now have a congeries of complex, sophisticated, and astonishingly accurate probability algorithms — the standard model[1] — and we are witnessing rapid growth in the exciting fields of quantum information and quantum technology. By contrast, the contemporaneous progress in quantum theory's philosophical foundations mainly consisted in finding out what does *not* work, such as the countless attempts to transmogrify statistical correlations between observations into physical processes that take place between and give rise to observations.

10. QBism: Wigner's friend

To make the centrality of human experience duly and truly stick, QBism emphasizes the *individual* subject. To a QBist, all probabilities are of the

[1] "Standard model is a grotesquely modest name for one of humankind's greatest achievements." [64]

subjective, personalist Bayesian kind. The so-called quantum state is something the individual user (of quantum mechanics) or agent (in a quantum world) assigns on the basis of her own experiences,[m] and it is used by her to assign probabilities to a set of possible personal experiences, which are determined by the action she takes to elicit one these experiences. Such an action does not have to take place in a physics laboratory. It "can be anything from running across the street at L'Étoile in Paris (and gambling upon one's life) to a sophisticated quantum information experiment (and gambling on the violation of a Bell inequality)" [65]. The only thing a QBist "does not model with quantum mechanics is her own direct internal awareness of her own private experience" [66].

Two of the pseudo-problems that quantum-state realists have to contend with are thereby taken care of: the matter of Wigner's friend [21] and the matter of Bell's shifty split [22]. In Wigner's scenario, Wigner's friend F performs a measurement on a system S using an apparatus A. Treating F as a quantum system, and treating quantum states as ontic states evolving unitarily between measurement-induced state reductions, Wigner concludes that a reduction of the combined system $S+A$ occurs for F when she becomes aware of the outcome, while a reduction of the combined system $S+A+F$ occurs for him when he is informed of the outcome by F. This scenario led Wigner to conclude that the theory of measurement was logically consistent only "so long as I maintain my privileged position as ultimate observer." QBism, on the contrary, maintains that Wigner's state assignment, which is based on his actual past and possible future experiences, is as valid as his friend's, based as that is on a different set of actual past and possible future experiences. This point, however, can be made without envisioning Wigner's friend in a coherent superposition of two distinct cognitive states:

> Wigner's quantum-state assignment and unitary evolution for the compound system are only about his *own* expectations for his *own* experiences should he take one or another action upon the system or any part of it. One such action might be his sounding the verbal question, "Hey friend, what did you see?," which will lead to one of two possible experiences for him. Another such action could be to put the whole conceptual box into some amazing quantum interference experiment, which would lead to one of two completely different experiences for him. [65]

[m] While Fuchs and Schack prefer the term "agent," Mermin prefers the term "user," in order to emphasize that QBists regard quantum mechanics as a "user's manual" [3].

QBists distinguish between (subjective) agent-dependent realities and a common body of (objective) reality:

> What is real for an agent rests entirely on what that agent experiences, and different agents have different experiences. An agent-dependent reality is constrained by the fact that different agents can communicate their experience to each other, limited only by the extent that personal experience can be expressed in ordinary language.... In this way a common body of reality can be constructed. [66]

What do we know about this common body of reality? Because we construct it from our experiences, and because our experiences are definite and irreversible, it is constructed from experiences that are definite and irreversible. I may be ignorant of your experiences and you may be ignorant of mine, but we cannot doubt the definiteness and irreversibility of our respective experiences. It is therefore inadmissible to assign to any (sane and healthy) subject a coherent superposition of distinct cognitive states. Wigner is not only perfectly justified but *required* to assign to the system that includes his friend an incoherent mixture reflecting his ignorance of the outcome that his friend has obtained. To treat his own experiences as definite but not those of his friend — that would be the solipsism which Wigner feared and sought to avoid by proposing "that the equations of motion of quantum mechanics cease to be linear, in fact that they are grossly non-linear if conscious beings enter the picture."

QBists are united in rejecting "the silly charges of solipsism" [3]. In order to avoid these charges, however, they need to do more than acknowledge the fact that "[m]y experience of you leads me to hypothesize that you are a being very much like myself, with your own private experience." They need to stop fantasizing about coherent superpositions involving distinct experiences.

11. QBism: Bell's shifty split

> *There is a straight ladder from the atom to the grain of sand, and the only real mystery in physics is the missing rung. Below it, particle physics; above it, classical physics; but in between, metaphysics.*
>
> — *Tom Stoppard,* Hapgood

Bell's "shifty split," a.k.a. the Heisenberg cut, is the mysterious boundary separating the system under investigation from the means of investigation.

While for Heisenberg its location was more or less arbitrary, for Bohr it was determined by the measurement setup.[n] In QBism, the experience of the individual user takes the place of the measurement setup. Accordingly there are as many splits as there are users, and there is nothing shifty about them. Mermin [68] explains:

> Each split is between an object (the world) and a subject (an agent's irreducible awareness of her or his own experience). Setting aside dreams or hallucinations, I, as agent, have no trouble making such a distinction, and I assume that you don't either. Vagueness and ambiguity only arise if one fails to acknowledge that the splits reside not in the objective world, but at the boundaries between that world and the experiences of the various agents who use quantum mechanics.

Let us disregard the ambiguity of "awareness of one's own experience," which could mean either awareness of something one is experiencing or awareness of one's experiencing something. The question is: what is meant by "objective world"? First guess: what is meant is "the common external world we have all negotiated with each other" or, equivalently, "a model for what is common to all of our privately constructed external worlds" [3]. In this case the split occurs between this world or model and the private experiences in which it originates.

Second guess: what is meant is something that induces experiences in conscious subjects. This interpretation is suggested by Mermin's statement that, according to QBism, "my understanding of the world rests entirely on the experiences that the world has induced in me throughout the course of my life" [3], or by the equivalent statement that "[t]he world acts on me, inducing the private experiences out of which I build my understanding of my own world" [5]. Judging by *these* statements, the split occurs between my private experiences and a world that induces them, rather than between my private experiences and the world as I understand it on the basis of my private experiences. (On the relation between these two worlds see Note g.)

What we are faced with here is an attempt to throw the baby out with the bathwater. The bathwater is the shifty split; the baby is the measuring apparatus. And not only the measuring apparatus. If QBism, as Fuchs and Schack affirm, treats "all physical systems in the same way, including atoms, beam splitters, Stern-Gerlach magnets, preparation devices, measurement

[n] See Camilleri and Schlosshauer [67] for a discussion of Bohr's and Heisenberg's divergent views on this matter.

apparatuses, all the way to living beings and other agents" [69], then Bohr's crucial insight that the properties of quantum systems are *contextual* — that they are defined by experimental arrangements — is lost.

For Bohr, the measurement apparatus was needed not only to indicate the possession of a property (by a system) or a value (by an observable) but also, and in the first place, to make a set of properties or values available for attribution to a system or an observable. The sensitive regions of an array of detectors *define* the regions of space in which the system can be found. In the absence of an array of detectors, the regions of space in which the system can be found do not exist. The orientation of a Stern-Gerlach apparatus *defines* the axis with respect to which a spin component is measured. In the absence of a Stern-Gerlach apparatus, the axis with respect to which a spin component can be up or down does not exist. What physical quantity is defined by running across the street at L'Étoile in Paris?

From a different QBist point of view, espoused by Fuchs and Schack, the measurement apparatus should be understood as an extension of the agent, and quantum mechanics itself should be regarded as a theory of stimulation and response:

> A quantum measurement device is like a prosthetic hand, and the outcome of a measurement is an unpredictable, undetermined "experience" shared between the agent and the external system. [65]
>
> The agent, through the process of quantum measurement stimulates the world external to himself. The world, in return, stimulates a response in the agent that is quantified by a change in his beliefs — i.e., by a change from a prior to a posterior quantum state. Somewhere in the structure of those belief changes lies quantum theory's most direct statement about what we believe of the world as it is without agents. [70]

This invites two comments. The first is that the question where the apparatus ends and the rest of the world begins is once more open to dispute. It appears that one shifty split has been traded for another. Fuchs responds by pointing out that the physical extent of the agent is up to the agent[o]:

> The question is not where does the quantum world play out and the classical world kick in? But where does the agent expect his own autonomy to play out and the external world, with its autonomy and its capacity to surprise, kick in? The physical extent of the agent is a judgment he makes of himself.

[o]Private communication (March 28, 2019).

By placing the dividing line — wherever the agent chooses to place it — between the agent-cum-instrument and the rest of the physical world, Fuchs does precisely what Mermin objects to when he writes that "[v]agueness and ambiguity only arise if one fails to acknowledge that the splits reside not in the objective world, but at the boundaries between that world and the experiences of the various agents who use quantum mechanics" [68].

The second comment concerns "the world as it is without agents." The phrase *could* refer to the unspeakable domain beyond the reach of our concepts, which only becomes speakable through the manner in which it is stimulated (i.e., by saying in ordinary language what the agent has done) and through the manner in which it responds (i.e., by saying in ordinary language what the agent has learned). Mermin,[p] however, rejects this interpretation: "QBists (at least this one) attach no meaning to 'the world as it is without agents.' It only means 'the common external world we have all negotiated with each other'."

Regardless, there is another boundary at which the Heisenberg cut can be placed. As our common external world has a "near" boundary (between it and the private experiences in which it originates), so it has a "far" boundary (between it and the unspeakable domain beyond the reach of our concepts). My contention is that the cut ought to be placed there. I therefore agree with Mermin that the cut does not reside in the objective world — the world of sense impressions organized into objects, the world we have all negotiated with each other. But instead of placing it at its near boundary, I maintain that it should be placed at its far boundary, and I take it that, to all intents and purposes, Bohr did the same. What is definite in this case is not just the measurement apparatus but the entire objective world. "Indefinite" and "objective" are mutually exclusive terms.

12. QBism and Bohr

As it is an irony that Bohr drew the battle lines in a way which put Kant and himself on opposing sides, so it is an irony that QBists draw their battle lines in a way which puts Bohr and themselves on opposing sides, notwithstanding that "QBism agrees with Bohr that the primitive concept of *experience* is fundamental to an understanding of science" [66]. Thus Fuchs *et al.*:

[p]Private communication (May 23, 2019).

> The Founders of quantum mechanics were already aware that there was a problem. Bohr and Heisenberg dealt with it by emphasizing the inseparability of the phenomena from the instruments we devised to investigate them. Instruments are the Copenhagen surrogate for experience.... [They are] objective and independent of the agent using them. [66]

And thus Mermin [3]:

> Those who reject QBism ... reify the common external world we have all negotiated with each other, purging from the story any reference to the origins of our common world in the private experiences we try to share with each other through language. ...by "experience" I believe [Bohr] meant the objective readings of large classical instruments.... Because outcomes of Copenhagen measurements are "classical," they are *ipso facto* real and objective.

While QBists are generally aware of the important distinction between what Mermin calls "reification" and what Schrödinger [36] called "objectivation," they share the now prevailing misappreciation of Bohr's thinking. Bohr was concerned with objectivation, the representation of a shared mental construct as an objective world, not with reification, which ignores or denies the origins of the objective world in our thoughts and perceptions. Objectivation means purging from the story any reference to these origins without ignoring or denying them, so that science may deal with the objective world as common-sense realism does — *as if* it existed independently of our thoughts and perceptions. Reification is the assertion that the world we perceive does in fact exist independently of our perceptions, or that the world we mentally construct does in fact exist independently of our constructing minds, or that the world we describe does in fact exist — just as we describe it — independently of our descriptions.

To Bohr, measurement outcomes are "classical" (i.e., definite and irreversible) and instruments are objective, not because (or in the sense that) they are reified, but because (or in the sense that) they are situated in an intersubjectively constituted world — like everything else that is directly accessible to human sensory experience. Instead of being a "surrogate of experience," instruments — like everything else in our common external world — are experiences that lend themselves to objectivation. They make it possible not only to apply classical concepts to quantum systems but also to extend their reach into the non-classical domain via principles of correspondence.[q]

The statement that those who reject QBism reify the common external world we have all negotiated with each other, also rings false. One can certainly reject some of the (sometimes mutually inconsistent) claims made by QBists without reifying the objectivized world. What is true is the converse: those who reify the objectivized world will have to reject QBism.

Admittedly, Bohr obscured his original thinking by compounding the incontestable irreversibility of human sensory experience with amplification processes or apparatus features like being sufficiently large or heavy. To invoke such processes or features was for him the price that one had to pay for achieving complete objectivation (i.e., for banishing every reference to experience). He himself, however, was clearly not inclined to pay this price. It bears repetition: for him the description of atomic phenomena had "a perfectly objective character, *in the sense that* no explicit reference is made to any individual observer and that therefore ... no ambiguity is involved in the communication of information" [BCW 7:390, emphasis added] — and thus *not in the sense that* no reference was made to the community of communicating observers or to the incontestable irreversibility of their experiences.

To Bohr, objectivity meant "a description by means of a language common to all ... in which people may communicate with each other in the relevant field" [BCW 10:XXXVII]. What QBists mean by objectivity is less clear, though arguably they mean the same, to wit: "language is the only means by which different users of quantum mechanics can attempt to compare their own private experiences," and it is only by communicating "that we can arrive at a shared understanding of what is common to all our own experiences of our own external worlds" [3].

There is just one detail in Mermin's argument to which Bohr would probably have objected, namely the idea that each of us first constructs a

[q]According to Falkenburg [50, pp. XII, 191], "quantum mechanics and quantum field theory only refer to individual systems due to the ways in which the quantum models of matter and subatomic interactions are linked by semi-classical models to the classical models of subatomic structure and scattering processes. All these links are based on tacit use of a generalized correspondence principle in Bohr's sense (plus other unifying principles of physics)." This generalized correspondence principle serves as "a semantic principle of continuity which guarantees that the predicates for physical properties such as 'position', 'momentum', 'mass', 'energy', etc., can also be defined in the domain of quantum mechanics, and that one may interpret them operationally in accordance with classical measurement methods. It provides a great many inter-theoretical relations, by means of which the formal concepts and models of quantum mechanics can be filled with physical meaning."

private external world, and that language comes in only after this is done, as a means of figuring out what is common to all our privately constructed external worlds. One cannot construct a private external world before being in possession of a language providing the concepts that are needed for its construction. At any rate, Mermin's claim that "[o]rdinary language comes into the QBist story in a more crucial way than it comes into the story told by Bohr," appears to me wholly unjustified.

The great merit of QBism is that it puts the spotlight back on the role that human experience plays in creating physical theories. One this is recognized, the mystery as to why measurements are irreversible and outcomes definite vanishes into thin air: it is because our experiences are irreversible and definite. Bohr could have said the same, and arguably did, albeit in such elliptic ways that the core of his message has been lost or distorted beyond recognition. The fundamental difference between Bohr and QBism is that the former was writing before interpreting quantum mechanics became a growth industry, while the latter emerged in reaction to an ever-growing number of futile attempts at averting the disaster of objectification in the same realist framework in which it arose.

There are several ways in which QBism goes beyond Bohr, but this in no wise affects my claim that Bohr was a QBist — what with his contention that "in our description of nature the purpose is not to disclose the real essence of the phenomena but only to track down, so far as it is possible, relations between the manifold aspects of our experience" [BCW 6:296]. And no, by "our experience" he did *not* mean the objective readings of large classical instruments. What qualifies him as a QBist, if not as the unwitting *founder* of QBism, is his insistence on *individual* experience, which only through communication becomes *objective* knowledge.

One of the ways in which contemporary QBism goes beyond Bohr consists in replacing the standard projector-valued probability measures by positive-operator-valued measures (POVMs). This makes it possible to formulate the Born rule entirely in terms of probabilities (as outlined in the Appendix) and to view quantum mechanics as nothing but a generalization of the (personalist) Bayesian theory of probability. In their justifiable enthusiasm, however, QBists also overshoot their mark, as when they permit incoherent superpositions to be assigned to possibilities involving distinct cognitive states, or when they claim that quantum mechanics is "*explicitly* local in the QBist interpretation" [66]. According to Fuchs [71],

> [QBism] gives each quantum state a home. Indeed, a home localized in space and time — namely, the physical site of the agent who assigns

> it! By this method, one expels once and for all the fear that quantum mechanics leads to "spooky action at a distance."

A quantum state has its home in an agent's mind, not at any physical site — which would have to be a site in our common external world, since there are no agents in "the world as it is without agents." Why inserting minds into our common external world is a bad idea has been explained by Schrödinger, who according to Fuchs *et al.* [66] took "a QBist view" of science. By the very fact that we treat some of our experiences as aspects of a shared external world, Schrödinger [37, 118–121] argued, we exclude ourselves from this world: "We step with our own person back into the part of an onlooker who does not belong to the world, which by this very procedure becomes an objective world." If we then reify the mind's creation, we are left with no choice but to insert the mind into its creation: "I so to speak put my own sentient self (which had constructed this world as a mental product) back into it — with the pandemonium of disastrous logical consequences" that flow from this error, such as "our fruitless quest for the place where mind acts on matter or vice-versa." Locating experiences in our common external world is therefore not an option. Instead of asserting that QBism is explicitly local, QBists ought to assert that QBism is neither local nor nonlocal in any realist sense of these terms.

It is strange indeed to see a QBist look upon spooky action at a distance as something to be feared. To banish it by claiming that quantum mechanics is local is to concede way too much to those who see themselves as modeling a reality not of their own making. Something fearsome is implied only if one forgets that physics deals with our common external world. One then has to worry how, in the absence of a common cause, measurement outcomes in spacelike relation can be so spookily correlated. Keeping in mind that measurement outcomes are responses from a reality beyond the reach of human sensory experience, one realizes (as Bohr did) that the answer to this question is beyond the reach of our concepts. After all, diachronic correlations (between successive experiences of the same agent) are no less inexplicable than synchronic correlations (between simultaneous experiences of different agents).

All in all, QBism, through its emphasis on the individual experiencing subject, brings home the intersubjective constitution of our common external world more forcefully than Bohr ever did. (The time wasn't ripe for this then. Perhaps it is now.) Bohr's insights, on the other hand, are eminently useful in clarifying the QBist position, attenuating its excesses, and enhancing its internal consistency.

In concluding, I want to extend my gratitude to the QBists for making me finally come round to seeing that there is *no* difference between observations qua experiences and observations qua measurement outcomes (which explains why measurements are irreversible and outcomes definite).

Appendix: The Born rule according to QBism

If quantum theory is so closely allied with probability theory, why is it not written in a language that starts *with probability, rather than a language that ends with it? Why does quantum theory invoke the mathematical apparatus of Hilbert spaces and linear operators, rather than probabilities outright? — Christopher A. Fuchs* [71]

For QBists, quantum mechanics is a generalization of the Bayesian theory of probability. It is a calculus of consistency — a set of criteria for testing coherence between beliefs. As there are no external criteria for declaring a probability judgment right or wrong, so there are no external criteria for declaring a quantum state assignment right or wrong. The only criterion for the adequacy of a probability judgment or a state assignment is internal coherence between beliefs. The Born rule thus is not simply a rule for updating probabilities, for getting new ones from old. It is a rule for relating probability assignments and constraining them. As such, it can be expressed entirely in terms of probabilities.

The proof of the last claim requires the use of POVMs, which generalize the standard projector valued measures used by von Neumann [53] and Dirac [57]. It goes like this: While a density operator ρ determines a potentially infinite number of probabilities, these cannot all be independent. On a d-dimensional Hilbert space, ρ is completely determined by the d^2 probabilities it assigns to the outcomes (represented by linearly independent positive operators E_i) of an *informationally complete* measurement. Any density operator ρ therefore corresponds to a vector whose d^2 components are the Born probabilities

$$p_i = \mathrm{Tr}(\rho E_i)\,, \tag{1}$$

and any POVM $\{F_j\}$ corresponds to a matrix whose elements are the conditional probabilities

$$R_{ji} = \mathrm{Tr}(\Pi_i F_j)\,, \tag{2}$$

where the Π_i are 1-dimensional projectors proportional to E_i. This makes it possible to write the Born rule in the generic form

$$q(F_j) = f\big(\{R_{ji}\}, \{p_i\}\big)\,, \tag{3}$$

where f depends on the details of the informationally complete measurement $\{E_i\}$.

The function f takes a particularly simple form if the positive operators E_i constitute a *symmetric* informationally complete (SIC) measurement. In this case, one of the ways in which the Born rule can be written turns out to be [71,72]:

$$q(F_j) = \sum_{i=1}^{d^2} \left[(d+1)\, p_i - \frac{1}{d}\right] R_{ji}\,. \tag{4}$$

If the positive operators F_j are mutually orthogonal projectors representing the outcomes of a complete von Neumann measurement, the Born rule takes the even simpler form

$$q(F_j) = (d+1) \sum_{i=1}^{d^2} R_{ji} p_i - 1\,. \tag{5}$$

While the probabilities (4) and (5) are expressed in terms of (i) the probabilities p_i that an agent assigns to the possible outcomes of the SIC measurement, and (ii) the conditional probabilities R_{ji} that the agent assigns to the possible outcomes of a subsequent measurement if the SIC measurement is actually made, they pertain to a situation in which the SIC measurement is *not* made. If it *is* made, the law of total probability applies, and we have

$$q(F_j) = \sum_{i=1}^{d^2} R_{ji} p_i\,. \tag{6}$$

Comparing Eqs. (4) and (5) with Eq. (6), one can see that "[t]he Born Rule is nothing but a kind of Quantum Law of Total Probability! No complex amplitudes, no operators — only probabilities in, and probabilities out" [71].

QBists hope to eventually be in a position to derive the standard Hilbert space formalism from the Born rule. And they hope so to distill the essence of quantum mechanics and the essential characteristic of the quantum world.[r] This a fascinating, highly ambitious, and seriously challenging project. Do SIC measurements even exist? Unfortunately, proofs of their existence are elusive. As of May 2017, such proofs have been found for

[r]While the Born rule is normative — it guides an agent's behavior in a world that is fundamentally quantum — it is also an empirical rule. It is a statement about the quantum world, indirectly expressed as a calculus of consistency for bets placed on the outcomes of measurements.

all dimensions up to $d = 151$, and for a few others up to 323 [65]. The mood of the QBist community nevertheless is that a SIC measurement should exist for every finite dimension. That said, it must be stressed that the general form of the Born rule, Eq. (3), does not depend on the existence of SIC measurements; it only presupposes informationally complete POVMs, and these are known to exist for all finite dimensions.

References

1. C.A. Fuchs (2010). *Coming of Age With Quantum Information: Notes on a Paulian Idea*, p. 192. New York: Cambridge University Press.
2. C.M. Caves, C.A. Fuchs, and R. Schack (2002). Quantum probabilities as Bayesian probabilities. *Physical Review A* **65**, 022305.
3. N.D. Mermin (2017). Why QBism is not the Copenhagen interpretation and what John Bell might have thought of it. In: R. Bertlmann and A. Zeilinger (Eds.), *Quantum [Un]Speakables II: 50 Years of Bell's Theorem* (83–93). Switzerland: Springer International Publishing.
4. N.D. Mermin (2013). QBism as CBism: Solving the Problem of "the Now," arXiv: 1312.7825 [quant-ph].
5. N.D. Mermin (2019). Making better sense of quantum mechanics. *Reports of Progress in Physics* **82**, 012002.
6. M.D. Howe (1961). *Holmes Pollock Letters*, p. 252. Cambridge, MA: Belknap Press.
7. M. Bitbol, P. Kerszberg, and J. Petitot (2009). *Constituting Objectivity: Transcendental Perspectives on Modern Physics.* Springer Science + Business Media.
8. B. Falkenburg (2009). A critical account of physical reality. In [7], 229–248.
9. S. Brock (2009). Old wine enriched in new bottles: Kantian flavors in Bohr's viewpoint of complementarity. In [7], 301–316.
10. B. d'Espagnat (2009). A physicist's approach to Kant. In [7], 481–490.
11. J. Faye and H.J. Folse (1994). *Niels Bohr and Contemporary Philosophy.* Dordrecht: Springer.
12. C. Chevalley (1994). Niels Bohr's words and the Atlantis of Kantianism. In [11], 33–55.
13. H. Folse (1994). Bohr's framework of complementarity and the realism debate. In [11], 119–139.
14. C.A. Hooker (1994). Bohr and the crisis of empirical intelligibility: An essay on the depth of Bohr's thought and our philosophical ignorance. In [11], 155–199.
15. M. Bitbol (2010). Reflective metaphysics: Understanding quantum mechanics from a Kantian standpoint. *Philosophica* **83**, 53–83.
16. M.E. Cuffaro (2010). The Kantian framework of complementarity. *Studies in History and Philosophy of Modern Physics* **41**, 309–317.
17. J. Honner (1982). The transcendental philosophy of Niels Bohr. *Studies in History and Philosophy of Science: Part A* **13**, 1–29.

18. D. Kaiser (1992). More roots of complementarity: Kantian aspects and influences. *Studies in History and Philosophy of Science: Part A* **23**, 213–239.
19. E. MacKinnon (2012). *Interpreting Physics: Language and the Classical/ Quantum Divide*. Dordrecht: Springer.
20. J.R. Searle (2004). *Mind: A Brief Introduction*. New York: Oxford University Press.
21. E.P. Wigner (1961). Remarks on the mind–body question. In: I.J. Good (Ed.), *The Scientist Speculates* (284–302). London: Heinemann.
22. J.S. Bell (1990). Against "measurement." In: A.I. Miller (Ed.), *62 Years of Uncertainty* (17–31). London: Plenum.
23. N. Bohr (1985). *Niels Bohr: Collected Works*, Vol. 6. Amsterdam: North-Holland.
24. N. Bohr (1996). *Niels Bohr: Collected Works*, Vol. 7. Amsterdam: Elsevier.
25. N. Bohr (1999). *Niels Bohr: Collected Works*, Vol. 10. Amsterdam: Elsevier.
26. N.D. Mermin (1990). *Boojums All the Way Through: Communicating Science in a Prosaic Age*, p. 189. Cambridge, UK: Cambridge University Press.
27. C. Chevalley (1999). Why do we find Bohr obscure? In D. Greenberger, W.L. Reiter, and A. Zeilinger (Eds.), *Epistemological and Experimental Perspectives on Quantum Physics* (59–73). Dordrecht: Springer.
28. C.A. Hooker (1972). The nature of quantum mechanical reality: Einstein versus Bohr. In: R.G. Colodny (Ed.), *Paradigms & Paradoxes: The Philosophical Challenge of the Quantum Domain* (67–302). Pittsburgh, PA: University of Pittsburgh Press.
29. D. Bohm (1952). A suggested interpretation of the quantum theory in terms of "hidden" variables. *Physical Review* **85**, 166–179, 180–193.
30. I. Kant (1998). *Critique of Pure Reason*. Cambridge, UK: Cambridge University Press.
31. M. Bitbol (1998). Some steps towards a transcendental deduction of quantum mechanics. *Philosophia Naturalis* **35**, 253–280.
32. J.S. Bell (1964). On the Einstein Podolsky Rosen paradox. *Physics Physique Fizika* **1** (3), 195–200.
33. S. Kochen and E.P. Specker (1967). The problem of hidden variables in quantum mechanics. *Journal of Mathematics and Mechanics* **17**, 59–87.
34. A.A. Klyachko, M.A. Can, S. Binicioğlu, and A.S. Shumovsky (2008). Simple test for hidden variables in the spin-1 system. *Physical Review Letters* **101**, 020403.
35. M. Friedman (2009). Einstein, Kant, and the relativized *a priori*. In [7], 253–267.
36. E. Schrödinger (1964). What is real? In: *My View of the World* (Part 2). Cambridge, UK: Cambridge University Press.
37. E. Schrödinger (1992). *What Is Life? With: Mind and Matter & Autobiographical Sketches*. Cambridge, UK: Cambridge University Press.
38. E. Schrödinger (1992). The Arithmetic Paradox: The Oneness of Mind. In [37], 128–139.
39. E. Schrödinger, The Principle of Objectivation. In [37], 117–127.

40. J. Levine (2001). *Purple Haze: The Puzzle of Consciousness*, p. 78. New York: Oxford University Press.
41. D. Welton (1999). *The Essential Husserl.* Bloomington and Indianapolis: Indiana University Press.
42. H. Putnam (1981). *Reason, Truth, and History*, p. 61. Cambridge, UK: Cambridge University Press.
43. H. Putnam (1987). *The Many Faces of Realism*, p. 36. LaSalle, IL: Open Court.
44. B. d'Espagnat (2003). *Veiled Reality: An Analysis of Present-Day Quantum Mechanical Concepts.* Boulder, CO: Westview Press.
45. E. Schrödinger (2014). *Nature and the Greeks, and Science and Humanism*, p. 162. Cambridge, UK: Cambridge University Press (Canto Classics).
46. N.D. Mermin (2016). What's wrong with those epochs. In: *Why Quark Rhymes With Pork And Other Scientific Diversions*, Chap. 9. Cambridge, UK: Cambridge University Press.
47. N. Bohr (1949). Discussion with Einstein on the epistemological problems in atomic physics. In: P.A. Schilpp (Ed.), *Albert Einstein: Philosopher–Scientist* (201–241). New York: MJF Books.
48. C.F. Von Weizsäcker (2006). *The Structure of Physics*, p. 257. Dordrecht: Springer.
49. W. Heisenberg (1958). *Physics and Philosophy.* New York: Harper and Brothers.
50. B. Falkenburg (2007). *Particle Metaphysics: A Critical Account of Subatomic Reality.* Berlin: Springer.
51. O. Ulfbeck and A. Bohr (2001). Genuine Fortuitousness. Where Did That Click Come From? *Foundations of Physics* **31**, 757–774.
52. N.D. Mermin (2009). What's bad about this habit? *Physics Today* **62** (5), 8–9.
53. J. von Neumann (1955). *Mathematical Foundations of Quantum Mechanics.* Princeton, NJ: Princeton University Press.
54. H. Everett III (1973). The theory of the universal wave function & "Relative state" formulation of quantum mechanics. In: B.S. DeWitt and R.N. Graham (Eds.), *The Many-Worlds Interpretation of Quantum Mechanics* (3–140, 141–149). Princeton, NJ: Princeton University Press.
55. N. Bohr (1935). Can quantum-mechanical description of physical reality be considered complete? *Physical Review* **48**, 696–702.
56. A. Einstein, B. Podolsky, and N. Rosen (1935). Can quantum-mechanical description of physical reality be considered complete? *Physical Review* **47**, 777–780.
57. P.A.M. Dirac (1958). *The Principles of Quantum Mechanics*, 4th Edition, p. 46. London: Oxford University Press (Clarendon).
58. B.C. Van Fraassen (1990). The problem of measurement in quantum mechanics. In: P. Lahti and P. Mittelstaedt (Eds.), *Symposium on the Foundations of Modern Physics* (497–503). Singapore: World Scientific.
59. D. Wallace (2008). Philosophy of Quantum Mechanics. In: D. Rickles (Ed.),

The Ashgate Companion to Contemporary Philosophy of Physics (16–98). Burlington, VT: Ashgate.

60. R. Penrose (1986). Gravity and state vector reduction. In: R. Penrose and C.J. Isham (Eds.), *Quantum Concepts in Space and Time* (129–146). New York: Oxford University Press (Clarendon).
61. M. Kaku (1995). *Hyperspace*, p. 263. New York: Oxford University Press.
62. D.F. Styer, M.S. Balkin, K.M. Becker, M.R. Burns, C.E. Dudley, S.T. Forth, J.S. Gaumer, M.A. Kramer, D.C. Oertel, L.H. Park, M.T. Rinkoski, C.T. Smith, and T.D. Wotherspoon (2002). Nine formulations of quantum mechanics. *American Journal of Physics* **70**, 288–297.
63. U. Mohrhoff (2009). Quantum mechanics explained. *International Journal of Quantum Information* **7**, 435–458.
64. F. Wilczek (2008). *The Lightness of Being: Mass, Ether, and the Unification of Forces*, p. 164. New York: Basic Books.
65. C.A. Fuchs (2017). Notwithstanding Bohr, the reasons for QBism. *Mind and Matter* **15**, 245–300.
66. C.A. Fuchs, N.D. Mermin, and R. Schack (2014). An introduction to QBism with an application to the locality of quantum mechanics. *American Journal of Physics* **82**, 749–754.
67. K. Camilleri and M. Schlosshauer (2015). Niels Bohr as philosopher of experiment: Does decoherence theory challenge Bohr's doctrine of classical concepts? *Studies in History and Philosophy of Modern Physics* **49**, 73–83.
68. N.D. Mermin (2012). Quantum mechanics: Fixing the shifty split. *Physics Today* **65** (7), 8–10.
69. C.A. Fuchs and R. Schack (2015). QBism and the Greeks: Why a quantum state does not represent an element of physical reality. *Physica Scripta* **90**, 015104.
70. C.A. Fuchs and R. Schack (2004). Unknown quantum states and operations, a Bayesian view. In: M. Paris and J. Řeháček (Eds.), *Quantum State Estimation* (151–190). Berlin: Springer.
71. C.A. Fuchs (2010). Quantum Bayesianism at the Perimeter. *Physics in Canada* **66**, 77–81.
72. C.A. Fuchs (2004). On the Quantumness of a Hilbert Space. *Quantum Information & Computation* **4**, 467–478.

GREEK PHILOSOPHY FOR QUANTUM PHYSICS. THE RETURN TO THE GREEKS IN THE WORKS OF HEISENBERG, PAULI AND SCHRÖDINGER

RAIMUNDO FERNÁNDEZ MOUJÁN

CONICET,
Institute of Philosophy "Dr. A. Korn",
Buenos Aires University, Argentina
Center Leo Apostel and Foundations of the Exact Sciences,
Brussels Free University, Belgium
E-mail: raifer86@gmail.com

Werner Heisenberg, Erwin Schrödinger and Wolfgang Pauli exhibited in their works a strong and insistent interest in Greek philosophy. And this interest — they claimed — was not at all separated from their investigations into the new quantum theory. Heisenberg directly affirmed that "one could hardly make progress in modern atomic physics without a knowledge of Greek natural philosophy". What does this claim mean? Why do these central figures in the development of quantum mechanics saw the Greeks as their main inspiring source? Why do they took them as a model for contemporary science, even over their modern predecessors of Enlightenment? This work attempts to answer these questions focusing on three main reasons: the revision of atomism, the recasting of the meaning of "understanding" in physics, and the critique of separations in science.

Keywords: Heisenberg; Pauli; Schrödinger; Greek philosophy; quantum mechanics.

Never once did it occur to me to consider the science and technology of our times as belonging to a world basically different from that of the philosophy of Pythagoras and Euclid

Werner Heisenberg

I hope no one still maintains that theories are deduced by strict logical conclusions from laboratory-books, a view which was still quite fashionable in my student days

Wolfgang Pauli

Introduction

The question, expressed in the title of this work, about the importance or value of Greek philosophy for quantum physics, about the importance of the *oldest* philosophy for the *newest* science, seems perhaps a bit extravagant, maybe far-fetched. This impression is however rapidly dissipated when one reads the works of three of the most important physicists of the early XXth century, and three of the "founders" of the quantum theory. Werner Heisenberg, Erwin Schrödinger and Wolfgang Pauli exhibited a strong and insistent interest in Greek philosophy. And this interest — they claimed — was not at all separated from their investigations into the new quantum theory. The subject of this work can then be paraphrased, for further precision, as the possible answers to the following questions: Why was the revision of Greek philosophy so important for them? Why were Schrödinger, Pauli and Heisenberg so interested in ancient Greek thought while working on the new quantum physics? But even before answering *why*, we can certainly say that they were not wrong in that interest. It will never be a mistake or a waste of time to go back to the ancient Greeks. They will always be able to teach us the art of thinking, the meaning of understanding, and they will once again remind us what a peculiar and prodigious thing this human endeavour of knowledge is. This rediscovery will most probably instill in us a new freedom of thought, while at the same time rendering us loyal to the rigors of that recovered original art of thinking and understanding. "Study the Greeks": that old advice is far from representing a conservative spirit, this as much is proved by the fact that the most innovative among physicists were the ones who saw the necessity to investigate what the Greeks could show them.

These first thoughts, however allusive, already lead us to a first, introductory answer to the questions about the importance of revisiting Greek philosophy for these physicists: the challenges quantum physics had set in front of the physicists from the early XXth century couldn't be tackled with the known tools of classical physics, they needed to recuperate a more flexible — yet always disciplined — way of thinking, and an approach to the question of understanding nature less burdened by the customs of their immediate scientific context. In order to achieve that, a jump back beyond the habits of modernity and into the lively Greek source, seemed to them a proper means. As Schrödinger puts it: "By the serious attempt to put ourselves back into the intellectual situation of the ancient thinkers (...) we may regain from them their freedom of thought" [11, p. 19].

Only if we approach things in this manner we can understand what, at first sight, seems like a persistent contradiction in the works of physicists like Heisenberg, Pauli and Schrödinger: the relation between the conscience of being faced with the necessity to take a jump forwards, to a new theory radically different from the known physics, and the fact that, in this task, they showed so much interest in what lay so far in the past, in the most ancient scientists of them all, in the source of western science and philosophy. Confronted with the strange indications of the quantum formalism, it may seem that the only thing we can do is to leap forwards into the void, to do something completely and radically new, or to abandon all hope of understanding and turn to instrumentalism. We tend to think that, given the fact that what QM shows us does not fit at all with the image of the physical world developed during the last five centuries, there is no other support or precedent to rely on. This is for instance Bohr's way of thinking, for whom the concepts of classical physics would necessary remain the concepts of physics for all times — and the reality expressed by quantum physics would remain unknowable beyond metaphorical allusions. But the fact is that there are other available associates that can, at least, help shake off some habits of modern thought while remaining faithful to a scientific spirit — and some of the founders of quantum physics understood this. Aware of the apparent oddness of the attempt, Schrödinger tells us that, when speaking about the subjects he later developed in his book *Nature and the Greeks*, he felt the need to explain himself:

> There was need to explain (...) that in passing the time with narratives about ancient Greek thinkers and with comments on their views I was *not* just following a recently acquired hobby of mine; that it did not mean, from the professional point of view, a waste of time, which ought to be relegated to the hours of leisure; that it was justified by the hope of some gain in understanding modern science and thus *inter alia* also modern physics [11, p. 3].

Heisenberg, less preoccupied by appearances, goes even further in claiming that "one could hardly make progress in modern atomic physics without a knowledge of Greek natural philosophy" [8, p. 61]. One hopes, the other is certain, but they both point to the same idea: this "progress", this advancement, the radically *new*, is conditioned or, in any case, assisted, by the *old*, by a knowledge of the original source. Without the ability to pose again questions of principle, to recuperate a broader view of the task of understanding nature, without the ability to shake off some modern presuppositions, gained by the knowledge of the Greeks, it is difficult — they believe — to be able to tackle the new atomic physics. For Pauli, the Greeks also allowed them to acquire a more profound view of the problems they were facing. Problems that seemed entirely new, but that could be in fact related to the great old problems and dilemmas, and therefore placed in a different perspective:

> The critical scientific spirit however reached its first culmination in classical Hellas. It was there that those contrasts and paradoxes were formulated which also concern us as problems, though in altered form: appearance and reality, being and becoming, the one and the many, sense experience and pure thought, the continuum and the integer, the rational ratio and the irrational number, necessity and purposefulness, causality and chance [9, p. 140].

As we revise those quotes, we are made aware of the striking contrast between the common views of Schrödinger, Pauli and Heisenberg, and the widely accepted vision today in scientific contexts — and certainly also among philosophers of science —, according to which science, properly speaking, originated in modern times, in the XVIIth century. These physicists from the early XXth century did not see what began in the XVIIth century and was developed over the following centuries as the origin of science, but rather as the origin of the specific parameters of modern science that quantum mechanics, among other developments, was rejecting. Enlightenment, however admirable, was for them the establishment of a certain view of science that had become problematic, and they saw better allies for their task among the ancient Greeks. They found Plato, Pythagoras or Heraclitus more useful than the modern philosophers and physicists. They were insistent critics of the parameters of science established in the XVIIth, XVIIIth and XIXth centuries. In their works they frequently make this contrast explicit, they distinguish between the aspects of science that had their origin in the XVIIth century and that seemed now problematic (the separation of the world in *res extensa* and *res cogitans*, the limiting of natural science to a narrow materialism, the sharp separations among disciplines and faculties, etc.) and the ancient Greek model, which could help them overcome those obstacles.

Quantum mechanics was born in a critical moment of modernity (that stretches still to our times), it should not surprise us then that the most lucid minds among those involved in the new physics saw the ancient more fruitful and inspiring than the modern. Schrödinger pointed that this critical moment had created a sort of *Zeitgeist* leading him to the Greeks: "Far from following an odd impulse of my own, I had been swept along unwittingly, as happens so often, by a trend of thought rooted somehow in the intellectual situation of our time." [11, p. 14]. This *Zeitgeist* was expressed in "the inordinately critical situation in which nearly all the fundamental sciences find themselves ever more disconcertingly enveloped" [11, p. 5]. Among those fundamental sciences in crisis, physics of course stood out:

> The modern development [relativity and quantum mechanics], which those who have brought it to the fore are yet far from really understanding, has intruded into the relatively simple scheme of physics which towards the end of the nineteenth century looked fairly stabilized. This intrusion has, in a way, overthrown what had been built on the foundations laid in the seventeenth century, mainly by Galileo, Huygens and Newton. The very foundations were shaken [11, p. 17].

Of course, this *Zeitgeist* pushing him to the Greeks arouse other responses as well. Ernst Mach's response is, as Schrödinger reminds us, exemplary: "he recommends a quaint method of getting beyond antiquity, namely to neglect and ignore it" [11, p. 21]. But Pauli, Schrödinger and Heisenberg, unconvinced by Mach's *tabula rasa*, went on to express their interest in Greek philosophy in relation to particular issues, that will be here divided into three. One was the question of *understanding*: they turned to the Greeks to recuperate a broader perspective on what it means to understand nature through science, and they hoped this perspective would allow them new room to maneuver. Another issue leading them to the Greeks was *separation*: they turned to the ancients to try to find a way out of the extremely compartmentalized and separated scientific landscape. And another very different reason to read Greek philosophy was for them related — as we shall see in the first section — not to its role in inspiring the new physics, but to the critical analysis of the basis of modern science, which in fact took much from some Greek philosophers. Schrödinger reminds us that "the thinkers who started to mould modern science did not begin from scratch. Though they had little to borrow from the earlier centuries of our era, they very truly revived and continued ancient science and philosophy" [11, pp. 17-18]. And this "is a further incentive for us to return once again to an assiduous study of Greek thought. There is not only (...) the hope of unearthing obliterated wisdom, but also of discovering inveterate error at the source" [11, p. 18]. As we shall see right away, the main Greek influence over modern physics, and the one that appeared as most problematic in the XXth century, was atomism.

1. Versions of atomism

Explaining the different motives for the necessary study of the Greeks, Schrödinger distinguished specially two reasons: on one hand, he hopes they can inspire us to reunite what is now separated (philosophy and science) and that shined as a unity among them; but the second reason aims at producing, at the same time, a critical analysis of what turned out to be problematic in the influence of Greek thought over modern physics. Undoubtedly, for Schrödinger, but also for Heisenberg and Pauli, the Greek philosophy that most strongly and effectively informed the basis of modern physical thought and that — given what QM was showing — appeared to them as extremely problematic, was atomism. As usual, among them it is Heisenberg who describes this philosophy best:

> The antithesis of Being and Not-being in the philosophy of Parmenides is here secularized into the antithesis of the 'Full' and the 'Void'. Being is not only One, it can be repeated an infinite number of times. This is the atom, the indivisible smallest unit of matter. The atom is eternal and indestructible, but it has a finite size. Motion is made possible through the empty space between the atoms. Thus for the first time in history there was voiced the idea of the existence of smallest ultimate particles — we would say of elementary particles, as the fundamental building blocks of matter. According to this new concept of the atom, matter did not consist only of the 'Full', but also of the 'Void', of the empty space in which the atoms move. The logical objection of Parmenides against the Void, that not-being cannot exist, was simply ignored to comply with experience [6, pp. 65-66].

Quite rightly, Heisenberg describes atomism in contrast to Parmenidean philosophy, and shows what, for many among the Greeks, seemed weak about the atomist proposal. Let's say a word about Parmenides to see if we can better understand Heisenberg's quote. In his poem, Parmenides aims at indicating the way which can lead us to true knowledge, if properly followed. This path is supported by a main certitude, that Parmenides puts in a famously brief manner: "*is* and not being is impossible" [DK 28 B]. Strangely, the conjugate verb *is* (*estín* in ancient Greek) appears without subject. *Who* or *what* is? Parmenides seems to be purposely leaving the verb without subject to differentiate his philosophy from those common already in Greek thought. If the previous or contemporary philosophers gave privilege to one or several 'elements' or substances as origin and foundation of nature (of which they primordially predicated being), or dedicated their thought to decipher the hidden order that governs reality (the case of Heraclitus), Parmenides starts by reflecting on a previous, humbler truth: any 'element' we may choose, any 'order', anything of any nature, must be, first, something that is, must share being. The simple yet universal, all-encompassing *fact of being* (as Néstor Cordero puts it [1]) is the origin of the Parmenidean wonder. Cordero's proposition (that we take Parmenides' being as the "fact of being") is useful for it shows that Parmenides wishes above all not to substantialize being. Parmenides' being is not a transcendent principle, is not The Being, The One, but the all-encompassing, irreducible, fact of being — that he will later on in the poem refer to as "what is". But, as we saw, Parmenides thesis adds something else, it also affirms the impossibility of not-being. In fact, it is this part of the thesis that is most extensively and effectively developed throughout the poem. According to Parmenides, when we understand being in this manner, its opposite, not-being, appears as impossible. And one of the most insisted upon ways in which Parmenides phrases this impossibility in his poem is the one that identifies not-being with *separation*: "you will not sever what is from holding to what is" [DK 28 B4]; "it is wholly continuous; for what is, is in contact with what is" [DK 28 B8.25]; "Nor is it divisible, since it is all alike" [DK 28 B8.22]. In this sense, "not-being is impossible" means that there is no cut, no strip, no ditch, inside being, through which not being would pass. Being has no cracks within, no interstices. There may be differences, but there are certainly no ontological separations. It is this impossibility that atomism, as Heisenberg puts it, simply ignores. They take being not in a Parmenidean manner as the irreducible totality of existence, but as a kind of indivisible and extremely small body, many times repeated, as a small substance infinitely combined to construct reality, and so they postulate that not-being (for this construction to happen) must also *be*.

Schrödinger, on his part, is satisfied with a more general description, one that does not pass through Parmenides, but that aims at rendering explicit the fact that this ancient theory is still essentially our own: "the bodies consist of discrete particles, which themselves do not change, but recede from each other or come closer together, leaving more or less empty space between them. That was their, and that is our, atomic theory." [11, p. 64]. But what interests us now is that he later explains how this ancient and quite strange theory came to be accepted in modern times and cemented in our common sense:

> From the lives and writings of Gassendi and Descartes, who introduced atomism into modern science, we know as an actual historical fact that, in doing so, they were fully aware of taking up the theory of the ancient philosophers whose scripts they had diligently studied. Furthermore, and more importantly, all the basic features of the ancient theory have survived in the modern one up to this day, greatly enhanced and widely elaborated but unchanged, if we apply the standard of the natural philosopher, not the myopic perspective of the specialist [11, pp. 82-83].

Those who established the basis of modern science found this ancient theory useful, and so it became a fundamental part of the great intellectual edifice they constructed, one that, in many essential aspects, lasts to our days. Heisenberg also describes the historical path through which atomism passed to us, specifically how it was united with the materialistic view that began to take shape in modern times and that, dividing reality in *res extensa* and *res cogitans* (a division to which we will go back later), limited physics to the study of the mechanics of this now merely material *res extensa*:

> Matter was thought of in terms of its mass, which remained constant through all changes, and which required forces to move it. Because, from the eighteenth century onwards, chemical experiments could be classified and explained by the atomic hypothesis of ancient times, it appeared reasonable to take over the view of ancient philosophy that atoms were the real substance, the immutable building-stones of matter. Just as in the philosophy of Democritus, the differences in material qualities were considered to be merely apparent; smell or colour, temperature or viscosity, were not actual qualities of matter but resulted from the interaction of matter and our senses, and had to be explained by the arrangements and movements of atoms (...). It is thus that there arose the over-simplified world-view of nineteenth-century materialism: atoms move in space

> and time as the real and immutable substances, and it is their arrangement and motion that create the colourful phenomena of the world of our senses [8, p. 12].

Schrödinger too sees this modern atomist worldview as naïve, he claims that the term "atom" has become "a misnomer" [12, p. 183], and states the fact that became evident for those involved in the development of the new quantum theory: "modern atomic theory has been plunged into a crisis. There is no doubt that the simple particle theory is too naïve. This is not altogether too astonishing, from the above speculations about its origin" [11, p. 87]. The origin Schrödinger refers to is precisely described in his book through a hypothesis that seems, in principle, well supported by the ancient sources. He sees the atomic postulate as originating from a widely known and discussed problem in Greek thought: the problem that arose for the understanding of the physical world from the mathematics of the continuum, mainly, the aspect of its infinite divisibility. This infinite mathematical divisibility, when applied to the understanding of reality, gave way to paradoxes (which were famously developed by Zeno of Elea, who Plato saw in fact as supporting Parmenidean philosophy). And so, a limit to this divisibility — for the physical world — was proposed by atomism as a postulate, and these small indivisible bodies, separated by not-being, were hypothesized. But this patch that once proved so effective seemed to be peeling off in the early XXth century and its weaknesses started to show. What was once a basic assumption for physical thought was now exposed as an inadequate postulate, one that represented an obstacle for the new theory. Schrödinger writes: "We have taken over from previous theory the idea of a particle and all the technical language concerning it. This idea is inadequate. It constantly drives our mind to ask information which has obviously no significance" [12, p. 188]. But perhaps it is Heisenberg who explains most clearly why the atomic postulate began to fail in quantum theory:

> Let us discuss the question: what is an elementary particle? We say, for instance, simply 'a neutron' but we can give no well-defined picture and what we mean by the word. We can use several pictures and describe it once as a particle, once as a wave or as a wave packet. But we know that none of these descriptions is accurate. Certainly the neutron has no colour, no smell, no taste. In this respect it resembles the atom of Greek philosophy. But even the other qualities are taken from the elementary particle. At least to

> some extent; the concepts of geometry and kinematics, like shape or motion in space, cannot be applied to it consistently. If one wants to give an accurate description of the elementary particle — and here the emphasis is on the word 'accurate' — the only thing which can be written down as description is a probability function [6, p. 70].

Quantum physics goes much further than atomism, which denied qualities as smell, colour or taste for their elementary particles. Now even shape or motion in space have to be excluded. What are we left with? The particle loses, in quantum theory, all of its defining characteristics. As Schrödinger writes, it even loses identity, "sameness": an "elementary particle", he says, "is not an individual; it cannot be identified, it lacks 'sameness'" [12, p. 183]. There's almost nothing left of it, and what's left, is certainly an aspect that doesn't seem too adequate to atomism, that could perhaps be better understood outside of the atomist frame: a strange probability that can't be interpreted in a classical manner, by ignorance, just as a mental calculation, as an abstract probability dependent on our degree of knowledge about a real, actual particle. On the contrary, this is a probability that, in fact, interacts. A new kind of probability, which seems to have physical existence independently of an actual particle, of which Schrödinger rightly points out: "Something that influences the physical behaviour of something else must not in any respect be called less real than the something it influences" [12, p. 185]. In one instance, trying to provide a new understanding of the reality of this quantum probability, Heisenberg finds a possible ally, among the ancient Greeks, in Aristotle and in his concept of *dynamis* (potency, possibility, capability: all possible translations) as a way of being; a concept modern physics, in its beginnings, ignored to focus only on the characterisation Aristotle did of the actual way of being — which of course in Aristotle's own philosophy could not be separated from the potential way of being, as modern scientists did (see [4]). We should add that this last example is quite representative of the conflict between the relation classical physics established with ancient Greek thought and the relation that quantum physics calls for. Exactly what was excluded, regarded as problematic (Aristotle's concept of potency) by classical physicists, became for those who were developing the quantum theory an interesting possibility; and what for classical physics functioned so well (the principles that for Aristotle defined the formal and actual aspects of entities) became for quantum theory naïve and inadequate parameters.

But apart from his brief speculations about Aristotle's concept of potency, Heisenberg actually attempts to develop a peculiar and radically different atomist view; a strange, non-materialistic atomism, based on the philosophy of one who, in fact, strongly opposed the atomists: Plato. We can safely say that the *Timaeus* was one of the philosophical works that most deeply influenced Heisenberg since a young age. Heisenberg himself, in *Physics and beyond*, describes his entire intellectual itinerary as different stations in his interpretation of Plato's *Timaeus*[a]. Focusing on this dialogue, where through the character of *Timaeus* it is proposed as plausible that pure geometrical forms are the building blocks of corporeal reality, Heisenberg sees a Plato that develops Pythagoreanism in a new direction, providing a different "atomism" — built up now from mathematical forms and not from material indivisible bodies — that can be better related to quantum mechanics. "Plato was not an atomist; on the contrary, Diogenes Laertius reported that Plato disliked Democritus so much that he wished all his books to be burned. But Plato combined ideas that were near to atomism with the doctrines of the Pythagorean school and the teachings of Empedocles" [6, p. 67]. Pauli shares with Heisenberg this Pythagorean interpretation of Plato: "As a reaction against the (...) atomists, Plato took over into his doctrine of ideas many of the mystical elements of the Pythagoreans. He shares with them his higher valuation of contemplation as compared with ordinary sense experience, and his passionate participation in mathematics, especially in geometry, with its ideal objects." [9, p. 141]. Both Pauli and Heisenberg see Plato as taking much from the Pythagoreans, but steering away from what appeared as mystical associations, which Plato wished rather to gain — without losing what was profound in them — for knowledge, constructing in this way a synthesis that Pauli regards — as we shall later see — as a model.

According to Heisenberg's Platonic atomism, "the smallest parts of matter are not the fundamental Beings, as in the philosophy of Democritus, but are mathematical forms. Here it is quite evident that the form is more important than the substance of which it is the form" [6, p. 69]. Heisenberg, through his peculiar Platonic atomism, is purposely disobeying the modern division between *res extensa* and *res cogitans* which confined physics to the material *res extensa*. And this is not done by incorporating the subject (his choice, his perspective, etc.); he is incorporating real intelligible, formal, elements in the explanation of the physical world as the elements that constitute the inherent order which untiredly unfolds in nature: "atoms are

[a]And he does this in a book of *dialogues*. Both the form and content of this book show the original and profound Platonism of Heisenberg.

not things. This was probably what Plato had tried to say in his *Timaeus*, and, seen in this light, his speculations about regular bodies were beginning to make more sense to me" [7, p. 11]. "Our elementary particles are comparable to the regular bodies of Plato's *Timaeus*. They are the original models, the ideas of matter. [As] Nucleic acid is the idea of the living being. These primitive models determine all subsequent developments. They are representative of the central order." [7, p. 241]. These mathematical forms, the ideas of matter, argues Heisenberg, should not however be interpreted as geometrical forms. This is the limit of the association with Plato:

> modern physics takes a definite stand against the materialism of Democritus and for Plato and the Pythagoreans. The elementary particles are certainly not eternal and indestructible units of matter, they can actually be transformed into each other. (...) But the resemblance of the modern views to those of Plato and the Pythagoreans can be carried somewhat further. The elementary particles in Plato's *Timaeus* are finally not substance but mathematical forms. (...) in modern quantum theory there can be no doubt that the elementary particles will finally also be mathematical forms, but of a much more complicated nature. (...) The constant element in physics since Newton is not a configuration or a geometrical form, but a dynamic law. The equation of motion holds at all times, it is in this sense eternal, whereas the geometrical forms, like the orbits, are changing. Therefore, the mathematical forms that represent the elementary particles will be solutions of some eternal law of motion for matter [6, pp. 71-72].

To further clarify the nature of these mathematical forms, it is useful to introduce another ancient Greek philosopher that Heisenberg sees as capable of inspiring an understanding of quantum mechanics: Heraclitus. Heisenberg repeats the traditional — yet highly questionable — interpretation of Heraclitus as a philosopher that, as some of his contemporaries, chose an "element" as origin and fundament of reality. In his case, this element is fire: "He regarded that which moves, the fire, as the basic element." [6, p. 62]. Heisenberg finds that what Heraclitus says about fire as the basic, transformable element, can now be applied to energy: "the views of modern physics are in this respect very close to those of Heraclitus if one interprets his element fire as meaning energy. Energy is in fact that which moves; it may be called the primary cause of all change, and energy can be transformed into matter or heat or light." [6, p. 71].

With the aim of reuniting these two Greek influences, Heisenberg theorizes the "elementary particles" — now thought of as mathematical forms — as formal stationary conditions of this same Heraclitean "stuff", as the elements of that central order (the dynamical law) that inform this basic energy/matter seen in analogy with Heraclitus fire:

> This state of affairs is best described by saying that all particles are basically nothing but different stationary states of one and the same stuff. Thus even the three basic building-stones [protons, neutrons and electrons] have become reduced to a single one. There is only one kind of matter but it can exist in different discrete stationary conditions. Some of these conditions, i.e., protons, neutrons and electrons, are stable while many others are unstable [8, p. 46].

As interesting as this Platonic atomism may seem, one can't help but to wonder why the insistence on developing an atomist view. Why does this obligation persist when there is in fact nothing left to support it? Why not leave atomism aside altogether? It is Schrödinger who finds an explanation for this fixation on atomism:

> atomism has proved infinitely fruitful. Yet the more one thinks of it, the less can one help wondering to what extent it is a true theory. Is it really founded exclusively on the actual objective structure of 'the real world around us'? Is it not in an important way conditioned by the nature of human understanding — what Kant would have called 'a priori'? [11, p. 88].

This worldview that seemed to describe the fundamental nature of physical reality started to appear — at least for Schrödinger — rather as a presupposition that constantly and inadvertently conditioned our way of perceiving the world — and one that was most inadequate. To follow the association with Kant's terminology, we can say that atomism appeared now also as dogmatism, as a metaphysical presupposition assumed without criticism. Criticism seemed then to be in order, a critical analysis of the worldview that determined the parameters of physical understanding. So why do we still today hear with such certitude that quantum mechanics talks about elementary particles? Because, for the most part, this critical task was prevented by the influence of Bohr and of positivist philosophy on physics and philosophy of science. The critical analysis of classical concepts and the development of different systems of concepts for this new physical theory was limited by Bohr, who claimed that the only language for physics could be the language of classical physics. At the same time, positivist philosophy evacuated metaphysics from its fundamental role in scientific theories, and turned it into just a mere storytelling one can — mostly inconsequentially — add, if one desires to do so, to an already functional, empirically adequate theory. So, in this way, atomism, instead of being the object of that criticism that Schrödinger alluded to, was crystalized, in a dogmatic manner, as a basic presupposition not to be questioned again and, at the same time, all other conceptual endeavours — which could be critical of that presupposition — were cast away or reduced to mere storytelling. One inadequate worldview was thus installed as a basic common-sense presupposition, and the — truly fundamental — intellectual endeavour which could favour criticism of that worldview and which could give rise to other possibilities, was reduced to inconsequential narratives. So physicists today claim to the general public that quantum mechanics talks about small elementary particles. Faced with interested questioning they will admit that these are not really small determined particles, but strange "quantum" particles, of which little can they say, and, ultimately, they will claim that, in any case, this talk of elementary particles is really just a "way of talking" we unimportantly adhere to an already empirically adequate theory. But actually, as Einstein insisted, "It is the theory which decides what we can observe" [7, p. 63]. Atomism is for physics not only — as we carelessly claim — just "a way of talking", but more profoundly, a way of thinking, imagining and even perceiving, of which we are many times unaware. This presupposed worldview is, as Schrödinger points out, the source of many of our missteps. It still makes us see what we are in fact not seeing (a particle when a "click" happens in a detector), it makes us ask the wrong questions, and forces us to give widely inadequate answers to those already problematic questions. The critical analysis that Schrödinger demands on the influence of certain aspects of Greek philosophy that appeared at the beginning of the XXth century as problematic — aspects that, through their reinterpretation in modernity, became a truly fundamental part of physical thought — remains still, for a great part, to be done. This is certainly one of the reasons why the task of investigating Greek philosophy for the sake of our own contemporary scientific theories has not grown old yet.

2. Revisiting *understanding*

Schrödinger writes: "the present crisis in modern basic science points to the necessity of revising its foundations" [11, p. 18]. It was not just a matter of filling some remaining gaps. They knew they were not faced only with new discoveries that could help complete the mosaic of modern physics. They were faced, on the contrary, with the evidence of a complete disconnection between what was being developed in quantum physics and the previous theoretical parameters as they were expressed not only in explicit scientific theories, but also — as we have seen with the example of atomism — in the presupposed worldview that gave sense to the perceived physical phenomena. For there to be understanding of the quantum phenomena, for them to be able to do physical science faced with these new developments, they were obligated to revise, rethink, the fundamental parameters of modern science. They had to ask again what doing physics meant and entailed, they had to take a look at the conditions — if they existed — of the human understanding of nature. The available modern answers were no longer enough. In all of these physicists we see this awareness: they take as a central matter they analyze, discuss, tackle, the question of what "understanding" means in physics. And in order to do that, they often turn to the Greeks. Of course, by that time there exists at their disposal a quite practical response to this question, an easy way out: the positivist renounce, the rejection of the possibility of understanding nature, and the redefinition of science as a merely coherent record of observations.

> One would in this context have to discuss the questions: what does comprehensibility really mean, and in what sense, if any, does science give explanations? David Hume's (1711-76) great discovery that the relation between cause and effect is not directly observable and enunciates nothing but the regular succession — this fundamental epistemological discovery has led the great physicists, Gustav Kirchhof (1824-87) and Ernst Mach (1838-1916), and others to maintain that natural science does not vouchsafe any explanations, that it aims only at, and is unable to attain to anything but, a complete and (Mach) economical description of the observed facts. This view, in the more elaborate form of philosophical positivism, has been enthusiastically embraced by modern physicists [11, pp. 90-91].

Before raising the obvious questions that this positivist stance provokes in us (are we to take these "observed facts" as innocent givens? What justifies "economy" as a parameter of this "description"? etc.), let us merely point out a fact that goes often unnoticed: this is an attitude towards knowledge that has as its sources not only modern empiricism, but also ancient Greek thought. We are referring, of course, to sophistry. The Manifesto of the logical positivists makes this influence explicit: "Everything is accessible to man; and man is the measure of all things. Here is an affinity with the Sophists, not with the Platonists; with the Epicureans, not with the Pythagoreans; with all those who stand for earthly being and the here and now" [3].

Contrary to what its fame indicates, it is not only the taste for controversy what lies behind sophistry, there are some originally sophistic positions — which evidently arose from an opposition to philosophy as it was developed among the first philosophers — that justify their praxis. Some strongly skeptical postures. These are positions that undermined the basis of the attempt to understand nature. For most of them, and specially for two of the greatest among them — Protagoras and Gorgias —, all we have is what we perceive. As radically as the modern empiricists, they discarded knowledge beyond individual perceptions. This aspect of their thought pushed them to relativism. The only remaining text pertaining to Protagoras, which, as we read, the logical positivist quoted in their Manifesto, makes this relativism explicit: "Man is the measure of all things, of the things that are, that they are, of the things that are not, that they are not" [DK 80 B1]. We don't possess more of Protagoras' text but we do have some comments about his philosophy that date back to antiquity, and they all seem to coincide in confirming a relativistic view. According to this stance, there is no such thing as 'a reality of things' — or at least, if there is one, we are not able to grasp it. We can only refer to our own perception. Although he fails to relate this with empiricism and positivism, Schrödinger takes the time to revisit Protagoras: "Protagoras regarded the sense perceptions as the only things that really existed, the only material from which our world-picture is made up. In principle all of them have to pass for equally true" [11, p. 30]. At the same time, since scientific and philosophical discourse could not be knowledge, could not relate to truth, the discursive aspect of these intellectual activities detached, in sophistry, from its function in knowledge and became a matter to be treated independently, and so formal characteristics, as mere coherency and economy, could be justified as ends in their own right.

What perhaps seems odd in this relation between sophistry, modern empiricism and positivism is that, while positivists claimed to enable by their position a scientific progress, the Greek sophists, as well as the British empiricists, concluded more reasonably in skepticism. But maybe the key to understand what this scientific "progress" means can still be found in these older sources, especially in sophistry and in the practical orientation they gave to scientific and intellectual activity (since true knowledge was discarded). They redirected the goal of discursive argumentations from knowledge to praxis, to the creation of effects. Since, among these effects, truth — if it existed — could not be differentiated from a mere rhetorical effect, it was this effect production, the ability to convince, act, produce changes by these discourses, the true purpose of what could no longer be regarded as knowledge of reality. And if one follows positivism until the end, both in arguments and in history, one arrives also to a practical justification for science: instrumentalism.

In any case, what is certain is that Schrödinger, Pauli, Heisenberg, as well as Einstein, coincide in rejecting the positivist "solution" to the fundamental question of understanding. Einstein, for example, sees the "economic" principle defended by Mach as way too naïve. But the main issue with positivism is, for all of them, its simplistic concept of observation. Heisenberg, Pauli and Einstein will respond to Mach what Kant added to Hume: observation is in itself theoretically laden. Observations cannot be taken as innocent givens. Causality — Kant argued — is surely not directly observable, but it *always* determines observation, and a priori determinations such as causality are what makes experience and objectivity possible. These a priori conditions are, quite fundamentally, part of all observations for all subjects. Mach wished to vanish a priori concepts and stick to observation but, by his blind *tabula rasa* attitude, he actually helped make the concepts that, in fact, determine observation, remain unconscious, unanalysed for many that, still today — willingly or not — continue to follow this naïve empiricism. In a conversation they held in 1925, Einstein explained to Heisenberg — as we commented earlier — that in fact "It is only the theory which decides what we can observe". Pauli also insists on this matter: "Personally I do not see how it is possible to give a definition of the phenomenon in physics which seeks to isolate the data of perception from rational and ordering principles. It seems to me rather that a separation of this sort is itself already the result of a special critical mental effort which removes the ever-present unconscious and instinctive ingredients of thinking" [9, p. 128].

But there is yet another thing that, according to Schrödinger, positivism leaves unexplained:

> For even if it be true (as they maintain) that in principle we only observe and register facts and put them into a convenient mnemotechnical arrangement, there are factual relations between our findings in the various, widely distant domains of knowledge, and again

> between them and the most fundamental general notions (as the natural integers 1, 2, 3, 4, ...), relations so striking and interesting, that for our eventual grasping and registering them the term 'understanding' seems very appropriate [11, p. 92].

This is of course the expression of one of the original and simplest amazements which always arouse the curiosity inherent in knowledge. These "factual relations" we apprehend between what appears in completely different domains, and between those discoveries and our general notions — a relation that, in fact, allows for the possibility of, among other things, a logical, coherent, even "economical" order to be applied to phenomena, as positivists want —, all of this must be explained. It seems, at least, that there is something like an order developed as well in our own activity as in different domains of reality. What is the nature of this order? Is it an illusion or does it point to an inherent order in nature? This is of course one of Kant's main questions. And his answer is well known: that "order" is originated in subjective — although universal, that is, transcendental — conditions. Transcendental conditions equally determine experience and understanding for every empirical subject. The order of nature (as experience) is for Kant in fact the order of the transcendental subject. Those "factual relations" have their origin in the a priori structure of the transcendental subject. But, unfortunately, Kant's proposal was not a possible solution for the founders of QM. In fact, Kant appeared to be rather part of the problem, and new allies were needed to rethink this fundamental question. The new discoveries of physics violated Kant's a priori conditions, there were in fact experiences which could not be subsumed by the pure concepts of understanding (the categories) and the pure forms of intuition (universal space and time) Kant believed to be universal:

> The theory of relativity has changed our views on space and time, it has in fact revealed entirely new features of space and time, of which nothing is seen in Kant's a priori forms of pure intuition. The law of causality is no longer applied in quantum theory and the law of conservation of matter is no longer true for the elementary particles [6, p. 88].

What was previously seen as the conditions for all possible experience and as the parameters of objectivity and understanding now appeared as the limited conditions of a particular worldview that no longer could be taken as universal. Kant's transcendental subject contained the conditions of modern science — it was in fact in part inspired by that science (specially by Newtonian mechanics) —, but that was no longer enough: "We agree with P. Bernays in no longer regarding the special ideas, which Kant calls synthetic judgements a priori, generally as the pre-conditions of human understanding, but merely as the special pre-conditions of the exact science (and mathematics) of his age" [9, p. 126]. Mach had also pushed for the rejection of Kant's a priori conditions, but without proposing a different solution, a different explanation for what those conditions allowed to understand. He rejected the Kantian a priori determinations as he rejected all possible a priori concepts, aiming to limit the scientific scope only to what is directly observable. But the question of the theoretical determination of observation and the question of those "factual relations" that arouse a sense of understanding nature remained then unanswered. Positivism denied the solution by also denying the problem. This was an attitude that Schrödinger, Heisenberg, Pauli and Einstein thought they could not allow themselves. Faced with the crisis of the modern parameters of knowledge and experience, some of these physicists turned instead, beyond the modern source, to the original source, they jumped back to ancient Greece, in order to rethink scientific understanding.

Heisenberg reconstructs the words of a young Pauli in a conversation they held in 1921, precisely about what "understanding" meant in physics when stripped down to its most essential meaning:

> knowledge cannot be gained by understanding an isolated phenomenon or a single group of phenomena, even if one discovers some order in them. It comes from the recognition that a wealth of experiential facts are interconnected and can therefore be reduced to a common principle. (...) 'Understanding' probably means nothing more than having whatever ideas and concepts are needed to recognize that a great many different phenomena are part of a coherent whole. Our mind becomes less puzzled once we have recognized that a special, apparently confused situation is merely a special case of something wider (...). The reduction of a colourful variety of phenomena to a general and simple principle, or, as the Greeks would have put it, the reduction of the many to the one, is precisely what we mean by 'understanding'. The ability to predict is often the consequence of understanding, of having the right concepts, but is not identical with understanding [7, p. 33].

Understanding is for Pauli — according to Heisenberg — not just piecing together multiple observations through coherent or economical logical propositions, it only comes with the recognition — conceptually elaborated — of their common principle, of the reality of the whole their interconnections point to. Understanding is being able to pass from a vision of isolated phenomena to a general reality in which those phenomena show their relation and their meaning (their place). Or, to take Schrödinger's words, it is to pass from the amazed yet uninterpreted vision of "factual relations" in reality to the common principle they express (and that explains their existence). As can be seen exemplified in almost every Platonic dialogue, an isolated individual "understanding" is not truly understanding and cannot help us gain knowledge. It is only through a wider interconnection that is able to point to a common principle, capable of explaining the subsumed experiences, that "understanding" happens, and, with it, knowledge. What those constellations indicate, what those general principles express, that is in fact the goal of knowledge for the Greeks philosophers (and not just what is "directly" observed).

It is also important to add that, according to Heisenberg's Pauli, this can only happen through concepts. There is no understanding without concepts: there is no possible representation of the principles the phenomena point to without concepts, and there is no understanding of the individual phenomena without that representation. Phenomena can be understood only through their principles, and these can only be represented with concepts. As remarked by Heisenberg [8, p. 264]: "For an understanding of the phenomena the first condition is the introduction of adequate concepts. Only with the help of correct concepts can we really know what has been observed." Of course, we need the ability to pose the right questions, to go against our own suppositions, to identify the decisive relations and arrive at an adequate system of concepts in order for that "reduction" Pauli speaks of to really occur (in Platonic terms: we need to learn the difficult art of dialectics). We cannot force its occurrence under inadequate concepts — as it widely happens today with QM. We cannot produce that unity by mere will, at out pleasure. We cannot force our presuppositions if they are not adequate. The Greeks propose we exercise an equilibrium: the fear of factual errors shouldn't make us renounce general principles for the sake of "exact" although unexplained observations, and, at the same time, the desire for grasping the general should not make us postulate no matter what inadequate principles too fast, or assume a detached mystical attitude. Although difficult, this "reduction" can be produced. And this ability is what Heisenberg refers to as the great heritage of Greek thought:

What always distinguished Greek thought from that of all other

> peoples was its ability to change the questions it asked into questions of principle and thus to arrive at new points of view, bringing order into the colourful kaleidoscope of experience and making it accessible to human thought. (...) Whoever delves into the philosophy of the Greeks will encounter at every step this ability to pose questions of principle, and thus by reading the Greeks he can become practised in the use of the strongest mental tool produced by western thought [8, pp. 52-53].

This fruitful strategy we take for granted, this disciplined yet broad, methodical yet artful ability to arrive at the unities manifested in multiples, and to proceed with those principles in a way that enables us to understand the world by ourselves, this is what the Greeks developed and what originated our intellectual attitude and strategy. According to Heisenberg, we are reminded of this each time we aim for the root of things in whatever scientific discipline:

> Those (...) who (...) wish to get to the root of things in their chosen vocation (...) are bound sooner or later to encounter the sources of antiquity, and their own work can only profit if they have learnt from the Greeks how to discipline their thoughts and how to pose questions of principle [8, p. 63].

The first expression of this ability directed Greek thinkers to the question of *physis*. This term is generally translated as "nature" and its meaning covers what we refer to when we talk about "the nature of reality" (its essence), as well as what we commonly, broadly and in an extensive way refer to as nature: the reality in which we take part. Nature is, for the first Greek philosophers, something dynamic, changing, which — at the same time — responds to some sort of internal order, substance or formula. This internal order is what they seek to understand. While for Kant reality's *readability*, its order, criteria, came from subjective conditions, for the Greeks those conditions were themselves real beyond the subject (although encompassing also the subject), they were the criteria, the formula, the order inherent to reality, which always rules reality. Those "factual relations" Schrödinger referred to were rather interpreted as signals pointing to an inherent order in reality that philosophy and science aimed to capture. Some of these first philosophers proposed an "element" (or a series of them) from which — and according to which — all reality develops and can be explained. Aristotle calls them the *physikoi*, the physicists or naturalists. Schrödinger also points to the fact that our scientific attitude finds its origin there: these *physikoi*, he says, "saw the world as a rather complicated mechanism, acting according to eternal innate laws, which they were curious to find out. This is, of course, the fundamental attitude of science up to this day" [11, p. 57].

Heraclitus redirected the search for the fundament of *physis* no longer to an "element" but to the description of a formula, an internal order that rules *physis*. He allusively described this formula and called it *lógos*. This denomination is very significant for the development of philosophy. Until Heraclitus' use of the word, *lógos* had a meaning almost exclusively related to language: discourse, argumentation, account, even tale. In all of those translations we can see already something that will be essential to all meanings and nuances of *lógos*, even when it doesn't refer to language: a significant combination, a reunion with criterion, a collection with purpose. *Lógos* never means an isolated word, or a meaningless sentence, or dispersed and ineffective ensembles of words. It always refers to a combination that has a reason, that is able to produce an effect or to exhibit a meaning[b]. We now begin to understand why Heraclitus chooses this specific word to name the internal order of *physis*. He sees in nature exactly that: a combination that is not meaningless dispersion, on the contrary, it responds to a formula, a criterion. This double meaning of *lógos* (order of *physis* and human discourse) also expresses for Heraclitus — and this is fundamental — an affinity between human discourse and reality, an affinity that allows for knowledge. Thus, there is a relation between the *lógos* of human beings and the *lógos* of *physis*. It is not a simple task to expose the true *lógos* since, as remarked by Heraclitus, "*physis* loves to hide" [DK 22 B 123], but it is none the less possible: in a particular *lógos* one can "listen" something that exceeds it, that is not only that personal discourse but the *lógos* of *physis*, or, as Heisenberg likes to call it, the central order: "Listening not to me but to the *lógos* it is wise to agree that all things are one" [DK 22 B 50].

Plato develops this further, he renders this *lógos* of *physis* more precise; it is no longer described only in a general manner, as the Heraclitean opposites in the tension of a peculiar harmony, and the path to its knowledge is not so obscurely alluded, but now methodologically developed. What Heraclitus calls the *lógos* of *physis* is for Plato the realm of Ideas. Ideas are the elements (the forms) that populate, constitute, the *lógos* of *physis*, the central order. These Ideas are different, each one is to be found and

[b]We follow Néstor Cordero, who thoroughly described in a recent book the transfigurations of the notion of *lógos* in Greek philosophy (see [2])

known through particular dialectical efforts; these Ideas have different relations among them and with phenomena. They allow for more than a general, allusive, assessment of unity and order in nature, they allow for the different unities, principles, expressed in phenomena to be, each of them, apprehended through methodical efforts, and related among them. In any case, it is this Platonic development, generally known as his "theory of Ideas", that Pauli reinterprets and, with Jungian accents, elaborates on. Specifically, what Pauli seems to emphasize in his reading of Plato is the question of *reminiscence*. For Plato, learning and understanding suppose necessarily a previous presence of Ideas in ourselves. Understanding comes then when, through methodical efforts, we "remember", we recognize in the investigation of the world something that is also essential to our thought. The dialectical endeavour of understanding reality is in Plato, undoubtedly, an effort that asks of us to reject what is only merely personal in our thinking, our opinions and judgments, but in doing so, we encounter that what is truly universal was in fact already in ourselves. An encounter with the universal in ourselves that only occurs through the methodical investigation of reality. As well as in the rest of *physis*, Ideas are expresses in us.

> With Plato's philosophy in mind, I should therefore like to suggest that the process of understanding nature, as well as the happiness that man feels in understanding, that is, in the conscious realisation of new knowledge, should be interpreted as a correspondence, a coming into congruence of inner images pre-existent in the human psyche with external objects and their behaviour. The bridge (...), which cannot be constructed by pure logic, rests, according to this conception, on a cosmic order independent of our choice — an order distinct from the world of phenomena, embracing psyche as well as physis [9, p. 125].

That "recognition" Heisenberg remembered in his friend's words is here specified. Understanding nature is the "coming into congruence" of pre-existent images in human psyche and external objects and their behaviour: this seems rather Kantian. But these convergent forms are not originated in the subject, projected by him, are not the property of human conscience, they are not restricted a priori to the limited parameters of modern subjectivity, they come from a greater cosmic order which is expressed as well in human psyche as in *physis*. Understanding is the result of a process by which the forms we read in phenomena — under certain methodically determined conditions — show their affinity, correspondence, with something we recognize in ourselves, and this is made possible by a cosmic order — an order which is, by this process, made available to our thought. The bridge that enables that recognition in understanding is not made of pure logic, cannot be developed from phenomena by pure logic. Logic alone cannot create that congruence, that moment of understanding (and happiness). Logic cannot create the pre-existent ideas, the forms that enable understanding (it can only suppose and eventually express them). A fundamental cosmic order contains the conditions of that affinity that is actualized in the congruence. For that understanding to occur — for us to recognize in ourselves the forms we perceive — there must exist for Pauli a common medium and origin that enables that real affinity. Understanding is thus the conscient correspondence between the human *lógos* and the *lógos* of *physis*, as it is rediscovered — or "remembered", to put it in Platonic terms — in its expression in phenomena. As Walter Benjamin was also attempting at the same time, Pauli is, in a way, Platonizing Kant, by broadening the conditions of understanding to allow for more than just the modern mechanical experience Kant's a priori determinations justified, and by relocating these conditions outside of the subject: "According to the conception here put forward, the a priori character of Kant's rationally formulated ideas, laid down once for all, is thus transferred to the pre-existent images (archetypes) present and operating outside of consciousness" [9, p. 126].

This "coming into congruence" is the astonishing fact the Greeks were always looking to show and explain, Pythagoreans with the magical correspondence between mathematics and reality, Plato with the experience of reminiscence, Heraclitus with the multiple meanings of *lógos*, etc. And it is by this understanding that theories come into being:

> Theories come into being through an understanding inspired by empirical material, an understanding which we may best regard, following Plato, as a coming into congruence of internal images with external objects and their behaviour. The possibility of understanding again demonstrates the presence of typical regulatory arrangements, to which man's inner as well as outer world is subject [9, p. 129].

By this realisation of knowledge, by this process of understanding nature, the typical regulatory arrangements, the ordering principles in *physis*, are demonstrated and represented. It is this central order that theories aim to represent. And it is by this order, as it is manifested in theories, that phenomena become meaningful, that they can be seen as belonging to a greater whole. They stop being isolated, or, as Plato said, they are saved.

3. Critics of separation

Another important motive for the return to the Greeks is easily found in the works of Heisenberg, Pauli and Schrödinger. They are quite insistent on this one. And perhaps it can be said that this reason subtends the other reasons, that it is expressed or supposed by them, although in different degrees. It is the problem with *separation.* It is — specially for Schrödinger, Heisenberg and Pauli, but also for Einstein — one of the main obstacles for the further development of physics in particular and science in general. A separation that they identified already in the XVIIth century, and that grew with time until it became untenable in the XXth century. Which separation? We actually find in their writings a multiplicity of them — although clearly related —: separation between scientific and speculative knowledge, between scientific and metaphysical attitudes, between experience and theory, between natural sciences and philosophy. Separation expressed also in the objects of knowledge: between *res extensa* and *res cogitans*; as well as more and more (and sharper and sharper) separations between disciplines. It was for instance the growing separation between scientific and philosophical endeavours that had caused, according to Schrödinger, a "grotesque" and childish spectacle he witnessed in the scientific world: "This produces the grotesque phenomenon of scientifically trained, highly competent minds with an unbelievably childlike — undeveloped or atrophied — philosophical outlook" [11, p. 12].

However problematic these separations can be in principle, they became in fact untenable due to the new scientific developments of the XXth century, especially in quantum physics. These authors couldn't help but notice that separations which may had been instrumental and effective for the physics of previous centuries appeared now as burdens. Starting from those separations, maintaining those separations, quantum mechanics was unintelligible. We already encountered some of the problems requiring they undo separations: how to recast the parameters of scientific understanding without revising natural philosophy? How to understand observation beyond modern presuppositions without rediscovering the irreducible and fundamental relation between theory and experiment, between metaphysics and experience? How to transform radically our atomist way at looking at things without critically analysing atomist natural philosophy? How to understand *a probability that interacts* from the point of view of a narrow materialism that only recognizes *res extensa* as physical reality? Einstein, Schrödinger, Heisenberg and Pauli saw very clearly that the critical analysis of those separations and the effort to overcome them were necessary tasks for the science of their time. The last three of them found, as they worked on the subject, an irreplaceable model for the reunion that they were seeking in the Greeks[c]:

> We look back along the wall: could we not pull it down, has it always been there? As we scan its windings over hills and vales back in history we behold a land far, far, away at a space of over two thousand years back, where the wall flattens and disappears and the path was not yet split, but was only one. Some of us deem it worthwhile to walk back and see what can be learnt from the alluring primeval unity. Dropping the metaphor, it is my opinion that the philosophy of the ancient Greeks attracts us at this moment, because never before or since, anywhere in the world, has anything like their highly advanced and articulated system of knowledge and speculation been established *without* the fateful division which has hampered us for centuries and has become unendurable in our days [11, pp. 13-14].

The Greeks were the model for an "advanced and articulated" effort of understanding that did not depend on separations, an uncompartmentalized way of thinking that they needed in order to face the challenges ahead. They were in fact confronted with problems that could not be tackled from the worldview and with the tools that modern physics presented them. They were in need of a broader perspective on the parameters of science and on the nature of physical reality. In their jump back to the Greeks they believed they could recuperate an understanding of the original scientific endeavour (which was one with natural philosophy), riding themselves of the habits and separations that had become common sense in their disciplines, and that were now blocking the way to the development of the new theory. Although developing different aspects of this process of separation in human knowledge, Schrödinger, Heisenberg and Pauli coincide in presenting the Greeks as the model for a non-separated, rigorous yet bold, scientific attitude.

---

[c]Einstein was more interested in a modern philosopher, one that was especially focused in tackling the problem of separation in our worldview: "I am fascinated by Spinoza's Pantheism. I admire even more his contributions to modern thought. Spinoza is the greatest of modern philosophers because he is the first philosopher who deals with the soul and the body as one, not as two separate things" (cited from [13])

Heisenberg concentrated especially on the critical analysis of the separation that determined for physics — and natural sciences in general — an over-simplified object of knowledge: the *res extensa*, a world defined and limited through a narrow materialism. Nothing beyond the mechanics of independent matter. A determinism of material bodies. A narrow materialism that, according to Heisenberg, enters inevitably into a crisis with the appearance of quantum mechanics. Indeed, how to account, for instance, for quantum probability, or for a principle of indeterminacy, in the context of this extreme form of materialism? He finds the basis of this separation were most clearly and definitely determined by Cartesian philosophy, which he presents in contrast with Greek thought:

> While ancient Greek philosophy had tried to find order in the infinite variety of things and events by looking for some fundamental unifying principle, Descartes tries to establish the order through some fundamental division [6, p. 78].

Let us start by the distinction that is made in Heisenberg's words between *finding* order and *establishing* order. It marks a fundamental difference between the Greek and the Cartesian way of looking at things; the difference between apprehending an order seen as existent, and establishing, forging, an order, to better determine and organize our knowledge. The second contrast presented by Heisenberg is between *unifying* and *separating*. While for the Greeks finding order meant understanding what *unifies*, for Descartes the aim was establishing order by being able to *separate*.

> This bases of the philosophy of Descartes is radically different from that of the ancient Greek philosophers. Here the starting point is not a fundamental principle or substance, but the attempt of a fundamental knowledge [6, p. 78].

For Heisenberg, Descartes' basic aim was not, as it was the case for the Greeks, to decipher the meaningful complexity of nature from its inherent principles or elements, but rather 'what and how can I know with certainty?'. It is the adaptation to the parameters that define certainty for the subject what is fundamental here, rather than developing a way of apprehending the inherent parameters of nature. The question is different and thus the answer is different, and Descartes' answer is undoubtedly quite a prodigious one; an answer that certainly transformed philosophy: as much as I doubt, I cannot deny that I think — since this doubting is thinking —, and since I am certain that I think, I certainly am (*dubito, cogito, sum*).

But Descartes finds himself locked in a solipsist certainty, he is only certain of the isolated *I*. How to advance beyond the *cogito*? How to recover reality? This is done through a version of the old ontological argument that functions as a proof of the existence of God: God must also exist, since I have an idea of God that I could have not produced by myself (since I don't have the amount of formal reality to produce the amount of objective reality that this idea entails); and given that God exists, the world in front of me, of the *I*, cannot be entirely a deceiving illusion. But, for Heisenberg, this reconstruction is problematic, since these added relations cannot disguise the fact that he sets a fundamental division as the bases for further reasoning:

> his starting point with the 'triangle' God-World-I simplifies in a dangerous way the basis for further reasoning. The division between matter and mind or between soul and body, (...) is now complete. God is separated both from the I and from the world. God in fact is raised so high above the world and men that He finally appears in the philosophy of Descartes only as a common point of reference that establishes the relation between the I and the world [6, p. 78].

Descartes is, in principle, only certain of himself, of the isolated *I*. Then through God he establishes the existence of the world — as what the *I* has in front of him. God sets up a narrow bridge between what is fundamentally an isolated *I* and a separated, strange, world. And at the same time, by the same movement, the world is impoverished: although God functions as guarantee of the existence of the world, his work or expression is not recognizable in it. The world is at the same time guaranteed and emptied. It is the mere world, an exterior landscape that affects my senses, that appears opposed to the *I*. God is evacuated from the world and the *I*, and represents only the guarantee of their relation. But this relation between the *I* and the world is the — weak — relation of two strangers, of two separated elements, and this reality now opposed to the subject and emptied of inherent meaning is defined as *res extensa*. Descartes establishes as the basis for further thinking, as the bases for knowledge, for the different scientific and philosophical endeavours, a fundamental separation. Starting from the separation of God, the I and the world, starting from the separation between a *res extensa* and a *res cogitans*, it seems we can better determine and organize knowledge. This is opposite in intention and in origin to the Greek way, which took as its most important object of knowledge the inseparable interrelation, interdependency, between *physis* and its *lógos*, between the multiplicity of the world and the Ideas that define its essence, between reality and its unifying principles. Surely, as Heisenberg points out, it is not completely fair to exclusively accuse Descartes for this sharp separation in human knowledge, but none the less — Heisenberg adds — it is certainly he who established more definitely the basis for such a separation, and for the narrow content of natural science:

> Of course Descartes knew the undisputable necessity of the connection, but philosophy and natural science in the following period developed on the basis of the polarity between the 'res cogitans' and the 'res extensa', and natural science concentrated its interest on the 'res extensa'. The influence of the Cartesian division on human thought in the following centuries can hardly be overestimated, but it is just this division which we have to criticize later from the development of physics in our time [6, pp. 78-79].

Where the Greeks saw the necessity of the interconnection between "sensible" and "intelligible", between becoming and *nous*, as they knew that neither of these could be understood without the other, modernity (at least an important part of it) was convinced by the necessity of division. And the merely "material", alienated from what was before related to its inherent meaning, was, with time, isolated as the business of independent natural science. It is a world emptied of its essential content, its inherent meaning, its principles. An emptied world which will understandably lead to the vision of its contingency, and, in empiricist philosophy, to skepticism (which will only be avoided by projecting objectivity and order into the world by means of the subject). The physical world was for the first time established as an object of knowledge completely separated as well from the subject as from God, from the Ideas, the *lógos*, etc. (although this separation will be rejected by later philosophers, as Spinoza for instance, for whom God and nature were synonyms). It is perhaps interesting to get back now, for a brief moment, to Heisenberg's atomism: takings as a starting point his critique of separation, we can perhaps better understand his attempt to reintroduce forms — in his case mathematical forms — as essential elements of physical reality — violating in this way the Cartesian separation, introducing something beyond *res extensa* as the object of physics, and attempting to understand nature from a worldview closer to the one the Greeks had. But, independently of his exploration of a Platonic atomism, what Heisenberg fails to point out — and, as we shall see, this will have a definitive influence on his conception of quantum physics — is that, while redefining the world as *res extensa*, Descartes is also redefining and narrowing *res cogitans*, thought, which is now — contrary to the Greek or Spinoza's worldview — identified exclusively with subjective conscience. In any case, what is certain is that this separation, with its modern redefinition of the object of physics, is something that, to Heisenberg, quantum mechanics calls into question. This is why, on this matter, Greek thought and not Enlightenment should be for him the inspiring source.

Although he shares some of Heisenberg's historical hypothesis, Pauli is the critic of another separation, between what he calls two types of knowledge or two attitudes towards knowledge, one related to mysticism, to the experience of "oneness", the denial of the world's multiplicity as illusion, and aimed — he argues — at salvation, and the other a scientific attitude, rational and methodical in its means, but "dispassionate" and incapable to see through multiplicity to the fundamental reality of unity. These represent for Pauli two poles, two attitudes presented as opposed for analytical pourposes, but with different relations in different times; sometimes reunited, even mixed, sometimes separated, related only by their mutual exclusion. It is the radical and sharp separation between the two poles that Pauli sees as problematic, the times when we witness an uncritical and superficial mysticism on one hand, and, completely separated, a narrow, short-sighted scientific view on the other. He believes to be living in one of those times.

He starts a conference from the mid 1950's reminding his listeners that many of the fundamental discoveries and bases of our science come from the synthesis of those attitudes, as is the case of the ancient discovery of the "enigmatical" possibility of applying mathematics to nature: "The possibility of mathematical proof, and the possibility of applying mathematics to nature, are fundamental experiences of humanity, which first arose in antiquity. These experiences were at once regarded as enigmatical, superhuman and divine, and contact was made with the religious atmosphere" [9, p. 139]. Pauli believes that "it is the destiny of the occident to keep bringing into connection with each other these two fundamental attitudes" [9, p. 139], to combine the oriental mystical inclination with a western scientific attitude, seeking surely to understand, but without sacrificing "oneness", rather gaining it for understanding, following, as he says, "the Greek spirit". But before entering the Greeks, let us remember first Pauli's characterization of the "mystical" attitude: "Mysticism seeks the unity of all external things and the unity of the inner man with them; this it does by seeking to see through the multiplicity of things as illusory and unreal" [9, p. 139].

Mysticism seeks an experience of unity that reveals multiplicity as illusory and unreal (as well as demonic). And its search is, according to Pauli, a search for salvation:

> Thorough-going mysticism does not ask 'why?' It asks 'how can man escape the evil, the suffering, of this terrible, menacing universe? How can it be recognised as appearance, how can the ultimate reality, the Brahman, the One (...) be seen?' It is however in keeping with the spirit of Western science — in a certain sense one might say with the Greek spirit — to ask, for instance, 'why is the One mirrored in the Many? What is it that mirrors, and what is mirrored? Why has the One not remained alone?' [9, p. 139].

The different questions show the transformation the Greeks, this is, philosophy and science in its origins, bring to the more mystical attitude. Their new scientific — or, as Pauli will say, "lucid" — mysticism, that inaugurates western philosophy and science. It is the transformation of a mysteric wisdom into an understanding in which, however, that "oneness" is not lost. The same that is for the mystic the object of a vision, is now, mostly striped from its esoteric aura, conceived and justified as a fundamental object of knowledge; and by this transformation the attitude towards multiplicity also changes: it is not denied, but, on the contrary, explained. The one is not lost, it is rather — more soberly — turned into a matter of understanding. Greek philosophy also seeks, as mysticism according to Pauli, "the unity of all external things and the unity of the inner man with them", but it does so in a different manner, where that unity can be an object of understanding and the goal of a somewhat methodical endeavour. They believe the experience of oneness can be rather its *understanding*. And, in this understanding, the unity is not gained by rejecting reality, by erasing multiplicity. On the contrary, there's no rigorous apprehension of the *one* without the specific *many* it encompasses, and conversely, it is only in the *one* that the *many* is interpreted and understood. It is a rephrasing that changes everything. The questions are no longer "how can I escape this deceiving multiplicity? how can I reject this apparent world to experience the true oneness beyond?" but rather turn into something like "what is the relation (which is not of mutual exclusion) between the one and the many? why is this multiple given? How does this one encompasses and determine this many?". As Heisenberg said, the Greeks rephrased the questions as questions about principles.

If the mystical attitude must be, according to Pauli, synthesized with the scientific, this also means that the former cannot be completely abandoned. The scientific attitude without the intention to understand what is object of the mystical attitude is insufficient, it cannot even justify and explain the nature of the multiplicity it concentrates on. It is a knowledge that eventually becomes short-sighted, dispersed and narrow, as it had occurred — Pauli believes — to the science of his time: "we can say that at the present time a point has again been reached at which the rationalist outlook has passed its zenith, and is found to be too narrow" [9, p. 147]. Pauli thinks that quantum mechanics, among other recent developments, demands a broadening of this separated scientific attitude, the recasting of what modernity had locked in too narrow limits, approaching and overcoming what lay separated: a too schematic scientific attitude and a too detached and exaggerated mysticism. To develop in further detail this synthesis that he sees as the task of the occident, Pauli names two attempts that were made in this direction, but concentrates rather on one of them:

> Among the attempts that have occurred in the course of history to effect a synthesis of the basic attitudes of science and of mysticism there are two which I would like particularly to stress. One of these originates with Pythagoras in the sixth century B. C., is then carried on by his disciples and developed further by Plato, appearing in late antiquity as Neo-Platonism and Neo-Pythagoreanism. Since much of this philosophy was taken over into early Christian theology, it continues thereafter in persevering association with Christianity, to blossom anew in the Renaissance. It was through (...) a return to Plato's doctrine of knowledge in Galileo's work, and through a partial revival of Pythagorean elements in that of Kepler, that the science of modernity, which we now call classical science, arises in the seventeenth century. After Newton it rapidly separates itself on rational-critical lines from its original mystical elements [9, p. 140].

As Heisenberg, Pauli also saw Plato in an essential continuity with the Pythagoreans: "Plato (428-348 B. C.) took over into his doctrine of ideas many of the mystical elements of the Pythagoreans. He shares with them his higher valuation of contemplation as compared with ordinary sense experience, and his passionate participation in mathematics, especially in geometry, with its ideal objects" [9, p. 141]. But Pythagoras represented still, according to Pauli, a clear mystical attitude: "He and his disciples founded an expressly mystical doctrine of salvation, which was most intimately tied up with mathematical thought, and was based on the earlier Babylonian number-mysticism." [9, p. 141]. It is Plato who takes a fundamental step forward:

> As a further development of Pythagorean teachings Plato's mysticism is a lucid mysticism, in which understanding, in its various degrees, from opinion (*dóxa*) through geometric knowledge (*diánoia*) to the highest knowledge of general and necessary truths (*episteme*) has found its place [9, p. 142].

In this continuity, it is in Plato that the transformation towards understanding settles, and, above all, that it does so with unequalled precision and awareness. Plato does more than suggesting that what was the object of a mystical approach can be the object of an understanding: he establishes the conditions of that understanding (as well as the mistakes or detours that commonly prevent understanding) and tries to develop the ways that can lead to it. And in that process of understanding he is able also to distinguish and specify its different degrees and modes, depending on their different objects, their distinct languages and capacities to encompass phenomena. What is only *dóxa*? What is *diánoia*? What is truly *episteme* — the knowledge of principles? What relations do they have, what dependencies among them? He even aims at distinguishing among the higher objects of knowledge, developing his theory about Ideas. In Plato we find the highest image of our faculties. In Plato, the mystical and the knowable, the different disciplines, the one and the many, are not separated. And this does not mean that they lie together in an undifferentiated and allusive unity. On the contrary, he is able to distinguish among what is reunited. And, most importantly, he teaches us how to develop the capacity to do it for ourselves.

After Plato, Pauli sees an emphasis on the mystical attitude gaining momentum in the work of the Neoplatonists:

> the mystical side of Plato's work gradually gave rise to Neo-Platonism, which achieves more or less systematic formulation in Plotinus (204 to 270 A. D.). Here we find the identity of the Good with the comprehensible carried to an extreme, as compared with Plato's own view, and coarsened by the doctrine that matter (*hyle*) is a simple lack (*privatio*) of ideas, that it is moreover the embodiment of Evil and that this is therefore a simple *privatio boni*, a lack of Good, which cannot be the object of conceptual thought [9, p. 141].

Pauli then jumps over the Middle ages, right to the Renaissance:

> The Renaissance was an epoch of extraordinary passion, of furor, which in 15th and 16th century Italy broke through the barriers between different human activities, and brought into the most intimate connection things formerly separated, such as empirical observation and mathematics, manual techniques and thought, art and science [9, p. 143].

This was not — Pauli argues — an epoch of "dispassionate science" [9, p. 143]. Science was not separated from debates about Greek philosophy, from Neoplatonism, not even from mysticism. It is only in the later XVIIth century, as expressed in Descartes' philosophy and Newton's mechanics, that he sees the scientific attitude beginning to take a separate path: "The later more dispassionate seventeenth-century way of looking at things led to Descartes' analytical geometry and to the absolute space of Newton's mechanics" [9, p. 144]. This leads, like Heisenberg also pointed out, to a separation of the world-picture, but also to the introduction of separations among different disciplines and faculties: "Among the general characteristic manifestation of the seventeenth century is the re-establishment of new boundaries between single disciplines and faculties, and the splitting of the world-picture" [9, pp. 144-145]. That path, Pauli believes, had found its limits and called for a renewed synthesis that would allow us to recast our relationship to knowledge. But how to answer that call? How to approach the poles? How to recuperate that original, broader, way of understanding nature?

> I believe that there is no other course for anyone for whom narrow rationalism has lost its force of conviction, and for whom also the magic of a mystical attitude, experiencing the external world in its crowding multiplicity as illusory, is not effective enough, than to expose himself in one way or another to these accentuated contrast and their conflicts. It is precisely by this means that the scientist can more or less consciously tread a path of inner salvation. Slowly then develop inner images, fantasies or ideas, compensatory to the external situation, which indicate the possibility of a mutual approach of poles in the pairs of opposites [9, p. 147].

Faced with a "narrow rationalism" on one side, and with the irrational, ineffective "magic of a mystical attitude", Pauli advices us not to escape this situation or to solve it dogmatically, but rather to honestly experience its conflicts and the limitations of both stances. He pushes us to experience and grasp how one lacks what the other has; how both seem unsatisfying in the end; how they turn superficial when isolated from each other; how both make good points and encompass things we recognize as meaningful but, in the end, remain so evidently narrow. It is important to understand why and how they fail, what they lack, what they have forgotten and, by these experiences, to develop the ideas which can effectively, precisely, lead to the needed synthesis. It is a truly Platonic advice, a dialectical advice. It is the kind of path that Platonic dialogues take, and it is for instance the one an old Parmenides advices a young Socrates to take in one of Plato's later dialogues[d].

Final remarks: Bohr's persuasion

It is puzzling: the developments we just investigated in Heisenberg's and Pauli's works coexist with some other affirmations that enter with the former, quite evidently, in clear and fundamental contradictions. This can only be explained by Niels Bohr's influence. We know, for instance, how those separations we talked about were "resolved" in some places of Heisenberg's and Pauli's works, and it was certainly not in line with the Greek spirit they took as a model. Instead of following his investigation into Greek philosophy as it pointed the way to a more complex and meaningful representation of the world, a world expressing its principles, its inherent meaning, Heisenberg chose frequently — following Bohr — to take the Cartesian concepts, to accept the reduction of the world realized in them, and simply collapse one over the other. Instead of following through with his critical analysis of the basic division and organization of the world in modern thought, instead of taking as a model what he pointed as exemplary in the Greeks, Heisenberg accepted the modern limitation of *res cogitans* to subjectivity and followed Bohr's "synthesis", according to which the subject directly creates physical reality. Faced with what he saw as naïve materialism, instead of following his own diagnostics, he accepted for the most part Bohr's proposal of an empowered relativism that subordinates physical reality to the subject. And contrary to the Platonic spirit, with its higher valuation of our intellectual capacities, Pauli curiously ended up, on some occasions, following Bohr's "solution" of a premature limitation of knowledge. Niels Bohr's power of persuasion will never cease to amaze. He disguised a renounce (the quantum realm is unknowable) as a new heroic breakthrough (the

[d]the *Parmenides*

subject creates physical reality); he simply rephrased the paradoxes that appeared when using classical concepts to represent quantum phenomena by calling them "complementary" — instead of critically analyzing those concepts — and yet somehow convinced many. He drew an arbitrary limit to our knowledge and the limit was incredibly respected. Rather than recasting the conditions of knowledge inspired by the broader Greek view, Pauli and Heisenberg accepted at times, following Bohr, the limitation of the concepts of physics to the concepts of classical physics: "the unambiguous interpretation of any measurement must be essentially framed in terms of classical physical theories, and we may say that in this sense the language of Newton and Maxwell will remain the language of physicists for all time." [14, p. 7].

It is certainly a perplexing aspect of their works. Heisenberg's Platonic atomism, with the introduction of forms as the real elements of the central order which is constantly determining reality, is very far from Bohr's proposal. Pauli's view about understanding, which take as its fundamental thesis a cosmic order that is expressed as well in *physis* as it is in psyche, seems also incompatible with Bohr's ideas. And yet, they both tended at times to embrace Bohr's turn. Schrödinger himself, even if he can't be accused of being close to Bohr, makes some of the same assumptions. This can be seen directly in some of his interpretations of Greek philosophers, as he projects in their philosophies a modern worldview. This is clear, for instance, in how he reads Parmenides, of whom he says: "The true reality he puts into thought, into the subject of cognizance as we should say. (...) The [real world] (...) resides in the subject, in the fact that it is a subject, capable of thinking, capable of some mental process at least" [11, p. 29]. And it is most especially clear in his final remarks about the "peculiar features" of our scientific view of the world, which he develops on similar lines. We have been thinking quantum mechanics mostly along Bohrian lines for quite some time now, and we still face the same unresolved paradoxes, we still regret our incapacity to complete the theory, we are still unable to represent the reality expressed in quantum physics. Maybe it is not a bad idea to revisit some of the less transited paths taken by those who first developed this theory, when, in search of a wider and flexible yet rigorous understanding of nature, they turned at times for inspiration, even beyond the Enlightenment, to the Greek source.

Acknowledgements

I would like to thank Dr. Christian de Ronde. An important part of what was here developed had its origin in our frequent discussions.

References

1. Cordero, N. L., 2005, *Siendo, se es. La tesis de Parménides*, Biblos, Buenos Aires.
2. Cordero, N. L., 2017, *El descubrimiento de la realidad en la filosofía griega. El origen y las transfiguraciones de la noción de lógos*, Colihue, Buenos Aires.
3. Carnap, H., Hahn, H. and Neurath, O., 1929, "The Scientific Conception of the World: The Vienna Circle", Wissendchaftliche Weltausffassung.
4. de Ronde, C., 2017, "Causality and the Modeling of the Measurement Process in Quantum Theory", Disputatio, Vol. IX, No. 47.
5. Diels, H., Kranz, W., 1922, *Die Fragmente der Vorsokratiker*, Berlin Weidmann.
6. Heisenberg, W., 1958, *Physics and Philosophy*, World perspectives, George Allen and Unwin Ltd., London.
7. Heisenberg, W., 1971, *Physics and Beyond*, Harper and Row, New York.
8. Heisenberg, W., 1973, *The Physicist's Conception of Nature*, J. Mehra (Ed.), Reidel, Dordrecht.
9. Pauli, W., *Writings on Physics and Philosophy*, Springer-Verlag Berlin Heidelberg GmbH, Berlin, 1994.
10. Plato, *Diálogos V*, Gredos, Madrid, 2015.
11. Schrödinger, E., *Nature and the Greeks*, Cambridge University Press, Cambridge, 2014.
12. Schrödinger, E., 1950, "What is an elementary particle?", Endeavor, VolIX, N35, July 1950. Reprinted in *Interpreting Bodies: Classical and Quantum Objects in Modern Physics*, pp. 197–210, Elena Castellani (Ed.), Princeton University Press, Princeton.
13. Viereck, G. S., 1930, *Glimpses of the Great*, Macauley, New York, pp. 372–373.
14. Wheeler, J., Zurek, W. (eds), 1983, *Theory and Measurement*, Princeton University Press, Princeton.

FUNDAMENTAL OBJECTS WITHOUT FUNDAMENTAL PROPERTIES: A THIN-ORIENTED METAPHYSICS GROUNDED ON STRUCTURE

VALIA ALLORI

Philosophy Department, Northern Illinois University,
DeKalb, IL, 60115, USA
E-mail: vallori@niu.edu
www.valiaallori.com

The scientific realist wants to read the metaphysical picture of reality through our best fundamental physical theories. The traditional way of doing so is in terms of objects, properties, and laws of nature. For instance, there are families of fundamental particles individuated by their properties of mass and charge, which determine how they move. One could call this view an *object-oriented metaphysics grounded on properties.* In this paper, I wish to present an alternative view that one can dub a *thin object-oriented metaphysics grounded on structure*: there are fundamental entities with no fundamental properties other than the one(s) needed to specify their nature. I argue that my view has several advantages over the received one. In particular, I compare my proposal to the traditional view in the quantum domain and I argue that my view provides a better fit for both approaches to quantum metaphysics, namely the primitive ontology program and wave-function realism.

Keywords: Scientific realism; laws of nature; fundamental properties; thin-objects; primitive ontology; wave-function realism; structural realism.

1. Introduction

Metaphysicians in the naturalized tradition as well as scientific realists have developed many views about the nature of reality. These views have in common a 'traditional core' according to which fundamental entities inhabit space and evolve in time according to the laws of nature. The fundamental entities could be for instance particles, as in the classical world, or particles and fields, as in classical electrodynamics. Their difference in nature is captured by their mathematical description: while particles are (exhaustively) mathematically described by points in three-dimensional space, fields are functions that assign to each point in space a field value. Laws of nature govern the behavior of such things: for instance, in a classical world Newton's law determines where the particles are at other times, given the description of the system at one time. To close the circle, one further assumes that the fundamental entities can be though as composed of different families. For instance, in the case of classical electrodynamics, particles divide into protons, electron and neutrons, and each family is identified by having specific masses and charges. That is, particles in different families move differently under the same circumstances because their fundamental properties are different. If one like labels, one can call this view an *object-oriented metaphysics grounded on properties*. In this paper I wish to explore an alternative approach that one can dub a *thin object-oriented metaphysics grounded on structure*: there are no intrinsic properties other than the one(s) necessary to identify the nature of the fundamental objects. That is, assuming the world is made of particles, they merely have the property of being located in space, which is the only property necessary to characterize them as particles (as opposed to, say, fields). Particles have no further properties: no mass and no charge. Their behavior is accounted for in terms of structural relations encoded in the laws of nature. In this paper I am going to argue that this view is better than the traditional view, especially when looking at the quantum domain, where it provides a nice fit for both of the main proposal for quantum metaphysics, namely the primitive ontology approach and wave-function realism. After discussing the traditional view in section 2, I present my view in section 3, where I also show its extension to the quantum domain. Before concluding in section 5, I present in section 4 the advantages of this approach over the traditional view and I respond to possible objections.

2. The traditional core

Restricting ourselves to the entities of the natural world, metaphysics aims at discovering what truly exists. Setting aside the so-called 'a priori' metaphysics, which is independent from empirical sciences, 'naturalized' metaphysics is informed by science. The debate between a priori and naturalized metaphysics is connected with the issues surrounding realism and anti-realism in the philosophy of science. Scientific realism fits well with naturalized metaphysics, for it maintains that we are justified in considering our best scientific theories as (approximately) true. Therefore, in this tradition one looks at natural sciences in order to individuate the nature of fundamental entities, their properties, and the laws that govern their motion. While there are several objections to scientific realism, in this paper I assume it to be tenable both in the classical and the quantum domain.

2.1. *Objects, laws and properties*

The various traditional object-oriented views in metaphysics have a common core. I call this core the traditional view. It comes down to four elements: space-time, fundamental things, their intrinsic properties, and laws of nature. First, there is space and time, or space-time: the arena in which the fundamental objects live. For instance, in a classical world the fundamental objects are point-particles. Mathematically, their nature as particles results in the fact that they are completely specified by their location in space. Newton's second law (i.e. the force F acting on an object generates an acceleration a proportional to its mass m, or $F = ma$) and Newton's law of gravitation (i.e., two objects attract each other in a way that depends directly on their masses, m_1 and m_2, and inversely to their relative distance r_{12} squared: $F_{grav} = G\frac{m_1 m_2}{r_{12}^2}$, where $G = 6.67408 \times 10^{-11} m^3 kg^{-1} s^{-2}$ is the gravitational constant) determine how the particles evolve in space through time. Other possible ontologies are material fields. In contrast to particles, they are extended objects that occupy the whole space, and whose intensity changes from point to point. Arguably (more on that in section 4.1), in classical electrodynamics one introduces the electromagnetic fields E and B as fundamental entities together with the particles. Both of these entities evolve in time in a way described by the fundamental laws of nature: Maxwell's equations determine the evolutions of the fields,[a] and the electromagnetic force ($F_{EM} = qE + qv \times B$, where q is the signed charges of the body, and v is its velocity) together with Newton's second law ($F_{EM} = ma$) account for the action of the fields on the particles. The temporal evolution of the particles can be described as a trajectory in space-time, and the evolution of the fields as a propagation of a wave in three-dimensional space. However, the traditional account needs more in order to successfully recover the empirical data: particles need to have mass to explain why a massive

[a] For completeness, Maxwell's equations are as follows: $\nabla \cdot E = \frac{\rho}{\epsilon_0}$; $\nabla \cdot B = 0$; $\nabla \times E = -\frac{\partial B}{\partial t}$; $\nabla \times B = \mu_0 (J + \epsilon_0 \frac{\partial B}{\partial t})$, where ρ is the charge density and J the current density (the rate of change of charge per unit area). They can be rewritten as $\frac{\partial^2}{\partial t^2} E - \frac{1}{\mu_0 \epsilon_0} \nabla^2 E = 0$, $\frac{\partial^2}{\partial t^2} B - \frac{1}{\mu_0 \epsilon_0} \nabla^2 B = 0$, where $\epsilon_0 = 8.85 \times 10^{-12} F m^{-1}$ (the vacuum permittivity) and $\mu_0 = 12.57 \times 10^{-7} N A^{-2}$ (vacuum permeability), which is an equation for a wave propagating with velocity $v = \frac{1}{\sqrt{\epsilon_0 \mu_0}}$, which is the velocity of light c.

body behaves differently from another massive body under the same circumstances. If body 1 and body 2 have respectively masses m_1 and m_2, given the law of gravitation F_{grav}, body 1 is accelerated toward body 2 with acceleration $a_{12} = F_{grav}/m_1$ and body 2 is accelerated toward body 1 with a different acceleration, namely $a_{21} = F_{grav}/m_2$.[b] In classical electrodynamics one needs also charge in order to account for the actual experiment outcomes. If particle 1 and particle 2 have the same mass but 1 turns left in a given magnetic field while 2 turns right, then this is explained by them having different charges. In the traditional account, therefore, one can divide particles as belonging to distinct families, for instance one is an electron and another is a proton, in virtue of their different masses and charges.[c] An electron, say, is defined in virtue of its properties, which therefore are intrinsic to it: its mass, $m_e = 9.11 \times 10^{-31} kg$, and its charge, $q_e = -1.60 \times 10^{-19} C$ determine its nature. In this sense, fundamental objects are thick: they are 'dressed up' with properties. Nonetheless, all fundamental entities follow the same laws of motion: the fields obey Maxwell's equations, the particles follow Newton's second law $F = ma$, supplemented by the laws of the forces given by F_{grav} and F_{EM} above, so that $F = F_{grav} + F_{EM}$.

So, given the ontology, there is *one fundamental law, and many fundamental properties*, and thus there are many families of the same kind of fundamental entities. It is as if whoever developed the traditional account followed this principle: let us minimize the number of fundamental laws (e.g. Newton's second law, the law of gravitation, Maxwell's laws of electrodynamics), and add fundamental properties as needed to allow the empirical adequacy of the theory. This would in turn multiply the number of fundamental kinds of entities.

Many debates are connected to the traditional core, each connected with its fundamental elements.[d] In particular, because in the traditional account

---

[b]I assume reductionism, so the mass of macroscopic objects is (suitably) the sum of the masses of their fundamental constituents. Notice that in classical mechanics as developed originally by Newton, who knew nothing of protons and electrons, all fundamental particles have the same mass and therefore macroscopic bodies behave differently only if they contain a different number of particles.

[c]The fact that protons are actually not fundamental particles is not relevant here. The example is just supposed to simplify the situation.

[d]First, we have the problem of the nature of space and time: is space a substance or does it merely consist of relations between objects? Then, we have the discussion over the nature of laws: one finds Humeans, who believe that laws are the axioms and postulates of the best theory of the world, (broadly understood) necessitarians, who argue that laws are over and above physical events, and primitivists or anti-reductionists, who maintain that laws are in their own metaphysical category.

properties play a mayor explanatory role, the debate about the nature of intrinsic properties has gained a lot of attention. The divide is between categorical properties and dispositional properties. However, even the definition of what counts as one or the other is controversial (see [18,25,51] for some attempts to characterize categorical properties, versions of sophisticated analyses of dispositions have been offered by [30,44,47,60], and [45]). For instance, one can say that a property is dispositional when its essence depends on what it prescribes its bearer to do. In contrast, a categorical property is defined as not connected to this. This could mean that a property is categorical when its essence does not depend on what its bearer is, or that they are the properties which remain the same in all possible worlds. Still, it is not clear whether the two definitions are compatible. An example of dispositional property is solubility: a substance has such a property if, when placed in water, will dissolve. On a more fundamental level mass and charge may be dispositional properties since they determine what the particle bearing a given mass, say, is disposed to do when subject to a force. However, mass and charge can also be seen as categorical properties: they identify what, say, an electron is in all possible worlds. Other examples of categorical properties are location and spatial relations (see [25]). Moreover, among dispositional properties one finds causal powers (such as mass), which allegedly intervene in causal processes, and propensities (such as half-lives of radioactive elements) which are such that the activities of their bearers do not depend on the circumstances (they are 'absolute' dispositions. Note that the debate over the nature of properties is connected with the debate over laws of nature: Humeanism requires a strong differentiation between dispositional and categorical properties, since Humean supervenience is expressed in terms of categorical properties ('perfectly natural properties') which fully determine dispositional properties.

2.2. *The traditional view in the quantum domain*

Aside from the problem of suitably defining properties, the most severe challenge to the traditional approach is its extension to the quantum domain. First, it is not obvious that quantum theory is compatible with scientific realism, and even so it is not obvious what its ontology is. In contrast with classical theories, it is not clear what the underlying metaphysical assumption is: what is matter made of, in this theory? Upon opening physics books, one finds a mathematical entity and an equation standing out: the wave-function and its temporal evolution equation, called the Schrödinger equation. So, one natural thought is to take this entity as describing the fundamental ontology. However, as emphasized by Schrödinger himself [61], this is problematical because of the so-called *measurement problem*, also known as the problem of the Schrödinger cat: given that the wave-function is supposed to represent every physical object, and given that it is a wave, it can superimpose also at the macroscopic level, which is something that we never observe. For instance, we have never seen a superposition of a dead-and-living cat.[e] Many physicists react to the cat problem conceding that quantum mechanics is contradiction-ridden, and thus it is unsuitable to describe reality. They therefore conclude that we have no choice but to become anti-realist: at best, quantum theory can describe measurement results. However, in the last 60 years many quantum theories compatible with a realist interpretation have been proposed, such as the pilot-wave theory [20,23], the spontaneous localization theory [33], and the many-worlds theory [29]. Mathematically speaking, the pilot-wave theory is one in which, in contrast with the common understanding, the complete description of a physical system is given by the wave-function and by the particles' locations. Accordingly, two equations define this theory: the Schrödinger equation for the wave-function, and the guidance equation for the particles, which describes how their velocity is governed by the wave-function. In this theory, only the part of the wave-function 'under which' the particle is matters for its future behavior, and thus getting rid of the superpositions for all practical purposes. In the spontaneous localization theory, the wave-function is usually thought as providing the complete description of any physical system, but it evolves according to the Schrödinger equation only until a random time, and then it instantaneously localizes around a random point, effectively eliminating the (macroscopic) superpositions. This so-called GRW-evolution involves two new constants: the rate of collapse (*per* 'particle') $\lambda = 10^{-16}s^{-1}$ and the width of the collapsed wave-function $\sigma = 10^{-7}m$. Finally, in the many-worlds theory the complete description of the system is given by the wave-function which evolves according to Schrödinger's equation. The basic idea is that the two terms of

[e]The reason this is often called the measurement problem has to do with the positivistic influences some of the founding fathers of quantum mechanics were under. The only moment they saw these superpositions to be problematic was in a measurement situation: we want to measure the position of a particle; the result is that it is the superposition of the-detector-on-the-right-having-clicked and the-detector-on-the-left-having-clicked; this is not what we observe because only one detector will click; so something has to happen that changes the evolution of the wave-function during the measurement process. However, a measurement process is like any other physical process, and therefore, the theory is not able to describe reality in a complete manner.

the superposition that describe a living and a dead cat both exists, even if they are never observed: the terms of the superpositions interact so little that they can be interpreted as living in distinct 'worlds' occupying the same region of space-time.

There is a sense in which the ontology of these theories is naturally given (at least partially) by the wave-function. That is, the most straightforward way of understanding the mathematics of these theories is to assume that the wave-function represents matter: in the pilot-wave theory, there are both particles and waves, while in the spontaneous localization theory and in the many-worlds theory the wave-function may evolve differently but equally describes physical objects. This view, dubbed *wave-function realism*, is (partly[f]) motivated by considerations arising from the comparison with Newton's theory. In classical mechanics, the fundamental equation is an equation for points in three-dimensional space. From this, we inferred that matter was made of point-like particles in three-dimensional space (or point-like particles in spatio-temporal relations between one another). Since in quantum mechanics (what is regarded as) the most fundamental equation, Schrödinger's equation or the GRW evolution, is an equation about the wave-function, similarly we should conclude that matter is represented by the wave-function (see [5], and then most notably [1,3,4,39–42,53–57], and [58]). Notice however that the wave-function by construction is an object that lives on configuration space, which classically is defined as the space of the configurations of all particles. As such, it is a space with a very large number of dimensions, while one would expect a physical object to live in three-dimensional space: call this the *configuration space problem* (or the macro-object problem). Wave-function realists therefore argue that what we take to be three-dimensional particles are instead emergent or derivative, and that configuration space is best seen as the space of the 'degrees of freedom' of the system. In this framework there are two main attempts to recover the macroscopic objects from the wave-function. The first is due to Albert [4] and based on a functionalist reduction of three-dimensional microscopic objects from the wave-function. The idea is that it is possible first to define functionally what it means to be a three-dimensional object, and then it is possible to show that the wave-function can play that role. This functional reduction can give rise to microscopic three-dimensional objects, which then can be understood as usual, in particular as compos-

[f]Ney [57] makes the case that another main motivation for this view is to have a separable and local metaphysics. See section 4.1 for a little more on this.

ing macroscopic objects. Ney [57] is critical of this approach, as she points out that in this reading there is no common three-dimensional space for inter-particle interactions. She therefore proposes her approach, in which she gives a privileged role to symmetries: she observes that only a three-dimensional 'decomposition' of the wave-function (as opposed to any other kind of decomposition) can explain symmetry properties, and because of this reason it makes sense that we represent our world three-dimensionally rather than (say) two-dimensionally. Thus, in her view three-dimensional objects exist, not as an additional postulate of the theory but as derivative from the wave-function when considering symmetry properties as fundamental facts about the world.

Another approach to quantum metaphysics is the so-called primitive ontology approach (see [9–11,24] and references therein) which follows the lead of scientists such as de Broglie, Lorentz, Einstein[g] and (interestingly enough) Heisenberg[h] in the 1920s, who, did not think that the wave-function should be thought of as material. They worry about the configuration space problem: the wave-function does not live in three-dimensional space, and as such it is not suitable to represent physical entities. The idea instead is that the most straightforward and direct way of making a connection between our observations and what is postulated by the theory is to assume that matter is made of three-dimensional stuff. For instance, it could be particles, (three-dimensional) fields, spatio-temporal events (sometimes called 'flashes'). Classically, as we have seen in the traditional core, we explain the behavior of macroscopic things by assuming that they are composed of microscopic things, and in the primitive ontology approach this straightforwardly follows also in the quantum domain. Note that in this approach the wave-function does not represent matter, but it is controversial what it actually is. Some have argued it is a property of matter (see [28,42,50,63,64] and references therein). However, its role in the theory does not seem to be straightforwardly the same as the one of properties: the role of properties in fact is to distinguish different kinds of fundamental entities, and this is not what the wave-function does; rather it helps generate the law of evolution of physical objects. Therefore, more fitting seems to be the view that the wave-function is nomological, or quasi-nomological: the wave-function is better understood as part of the law of nature (see [8,35]

[g]See [15] for an interesting discussion of the various positions about this issue and others at the 1927 Solvay Congress.

[h]Heisenberg very vividly said to Bloch [19], referring to configuration space realism: "Nonsense, [...] space is blue and birds fly through it."

and references therein for criticisms and replies). In fact the wave-function behaves similarly to other entities we label as 'law-like' in classical mechanics, like for example potentials: matter is not 'made of' them; rather, they help defining the law for matter, just like the wave-function in the quantum domain. If so, we are not really adding anything mysterious to the picture. The complicating feature is that the wave-function evolves in time, which however is not too concerning in a Humean account of laws (see. e.g. [17,21,49] and [27]). However, perhaps better is the idea that the wave-function should be understood structurally, which is what I am defending in this paper.

So, to summarize, under the assumption that quantum theory can be interpreted from a realist perspective, the situation seems to be as follows. On the one hand it seems that also in the quantum domain we need properties to explain why things behave differently in the same circumstances. Indeed, we add more properties, such as spin, to explain the behavior of more particle types: for instance, we talk about bosons (which share the properties of having integer spin) and fermions (which instead have half-integer spin). On the other hand, there are two approaches to quantum metaphysics, wave-function realism and the primitive ontology approach. In the case of the former, because of the ontological and explanatory continuity between the classical and the quantum domain, it seems that the traditional view may hold just as well, we just need to add more properties to the mix. However, the situation is more troublesome in the case of wave-function realism. In fact, while in the traditional view properties did the majority of the work in explaining the behavior of things, nothing like that happens here. While in the traditional view one explains why electrons go up in a magnetic fields and protons go down by invoking the fact that they have opposite charge, in this context matter is described by the wave-function and its fundamental property is (presumably) its field value. In Albert's approach three-dimensional objects are 'functional shadows' of the high dimensional wave-function, which does not naturally or obviously 'carve up' into different properties to assign to the different shadows. All matter is described by the same wave-function, with the same field value. So, Albert's theory explains the electrons going up in a magnetic field not by invoking the property of the wave-function, but rather by saying that there is something in the wave-function that plays the functional role of the electron, namely to go up in a magnetic field. In Ney's account something similar happens: electrons exist because, among all the decompositions of the wave-function, the one privileged by symmetries is the three-dimensional decomposition. This decomposition shows 'particle-like' entities which *appear* to have charges in the sense that they go up in a magnetic field: they go up because that is what that decomposition of the wave-function ends up doing. So, in both cases 'electrons' exist derivationally from the wave-function, and appear to have properties insofar as their behavior allows us to use that kind of talk. That is, what emerges derivationally from the wave-function (either structurally or using symmetries) is the object, not its properties. In other words, in this approach, property talk is *purely fictional*: we can use it, but it is not fundamental. The most fundamental explanation of the behavior of macroscopic objects is instead the one given by the wave-function and its temporal evolution.

So it seems to me that, while in the case of the primitive ontology approach the traditional view *may* still be viable, within wave-function realism it would be a stretch to think of explanation in terms of properties of fundamental objects as the traditional approach suggests. Instead, in the next section I propose an approach which I argue should be preferred to the traditional view for general reasons, but also because it provides, contrary to the alternative, a good fit for both the primitive ontology approach and wave-function realism.

3. Fundamental entities without fundamental properties

My approach is an alternative to the traditional account. I argue in this section that my view has several advantages over the traditional approach and over the approaches proposed by the wave-function realists. My view requires a fundamental re-interpretation of the role of laws of nature, which I think is best understood in structuralist terms. The basic idea is that, as in the traditional account, there is space-time and there are fundamental entities. Yet, these entities have only *one fundamental property*, namely the one that *uniquely characterizes their nature*. For instance, if the fundamental ontology is particles, their only property is their spatial location. If the world is instead made of fields, their only property is having a set of intensity values for every point in space. In contrast with the traditional account therefore there are no different families of fundamental entities. That is, *all fundamental entities are identical*: there is just one kind. That is, if the fundamental ontology is particles, all particles are identical: there are no different families of electrons, protons and neutrons.

A first reaction to this would be an immediate dismissal: we already tried and failed. One *has* to add properties to account for the different behaviors under the same circumstances. However, I think this rejection is too hasty:

one could suitably account for the observed different behaviors of particles *appearing* to belong to different families in terms of laws of nature. This can be done by introducing the notion of 'effective law:' fundamental entities behave differently in the same situation not because they have different properties (they do not), but because they are *governed by different effective laws*. Effective laws are different ways in which the same (kind of) law can be implemented, and there is one effective law for what appears to be a different family of ontology in the traditional approach. For instance, in a particle theory, there are as many effective laws as there seem to be families of particles. Thus, the main idea is the opposite of the traditional view: *let us minimize the fundamental properties, allowing for as many effective laws of nature as needed to make the theory empirically adequate*. In other words, there is a network of structures between the fundamental entities that can be captured in terms of general laws which are implemented differently through effective laws. Let me elaborate on this in the next section.

3.1. *Thin objects, effective laws, and symmetries*

Consider Newton's law of gravitation $F_{grav} = G\frac{m_1 m_2}{r_{1,2}^2}$. It contains the gravitational constant G, which is fixed and immutable. This is the case also for the other forces, including the electromagnetic force, where two constants ϵ_0 and μ_0 are 'hidden' in the equations of the evolution of the fields. These constants are part of the definition of the *law*. In addition, there are other parameters, like masses and charges. They are different from constants: they appear in laws, but *they are part of what defines matter*. In fact, constants remain identical independently of the objects the law applies to, while masses and charges do not. As we saw, in the traditional account one recovers the empirical data by postulating that particles have certain properties, over and above their spatial location, and accordingly there are different fundamental families. In contrast, in my view there is no such distinction, and empirical adequacy is obtained by introducing effective laws. That means that the parameters in the law, namely the values of the fundamental masses and charges, are seen as *part of the definition of the law* too, together with the constants. For example, assume that in the traditional view there are three families of particles: electrons, protons and neutrons. Thus, we have six parameters: the three masses m_e,m_p,m_n, and the three charges q_e,q_p,q_n. Now, forget the charges for the moment and focus on Newton's gravitational law. One can rewrite it for an 'electron' as follows: $F_{\text{`}electron\text{'}} = \frac{H_1}{r^2}$, where $H_1 \equiv Gm_eM$, where M is the mass of a reference body that generates the force. For all families, the form of the law (its kind, if you prefer) is the same: its intensity is proportional to $1/r^2$. However, there is a different constant H_i for each kind of 'fundamental particle.' In this way, to each 'fundamental particle' is associated an effective law $\mathit{Eff}_{law\,i} = \frac{H_i}{r^2}$, where $i = 1, 2, 3$ is associated to one of the three 'families.' There is one effective law for the 'electron' $\mathit{Eff}_{law\,1} = \frac{H_1}{r^2}$, where $H_1 = Gm_eM$, one for the 'proton,' $\mathit{Eff}_{law\,2} = \frac{H_2}{r^2}$, where $H_2 = Gm_pM$, and one for the 'neutron,' $\mathit{Eff}_{law\,3} = \frac{H_3}{r^2}$, where $H_3 = Gm_nM$.

In other words, and to sum up, in my view matter is represented by *stuff* in three-dimensional space whose nature is specified by our best theory as the simplest and most explanatory. In classical mechanics, this stuff is particles. As already said, in my approach there is just one kind of fundamental entity whose only property is the one that defines its nature. Classical particles have only one property, namely their position in three-dimensional space, since it determines their nature of particles (as opposed to fields, say). Moreover, two fundamental entities differ only because they differ in behavior under the same circumstances, and this is accounted for by assuming that they are governed by different effective laws $\mathit{Eff}_{law\,1}$ and $\mathit{Eff}_{law\,2}$. This is easy to see in classical electrodynamics. Assume that, pictorially, in the traditional view electrons are red dots in space-time, protons are yellow ones, and neutrons are green ones. Since their temporal evolution is described by the same law, they follow respectively the red, the yellow and the green trajectories. Instead in my approach an electron is a point which has no color which follows the red trajectories generated by $\mathit{Eff}_{law\,1}$, a proton is a point which has no color which follows the red trajectories generated by $\mathit{Eff}_{law\,2}$, and a neutron is a point which has no color which follows the red trajectories generated by $\mathit{Eff}_{law\,3}$.

Moreover, as we saw in section 2.1 in classical electrodynamics one introduces the fields as part of the fundamental ontology to make the theory empirically adequate: the fields appear in the force $F_{EM} = qE + qv \times B$ and affect the particle trajectory in virtue of its charge. However, in the spirit of this approach it seems more natural to think of the *fields as part of the law* too, just like the particle properties. Consider the example above, this time focusing on the electromagnetic force. For simplicity, consider only the Coulomb force: $F_{Coulomb} = k\frac{q_1Q}{r^2}$ (where $k = \frac{1}{4\pi\epsilon_0} = 8.99 \times 10^9 Nm^2C^{-2}$ is Coulomb's constant, generated by particle with charge Q and acting on the particle with charge q_1). One can define the electric field as $E = k\frac{Q}{r^2}$. Thus, $F_{Coulomb} = q_1E$. If particle 1 is an 'electron,' one can rewire this as follows: $F_{electron} = q_eE = \frac{K_1}{r^2}$, where $K \equiv kq_eQ$. As in the gravitational case, for all particles the form of the law is the same: it is an inverse square law. However, there is a different proportionality constant K_i for each kind of 'fundamental particle' so that to each is associated an effective law $Eff_{law\,i} = \frac{K_i}{r^2}$, where in this example $i = 1, 2, 3$. In this way, not only there are no particle families, *there are no electromagnetic fields either*! They are 'absorbed' into the definition of the law as well: one law for the 'electron' $Eff_{law\,1} = \frac{K_1}{r^2}$, where $K_1 = kq_eQ$; one for the 'proton,' $Eff_{law\,2} = \frac{K_2}{r^2}$, where $K_2 = kq_pQ$, and one for the 'neutron' $Eff_{law\,3} = \frac{K_3}{r^2} = 0$, because the neutron has no charge and thus $K_3 = kq_nQ = 0$.

One important reason, perhaps the most important one, to question that electromagnetic fields are part of the material constitution of the world has to do with symmetry properties (however, for other arguments against a field ontology, see [52] and [38]). In fact, as explained below, it turns out that they transform in a way that is *dissonant* with their apparent nature. Roughly put, symmetries are transformation that do not affect the applicability of the theory. For instance, Newtonian mechanics being translation invariant implies that it equally describes the actual world and a world exactly like this but translated one meter to the left, say. Among others symmetries, classical electrodynamics is taken to possess also time-reversal symmetry (for a discussion on the meaning of time-reversal symmetry and its implication for classical electrodynamics see [9] and references therein. For time reversal in the quantum domain, see [8]). A time-reversal transformation reverses time, and roughly a theory is said to be time-reversal invariant when the world 'running forward' and the world 'running backwards' are both described by the theory. Assume that in classical electrodynamics there are both particles and fields. Particles are defined as points in three-dimensional space, but what about the fields? One natural way to think about them is as vector functions: they associate to each point in three-dimensional space a field value in terms of a vector. As such, they do not change under a time-reversal transformation: why would a field value change depending on whether it is part of the forward-running movie of the world or the backward-running movie? It would be as if in "Back to the Future" projected forward Marty McFly is Michael J. Fox, and if we project the movie backward now Marty is played by Keanu Reeves. However, if we want the theory to keep time-reversal symmetry, the magnetic field in the backward-running world would have to change sign, and this is puzzling. One way of keeping the symmetry and explain the field transformation is to think of the electric and magnetic fields as respectively a polar vector (or merely a vector) and an axial vector (or pseudo-vector) which transform as needed to preserve invariance. A pseudo-vector transforms like a vector under rotation, but it changes sign under reflection, accounting for the behavior of the magnetic field changing under a time-reversal transformation needed to preserve invariance. Yet, one needs to provide additional independent reasons to believe that the true nature of the fields is the one proposed: why would the magnetic field be a pseudo-vector if it looks like a vector? An explanation is given by my approach: if one wants to keep the symmetry, one should not allow the fields to represent physical objects. And assuming that they do not, then we can allow the fields to transform as needed to allow the symmetry to be preserved. This amounts to define the electric and the magnetic fields respectively as a true vectors and a pseudo vector. Similar consideration will also be true in the case of the wave-function, as we will see below: only if one thinks of it as non-material, we can account for certain symmetries of quantum theory.

To summarize, in my view we have *thin* objects, individuated by their only natural property, namely by the property that uniquely characterizes their nature. In this sense therefore the objects in this approach are 'thin.' Then the law of nature determines how they evolve in time. For the theory to be empirically adequate, the law has to break down into a set of effective laws, each of which applies to a subset of the set of fundamental objects. The laws and the effective laws are naturally seen as a *structure*, a network of relations, between the thin objects.

Before we proceed, I wish to make a couple of remarks. The first is on the use of the term 'structure.' In my view one has identical thin objects, each governed by an effective law, which is a specific instantiation of the fundamental law of nature, which includes parameters that capture what we call in the traditional view 'properties' of the fundamental objects. There are many questions that immediately come to mind, some of which have to do with ontological priority: does the fundamental law exist over and above the thin objects? Do the effective laws exist over and above the fundamental law? These questions are the typical questions one asks when investigating the metaphysics of laws and have Humean and non-Humean perspectives: while the Humean maintains that laws supervene on the arrangements of matter, the non-Humean disagrees. I will bypass the whole debate by simply assuming that one can be neutral about the metaphysical status of laws and endorse a structural approach: regardless of whether one is Humean or not with respect to laws, the set composed of the fundamental law and the effective laws can be seen as a set of relations between the thin objects. They are a structure which, regardless of whether it supervenes of not on matter, describes how matter behaves. And this is sufficient for the purpose of this paper.

The second remark instead has to do with symmetries. In this approach *symmetries help identifying the true nature of objects*: if we allow electromagnetic fields to be objects then the theory loses an important symmetry, so they should be regarded as part of the law. To put it differently, there is a sense in which *symmetries give indications on what the law is*. If one also adds simplicity considerations, one may argue that symmetries allow to generate the law for the thin objects. While undoubtedly there is no logic of theory construction, one could notice that historically laws of nature were proposed in the context of a given theory using symmetry considerations. Take the case of classical electrodynamics. As we have seen, one needs fields in order for the theory to be empirically adequate. However, how should they transform? As we have seen, they should transform as to preserve certain symmetries of the motion of the particles. Similar considerations have been used also in the quantum domain: in the pilot-wave theory discussed in the next section the law of evolution of the particles is obtained entirely using symmetry (and simplicity) considerations.[i] Moreover, relativity theory was developed, together with the light postulate, using a symmetry principle, namely the principle of relativity, which was adopted by Einstein solely with the purpose of restricting the form of the law. In addition, Wigner and Weyl were among the first to recognize the great relevance of symmetry groups to quantum physics, which were (and still are) systematically used to construct new models and theories, similarly as how now in string theory we have dualities. This kind of considerations should be taken with a grain of salt, as epistemological considerations about how a theory has been proposed not necessarily track ontology. However, it seems to me that if one could show that symmetry considerations are enough to select one particular law of nature among the infinite possibilities as well as a type

[i]This is how Dürr Goldstein and Zanghì [24] infer the laws for the trajectories of particles in the pilot-wave theory. The consideration that a first order derivative of position is simpler than a second order one, leads to an equation for the velocities so that the k-th particle will move according to $\frac{dq_k}{dx} = v_k^{\psi}(q_1, \ldots, q_N)$, where is a Schödinger evolving wave-function. Moreover, the fact that two wave-functions that differ by a constant are physically equivalent leads to the fact that the velocity needs to be a homogeneous of degree 0 as a function of the wave-function, that is $v_k^{c\psi} = v_k^{\psi}$, where $c \neq 0$ is a constant. Asking for rotational invariance one gets $v_k^{\psi} = \alpha \frac{\nabla \psi}{\psi}$, where α is a scalar constant. In addition, time-reversal invariance is implemented for the wave-function as a transformation into its complex conjugate $\psi \to \psi^*$, which leads to $v_k^{\psi^*} = -v_k^{\psi}$, and thus $v_k^{\psi} = \alpha \Im \frac{\nabla \psi}{\psi}$. Finally, Galilean invariance requires $\alpha = \frac{\hbar}{m}$, given that the wave function transforms as $\phi \to e^{i \frac{m}{\hbar} v_0 q} \psi$.

of ontology, then one would have further simplification of the explanatory structure: from symmetries one would get the type of law and the type of objects, and from further considerations about empirical adequacy one would get the effective laws. In turn, from this one would then explain the 'apparent' properties of matter. For instance, in classical electrodynamics one would have a particle ontology, from simplicity and symmetry considerations: particles are simpler than fields, mathematically, and if we allow fields then we lose symmetries. Then we have a type of law for the particles, the inverse square law, based on considerations of empirical adequacy and simplicity. Effective laws, however, need to be postulated in addition to the law: there is noting that tells us, merely from using symmetries, how many particle types there are, in the traditional language, which means that there is nothing in the symmetries that tells us how many effective laws one needs in order for the theory to be empirically adequate.

3.2. *Extension to the quantum domain*

In this section I show how this view extends nicely in the quantum domain, regardless of whether one endorses the primitive ontology approach or wave-function realism. Let us see how that is first in the case of the primitive ontology approach. As we have already discussed, within this framework, matter is represented by (microscopic) variables in three-dimensional space (or four-dimensional space-time). Because of the more or less straightforward classical-to-quantum continuity, this view can naturally be read as saying that matter has no property other than the one that determines its nature. To close the circle, within quantum mechanics matter evolves in time according to a law that involves the wave-function, which is taken to be part of the definition of the laws. Every quantum theory can be analyzed in these terms, and each of them provides a picture of the world in terms of its (more or less natural) fundamental entities. For instance, consider the pilot-wave theory. Matter is constituted by particles, whose law of evolution involves constants, such as $\hbar = 1.05 \times 10^{-34} m^2 kg s^{-1}$ which is the reduced Planck's constant, and parameters such as masses.[j] Following the approach developed in this paper, this law breaks down into a series

---

[j]Assuming there are N particles with positions $Q_i, i = 1, \ldots, N$, then the k-th particle would evolve in time with a velocity given by $v_k = \frac{dQ_k}{dt} = \frac{\hbar}{m_k} \Im \frac{\psi^* \partial_k \psi}{\psi^* \psi}(Q_1, \ldots, Q_N)$, where I denotes the imaginary part, ψ is the Schrödinger evolving wave-function and m_k is traditionally intended mass of the k-th particle.

of effective laws by considering the parameters as part of the law too.[k] In the spontaneous localization theory one may have a particle primitive ontology whose evolution is implemented by a GRW-evolving wave-function (this theory was dubbed GRWp3 in [14], [7], while it is called GRWp in [6]). In this theory the law of nature is the law of the particles' motion, which is the same as the one of the pilot-wave theory (with qualifications[l]). Thus, one can similarly define the effective laws using the particles masses as new constant in addition to the GRW constants λ and σ.[m] Alternatively, one could assume that the fundamental entity is a scalar field in three-dimensional space representing the matter density of things, defined in terms of the wave-function (this theory was originally proposed by [32], and was dubbed GRWm in [12]). In the reading I am proposing, assuming that there are, say, three 'particle' kinds (electrons, protons and neutrons), each associated traditionally with mass m_i, $i = 1, 2, 3$ then the total matter density field is M^ψ such that $M^\psi(x,t) = M_1^\psi(x,t) + M_2^\psi(x,t) + M_3^\psi(x,t)$, located in three-dimensional space.[n] The matter density $M^\psi(x,t)$ has no property other than the field values assigned by the formula above. Also, the expression for this field defines the law of nature of the theory, namely the law of evolution of the matter density in terms of the wave-function, and

[k]One can in fact rewrite the guidance law of the particle as $v_j = W_j \Im \frac{\partial_k \psi^* \psi}{\psi^* \psi}(Q_1, \ldots, Q_N)$, where $W_j = \frac{\hbar}{m_j}$ is the new constant, and $j = 1, \ldots, n$, where n is the number of 'apparent families'. Similar considerations as the one I am providing here can be also found in [34].

[l]For particle k, the velocity is $v_k = \frac{dQ_k}{dt} = \frac{\hbar}{m_k} \Im \frac{\partial_k \psi^*_{GRW} \psi_{GRW}}{\psi^*_{GRW} \psi_{GRW}}(Q_1, \ldots, Q_N)$, where the wave-function evolves as in the GRW process but it collapses to a point centered where the particle is at that time and shifted at random.

[m]The effective laws is similar to the one in the pilot-wave theory: $v_j = W_j \Im \frac{\partial_k \psi^*_{GRW} \psi_{GRW}}{\psi^*_{GRW} \psi_{GRW}}(Q_1, \ldots, Q_N$, $j = 1, \ldots, n$, where n is the number of 'apparent families' of particles, $W_j = \frac{\hbar}{m_j}$ is the new constant, together with the rate per particle λ, and the wave-function width σ present in the GRW evolution. Notice that the wave-function in the law for the velocity will change at the time of collapse into the collapsed wave-function.

[n]In formulas: $M^\psi(x,t) = m_1 \int dq_1, dq_2, dq_3 \delta(q_1 - x)|\psi_{GRW}(q_1, q_2, q_3, t)|^2 + m_2 \int dq_1, dq_2, dq_3 \delta(q_1 - x)|\psi_{GRW}(q_1, q_2, q_3, t)|^2 + m_3 \int dq_1, dq_2, dq_3 \delta(q_3 - x)|\psi_{GRW}(q_1, q_2, q_3, t)|^2 = M_1^\psi(x,t) + M_2^\psi(x,t) + M_3^\psi(x,t)$. Visually, one has three distributions of matter: one, M_1, corresponding to the 'electrons' of 'mass' m_1; the other, M_2, corresponding to the 'protons' of 'mass' m_2; and the third, M_3, to the 'neutrons' of 'mass' m_3, each centered around where one would think they are located, that is respectively q_1, q_2, q_3. This generalizes to N 'particles' as follows: $M(x,t) = \sum_{i=1}^N m_i \int dq_1, \ldots, dq_N \delta(q_i - x)|\psi_{GRW}(q_1, \ldots, q_N, t)|^2$.

as such it contains only one parameter: m_i.[o] Therefore, the effective laws can be obtained by turning these parameters into constants.[p] Also in the case of the many-worlds theory, one could assume matter to be represented by the matter field like we have seen above but defined in terms of the Schrödinger evolving wave-function. This field inherits the superpositions of the wave-function, but since the various terms effectively do not interact they can be seen as describing different worlds superimposed in the same space-time (this theory was proposed by [13], who argued that it may have been what Schrödinger had in mind when he first proposed his wave equation). The law of nature and the effective laws are defined similarly to the case of the spontaneous localization theory with matter density.[q] Moreover, symmetry properties help us guiding in the choice of the physical objects: in order to preserve symmetries, one has to deny that they are represented by the wave-function, as in the case of electromagnetic fields. In this context, indeed, it is even easier to deny the materiality of the wave-function, given that, in contrast with the electromagnetic fields, it is not even living on three-dimensional space. The best way to think of the wave-function in this context is as quasi-nomological, for the same reasons we gave for the fields in classical electrodynamics.

Now let us explain how my view provides a nice fit for wave-function realism as well. My proposal drops properties altogether, and pass on their explanatory role to the laws, which are now the true explanatory entities

[o]So that, for an 'electron' with 'mass' m_1 it will be: $M_1(x,t) = m_1 \int dq_1, \ldots, dq_N \delta(q_1 - x)|\psi_{GRW}(q_1, \ldots, q_N, t)|^2$.

[p]For completion, one could imagine the fundamental objects not to be in three-dimensional space, evolving in time, but to be directly located in four-dimensional space-time. In this way, there are certain events in space-time, that one can call 'flashes,' that are non-empty, so to speak, and they represent matter. This theory was initially proposed by [16], and then dubbed GRWf by [12]. Every flash (X, T) corresponds to one of the spontaneous localizations of the wave-function, and its space-time location is just the space-time location of that collapse. Assuming there has been n collapses, the flashes will be the set $F_n = (X_1, T_1), \ldots (X_k, T_k), \ldots, (X_n, T_n)$, k being a progressive natural number indicating the time progression of the flashes. As one can see, the only property the flashes possess is their 'location' in space-time. Since T_k, X_k and n are random, the wave-function is also random. However, given the initial wave-function, the statistic of the future evolution of the flashes is determined. For simplicity reasons, I overlook the derivation of the effective laws in this case, which is however possible.

[q]Explicitly, the law of nature is:

$$M(x,t) = \sum_{i=1}^{N} m_i \int dq_1, \ldots, dq_N \delta(q_i - x)|\psi(q_1, \ldots, q_N, t)|$$

and the effective law for 'electrons' is:

$$M_1(x,t) = m_1 \int dq_1, \ldots, dq_N \delta(q_1 - x)|\psi(q_1, \ldots, q_N, t)|^2.$$

in that they govern the motion of fundamental stuff. As we discussed, both in the case of Albert and Ney, property talk is completely fictional: three-dimensional objects as we usually think of them are either functionally defined or symmetry-privileged 'shadows' of the wave-function, but no reference to their properties in the traditional sense is given or needed. In the case of Albert, what it is to be a system of three-dimensional objects is to have their behavior described by a given Hamiltonian. That is, to be a three-dimensional object is to have one's behavior depending on changes in position and on inter-particle distances in three dimensions (see [1,3] and [4]). In the case of Ney, symmetries select three-dimensional objects as privileged, and three-dimensional entities exist derivationally as a partially instantiated by the fundamental high dimensional wave-function, where, again, no mention of fundamental or derivative properties is given or needed. Of course, there is a sense in which this approach is not a thin object-oriented metaphysics grounded on structure in the same way as the primitive ontology approach. In the classical theory and in the primitive ontology approach the continuity is greater: the thin objects are three-dimensional, and the structure is the set of the relations among them given by the wave-function. In the wave-function realist framework instead the situation seems to be as follows. First, the wave-function is what physical objects are made of, and it has no fundamental property other than its amplitude in each point of configuration space: thus, it is itself a thin object. However, this is not the thin object one uses to explain our observations directly. Rather, they are accounted for in terms of the derivative three-dimensional objects given the structure provided by the wave-function. These three-dimensional objects are also thin: they have no fundamental property other than the fact that they are three-dimensional and localized somewhere. The approach is, in a sense, doubly structuralist: the structure of the wave-function allows for the derivative existence of three-dimensional objects, and it also explains their behavior. No such thing happens in the primitive ontology approach, given that the wave-function is not considered material. Finally, however, at least in the case of Ney, symmetries help us select what the objects are, just like in the case of the primitive ontology approach, with the qualification that here these objects are derivative rather than fundamental.

4. Comparison

In the previous section I have presented an alternative view to the traditional metaphysics grounded on properties. I have shown that an approach like mine, in which all fundamental entities are identical and guided by the same law, can be explanatory of different behaviors under the same circumstances without invoking properties but making use of effective laws. In this section I argue that it is also preferable to the traditional view because it is simpler, more explanatory and more compatible with quantum theory, regardless of whether one is a wave-function realist or a proponent of the primitive ontology approach. I discuss some objections in section 4.2, before concluding in section 5.

4.1. *Advantages of my view over the traditional view*

Presumably, one motivation for the traditional view is its simplicity: it is the most straightforward way of accounting for the behavior of the fundamental entities. But is it really so? In this account, most of the explanatory burden is carried by the properties. However, as Esfeld has pointed out [27] and as I briefly mentioned in section 2.1, we are far from having a satisfactory notion of categorical and dispositional property. In contrast, in my framework the entire debate over the nature of properties is completely eliminated: there are no fundamental properties other than spatio-temporal properties.

Moreover, my approach is ontologically more parsimonious than the traditional view. In the standard view, in addition to space-time, we have three categories: ontology, fundamental properties of the ontology, and laws of nature. Here we have only two: ontology, and laws of nature (loosely understood as structure). Since we have fewer categories to account for, Ockham's razor seems to favor the latter approach. Ockham's razor is the principle that, when presented with competing alternatives, one should select the one that makes the fewest assumptions, all other things being equal. It is widely used in theory selection but notoriously difficult to justify (see [62] for some history and some means of evaluating parsimony-based reasoning). Anyway, one could construct an optimistic meta-induction argument to support parsimony: my view postulates only one kind of stuff, and theories like that have been historically more successful than theories that have postulated more than one. Think for instance of ancient astronomy, biology pre-Darwinism, pre-genetic physiology: they respectively postulated the existence of ether to constitute the Heavens in addition to earth, of an Intelligent Designer to create the wonders of nature, and of an *Elan Vital* to account from the difference between living and death things. Now we know that we do not need to postulate these additional kinds in order to satisfactorily account for the evidence. In fact, more empirically adequate theories have each an ontology which is more parsimonious: respectively,

Newton's theory of gravitation, the theory of natural selection, and modern physiology. Therefore, it is likely that the traditional approach, which postulates more categories of entities than mine, is false too.

One could think that, from a Humean perspective with respect to laws, the claim that my approach is more parsimonious is unwarranted, as Humeans conceive of laws not as ontological additions. Even so, however, my view seems to be the natural extension of Humeanism, since it provides a better best system (see also [27,36]). In the context of the Humean account, in which laws are theorems and axioms of the best system, my view provides a better combination of simplicity and strength. In fact, to have laws and effective laws does not change the situation from the traditional picture: theorems are the same and so are the axioms; the only thing that changes is that some parameters are now constants. On the other hand, there is only one kind of matter, rather than many. In the context of a necessitarian, or a primitivist account of laws, the case may be not so compelling. However, one could still use the parsimony argument to argue my view is to be preferred. In any case, two things should be noted: one does not have to take a stand on the nature of laws to hold my view, and in any case laws are mysterious things, and in my approach one is pushing all the mysteries in one place (the laws, intended as structure) instead of two (laws *and* properties).

My view may also be compatible with Humean supervenience as long as the objects have as fundamental properties the only ones that specify their nature and that symmetry constraints allow generating the laws from the objects. Humean supervenience is the doctrine that all the facts about the world supervene on the distribution of fundamental properties of points in three-dimensional space. However, in wave-function realism, the locality condition in the definition of Humean supervenience (namely that there are properties for each point) does not hold. The wave-function is a nonlocal object, in fact by definition it is a function of the particles' configurations: $\psi(x_1, \ldots, x_N)$. Therefore, the particle's motion, through the wave-function, will be influenced by other particles far away. Because of this nonlocality, Teller [65] argued that Humean supervenience is false, and further criticisms followed (see for instance [26,37,46] and [31]). Nonetheless, some proponents of wave-function realism like Loewer [43] and Ney [57] modify the Humean supervenience thesis to be formulated in configuration space, so that the theory is local in that space. This is perfectly compatible with my view: in the case of wave-function realism one is naturally led to see the Humean supervenience thesis as formulated in configuration space, and then one needs to explain how it grounds the fact that for the derivative three-dimensional objects locality fails. Moreover, in the primitive ontology framework, the wave-function is not a physical entity but merely a part of the law. Assuming that enforcing symmetry constraints allows to select one law of nature over the infinite other possibilities, one can argue that the wave-function supervenes on the local matter of facts given by the thin objects (this is similar to what Esfeld [27] calls 'super-Humeanism.' See also [36]). Similar considerations however do not seem to hold also for wave-function realism, given that the wave-function represents what is fundamental, and the thin objects are derivative.

A clarification regarding parsimony: 'less is more' as long as one is able to satisfactorily explain. So, one could complain that even granting that my view is more parsimonious than the alternative it is less explanatory. However, I argue, my view turns out to be more explanatory than the traditional view. In fact, it has less things to account for. One of the main problems of the Standard Model is that it cannot explain why particles have they mass they do. Why does the proton and the electron have masses in the particular relation we observe? My view eliminates the mystery because there are no masses or charges, but merely objects and structure (laws and effective laws). One could complain that the situation has not changed since we have no explanation of why different effective laws exist and why they apply as they do. Nonetheless, I think this is not quite true: we do not know why the laws are what they are anyway, and in the traditional view, we *also* do not know why particles have the properties they do. In my view instead, we do not need to explain things about matter, we 'just' need to explain things about laws. And as I suggested already in section 4.1, laws may be generated by symmetry constraints as well as assuming simplicity as a guide to theory choice.

In addition, this approach also seems to be compatible with the fact that in relativity theory the mass changes with velocity. One could say that the rest mass, the one that does not change with velocity, is the property. Still, she would have to explain what the other portion of the mass is. Instead, if mass is not a property, we have no puzzle whatsoever.

Also, in the framework of classical electrodynamics, my account makes sense of certain asymmetries between particles and fields. The first asymmetry that comes to mind is that particles have properties and fields do not. However, why is it that one can have particles with different masses and charges, but only one field (the electromagnetic field)? In my approach, there is no problem: one particle type, no properties, and no fields. Moreover, particles generate fields, but fields do not generate particles. Then, the fields act on the particles in virtue of the particles' properties, however, particles do not act on fields in virtue of the field's 'properties' (since they haven't got any). In my approach, these puzzles disappear because there are no fields. Somewhat similarly, a common objection to the pilot-wave theory is that the wave-function acts on the particles but the particles do not act on the wave-function. One can respond to this objection easily from the point of view of the primitive ontology approach by pointing out that the wave-function is nomological: it acts on the particles as a law of nature would, and matter do not 'act' on the law. Form the point of view of wave-function realism, instead, there is no straightforward answer, which presumably means that this framework is not a good fit for the pilot-wave theory. Indeed, someone motivated to have the local and separable ontology arguably provided by wave-function realism would not tend to consider this theory as satisfactory.

Even if this consideration would not apply to wave-function realism, one could point put that my view could appeal to the more empiricist minded philosopher or scientist. In fact, to distinguish a proton from an electron we look at its track in the bubble chamber (or in the fancier, more modern version of it), and we see it curve one way rather than another: we measure its trajectory, not its mass or charge. *The only thing we see is stuff that moves*: we see positions that change; we do not see masses, charges, or spin. Thus, from a point of view in which one attributes reality to what is observed, this is the natural ontology, this is what we should take seriously: locations.

Perhaps more strikingly, the fact that in this approach there are no other properties than the spatio-temporal ones make contextuality in quantum mechanics go away. A more detailed discussion is probably needed, however let me note the following. In quantum mechanics 'spin' has usually been considered as a paradigmatic quantum property. However, it turns out that this property is contextual: its value depends on the experiment made to reveal it. That is, the value of the spin of a particle changes depending on whether it is measured together with one property or another. It is as if the value of the length of a square table when measured together with the weight of the table would be different from the value obtained when the length is measured together with the height. This is a serious problem for someone who thinks spin is a property: what kind of fundamental, intrinsic property changes its value like that? This kind of considerations have led many to think of spin as another example of the paradoxical nature of quantum mechanics, and spent oceans of ink to speculate about its possible meanings (see for instance [48]). Instead, if we follow the approach proposed here, there is no mystery left: spin is not a property; spin is part of the law of nature that governs the motion of matter. The fact that its value changes depending on the way in which it is 'measured' (the scary quotes indicate that nothing is truly measured here), is merely a consequence of the fact that experimental apparatuses can affect the system in ways that may destroy a 'property' one is trying to reveal. That is, if we measure a table's length by taking its legs apart and then comparing the size of the flat surface to another of known value, and then we measure its height afterwards, we find that it is very different from what we would have found had we measured it before disassembling the table. So, there is no mystery about contextuality and spin is not a property (see [22] for more on this point). Therefore, as we saw previously, the primitive ontology approach is compatible with the traditional view. However, as we just saw, if one wants to keep this approach she will end up with contextual properties, which are undesirable. What does it even mean that a property is contextual? If it is a property, it is some feature of the object, and not of the context. Therefore, I argue, one should endorse my view, which would make contextuality disappear. In wave-function realism, as we discussed previously, we were already departing from the traditional view given the fictional role of properties in this framework. Now, we are simply pointing out an additional reason to abandon the traditional view: if we keep the traditional framework our 'property talk,' which could so far be motivated for being familiar, simple, and informative, would now become far-fetched, convoluted, and non-explanatory. Because of this, property talk loses its only purpose in the approach, and thus, has no reason to be. So, no matter which your favorite quantum metaphysics is, my approach is more compatible with it than the traditional view.

4.2. *Possible objections to my view*

First, one could complain that this view is unnecessarily radical and revisionary: why someone would want to get rid of properties if they work so well in the standard schema? My reply is that the proposal is far from being unnecessary: properties are notoriously rough nuts to crack and, as already mentioned, there are severe problems in trying to spell out what fundamental properties are. Also, as noted in the previous section, my view has several advantages over the traditional core, which makes it at least worth exploring: it is compatible with Humean supervenience; it extends smoothly to the quantum domain; it is more explanatory and simpler.

Similarly, someone may think that asserting that there are no fields in classical electrodynamics or in general asserting that there are no properties is contrary to our intuitions and our ordinary beliefs. That is, this approach makes many of our beliefs false, and theories like that are not to be favored, all things being equal. However, one could respond by rejecting such a criterion, or maintaining that all things are not equal. After all, many of the modern scientific theories ask us to revise many of our ordinary beliefs, but we do not reject them because of it. In fact, they provide explanatory insight, and they are more empirically adequate than out ordinary beliefs. Our ordinary belief is that matter is continuous, but we are mistaken because atomic theory, which is more explanatory and empirically adequate than our ordinary beliefs, teaches that matter is mostly void. Similarly, our ordinary belief is that fields exist and that properties exist, but we are mistaken because my approach, which is more explanatory and empirically adequate than our ordinary beliefs, teaches that this is not the case.

In addition, one may think that in my view there are infinitely many laws, one law for every massive object, and this makes the claim that my view is more parsimonious mistaken. Nonetheless, this is not accurate: not one law for every massive object, but one law for every entity that in the traditional view we take to be fundamental. In my approach, there is just one kind of material entity, and one kind of law. Laws are implemented differently depending on what they are acting on. If one thinks structurally, the law and the effective laws are the network of relations between the fundamental objects. Macroscopic objects are made of the microscopic entities, and their behavior can be explained and accounted for in terms of them. Once we discuss the situation at the microscopic level, in the standard view we have, say, N fundamental particles identified by their fundamental properties (masses, for instance) and one law of nature; here we have one kind of particles, and N effective laws. Or, better, one kind of particle and a more complex structure.

Another worry could be that it is mysterious how different effective laws act on matter, if there is just one kind. The traditional idea is that the positive charge of the proton makes it go down rather than up in a given magnetic fields. In the view proposed here instead we have just matter. So, how is a material entity 'paired up' with its effective law? To respond, notice that in the traditional view it is a primitive fact that positive charge will result in 'going up' in a given magnetic field. In contrast, here it is a primitive fact that that effective laws act as they do on the various material entities. That is, every view needs to have primitives: in the traditional approach, fundamental properties are primitive; here it is the pairing between objects and their effective laws, or, if one wishes, to the thin objects and the structure. Instead of focusing on laws and effective laws, think about what they truly are: a structural network of relations that determines how matter moves. One may complain that not all primitive postulates are on the same footing and that mine are more radical. If so, we have to agree to disagree on that: the use of properties as in the traditional approach to explain the particle behavior just looks like magic to me. What is mass? Where is it? How does the law act on the particle mass and modifies its behavior? One does not ask these questions in my account but the last one: how does the law 'hook up' with the particle and modify its behavior?

Another related objection may be expressed by this question: what characterizes the difference between empty points and material points? The response of the traditional view, namely that they are massive points, is of course unavailable to me. However, I wonder whether this is even a satisfactory response, given that it is mysterious what masses are. In any case one could reply that, whatever masses are, they are something, while I have nothing. Nonetheless that is not true, as I can indeed claim the same. In fact, I can say that there is a primitive distinction between space points: those that are 'full of matter' and those that are not, and the former change their locations in time according to the laws.

One could think of other more technical problems. For instance, one could think that the relation $E = mc^2$, which establishes that the energy of the particle is associated to its mass, cannot fit in this approach because if mass is not a property, so energy is not a property, and that is wrong. However, there is nothing intrinsically wrong in this, and neither there are tragic consequences if energy turns out not to be a property. We are used to think of mass as a property, but we are mistaken; similarly, we are used to think of energy as a property, and we are mistaken as well.

5. Conclusion

In this paper, I have described a view that provides an alternative account to the traditional way which naturalized metaphysicians explain the behavior of physical objects. While the traditional view is in terms of laws of nature which guide objects in their motion and in which properties such as mass or charge play a fundamental role in the explanatory schema, I have proposed a view in which fundamental objects are without any fundamental property other than the one(s) which allow to determine their nature. While in the traditional view, there are families of particles, divided in virtue of the different properties these particles possess, in my account all matter is identical. The explanation of why two objects behave differently in the same situation is not given in terms of them having different properties. Rather it is provided by the fact that they obey different effective laws, namely different variants of the same fundamental law of nature which encode what we traditionally call properties. So, if in the traditional picture an electron going up in a magnetic field as opposed to a proton going down is explained by invoking that they have opposite charges, in my view there is no such difference between an electron and a proton: they are two particles which are however guided by different effective laws. This view can be dubbed a thin-oriented metaphysics grounded on structure as opposed to the traditional view which could be called an object-oriented metaphysics grounded on properties. In fact, in my approach objects are 'bare' or 'thin,' given that they have no fundamental property, and it is a structuralist view insofar as the effective laws provide structural relations between the fundamental objects. Also, symmetries are important, as they play an important role in selecting the ontology as well as the law of nature. I have argued that my view is more explanatory and simpler than the traditional account, that it is more compatible with quantum theory and Humean supervenience. In particular I have argued that in the quantum domain this is the best approach to metaphysics regardless of one endorsing the primitive ontology framework or wave-function realism.

Acknowledgments

I wish to thank the participants of the following meetings for helpful feedback on previous versions of this paper: *Bridging Metaphysics and Philosophy of Physics*, University of Rochester (2013); *Midwest Annual Workshop in Metaphysics (MAWM)*, University of Nebraska-Lincoln (2016); *Symposium on Radical Ontic Structural Realism*, Eastern APA (2017); *Workshop on Structural Realism and Metaphysics of Science*, Rutgers (2017); *Fundamentality in Physics and Metaphysics Workshop*, University of Geneva (2018).

References

1. Albert, D. Z., in: J. Cushing, A. Fine, and S. Goldstein (eds.) *Bohmian Mechanics and Quantum Theory: An Appraisal* (Kluwer, 1996).
2. Albert, D. Z., *Time and Chance* (Harvard University Press, 2000).
3. Albert, D. Z., in: D. Z. Albert, and A. Ney (eds.) *The Wave-function: Essays in the Metaphysics of Quantum Mechanics* (Oxford University Press, 2013).
4. Albert, D. Z., *After Physics* (Harvard University Press, 2015).

5. Albert, D. Z., and A. Ney, *The Wave-function: Essays in the Metaphysics of Quantum Mechanics* (Oxford University Press, 2013).
6. Allori, V., in: Allori, V., Bassi, A., Dürr, D. and N. Zanghì (eds.) *Do Wave Functions Jump? Perspectives on the Work of GianCarlo Ghirardi* (Springer, 2020).
7. Allori, V., in: J. Saatsi, S. French (eds.) *Scientific Realism and the Quantum* (Oxford University Press, 2020).
8. Allori, V., *International Studies in the Philosophy of Science* 31 (2): 177–188 (2018).
9. Allori, V., *Analytica* 1: 1–19 (2015).
10. Allori, V., in: S. Lebihan (ed.) *Precis de la Philosophie de la Physique* (Vuibert, 2013).
11. Allori, V., in: D. Albert, A. Ney (eds.) *The Wave-function: Essays in the Metaphysics of Quantum Mechanics* (Oxford University Press, 2013).
12. Allori, V., S. Goldstein, R. Tumulka, and N. Zanghì, *The British Journal for the Philosophy of Science* 59 (3): 353–389 (2008).
13. Allori, V., S. Goldstein, R. Tumulka, and N. Zanghì, *The British Journal for the Philosophy of Science* 62 (1): 1–27 (2011).
14. Allori, V., S. Goldstein, R. Tumulka, and N. Zanghì, *The British Journal for the Philosophy of Science* 65 (2): 323–352 (2014).
15. Bacciagaluppi, G., and A. Valentini, *Quantum Theory at the Crossroads* (Cambridge University Press, 2009).
16. Bell, J. S., *Speakable and Unspeakable in Quantum Mechanics* (Cambridge University Press, 1987).
17. Bhogal, H., and Z. Perry, *Nous* 1, 74–94 (2017).
18. Bird, A., *Nature's Metaphysics: Laws and Properties* (Oxford University Press, 2007).
19. Bloch, F., *Physics Today* (1976).
20. Bohm, D., *Physical Review* 85 (2): 166–179; 180–193 (1952).
21. Callender, C., *Synthese* 192, 3153–3177 (2015).
22. Daumner, M., S. Goldstein, D. Dürr, and N. Zanghì, *Erkenntnis* 45: 379–397 (1996).
23. de Broglie, L., in: *Solvay Conference, Electrons et Photons*, Translated in G. Bacciagaluppi and A. Valentini (2009).
24. Dürr, D., S. Goldstein, and N. Zanghì, *Journal of Statistical Physics* 67: 843–907 (1992).
25. Ellis, B., in A. Marmodoro (ed.) *The Metaphysics of Powers* (Routledge, 2010).
26. Esfeld, M. A., *International Studies in the Philosophy of Science* 23: 179–194 (2009).
27. Esfeld, M. A., *The Philosophical Quarterly* 64: 453–470 (2014).
28. Esfeld, M. A., D. Lazarovici, M. Huber, and D. Dürr, *The British Journal for the Philosophy of Science* 65, 773–796 (2014).
29. Everett, H., *Reviews of Modern Physics* 29: 454–462 (1957).
30. Fara, M., *Nous* 39: 43–82 (2005).
31. French, S., *The Structure of the World* (Oxford University Press, 2014).

32. Ghirardi, G. C., R. Grassi, and F. Benatti, *Foundations of Physics* 25 (1): 5–38 (1995).
33. Ghirardi, G. C., A. Rimini and T. Weber, *Physical Review* D: 34: 470 (1986).
34. Goldstein, S., J. Taylor, R. Tumulka, and N. Zanghì, *Journal of Physics A* 38 (7): 1567–1576 (2005).
35. Goldstein, S., and N. Zanghì, in: D. Albert, A. Ney (eds.) *The Wave-function: Essays in the Metaphysics of Quantum Mechanics* (Oxford University Press, 2013).
36. Hall, N., in: B. Loewer, and J. Schaffer (eds.) *A Companion to David Lewis*, John Wiley and Sons (2015).
37. Ladyman, J., and D. Ross, *Everything Must Go: Metaphysics Naturalized* (Oxford University Press, 2007).
38. Lazarovici, D., *European Journal for Philosophy of Science*: 1–26 (2017).
39. Lewis, P. J., *The British Journal for the Philosophy of Science* 55: 713–729 (2004).
40. Lewis, P. J., *Studies in History and Philosophy of Modern Physics* 36: 165–180 (2005).
41. Lewis, P. J., *Philosophy Compass* 1: 224–244 (2006).
42. Lewis, P. J., in: D. Albert and A. Ney (eds.) *The Wave-function: Essays in the Metaphysics of Quantum Mechanics* (Oxford University Press, 2013).
43. Loewer, B., *Philosophical Topics* 24: 101–127 (1996).
44. Malzkorn, W., *The Philosophical Quarterly* 50: 452–469 (2000).
45. Manley, D., and R. Wassermann, *The Philosophical Quarterly* 57: 68–75 (2007).
46. Maudlin, T., *The Metaphysics within Physics* (Oxford University Press, 2007).
47. Mellor, D. H., *Mind* 109: 757–780 (2000).
48. Mermin, D. H., *Review of Modern Physics* 65: 803–815 (1993).
49. Miller, E., *Australasian Journal of Philosophy* 92, 567–583 (2014).
50. Monton, B., in: D. Albert and A. Ney (eds.) *The Wave Function: Essays on the Metaphysics of Quantum Mechanics* (Oxford University Press, 2013).
51. Mumford, S., *Dispositions* (Oxford University Press, 1989).
52. Mundy, B., *The British Journal for the Philosophy of Science* 40 (1): 39–68 (1989).
53. Ney, A., *Nous* 46, 525–560 (2012).
54. Ney, A., in: D. Albert and A. Ney (eds.) *The Wave Function: Essays on the Metaphysics of Quantum Mechanics* (Oxford University Press, 2013).
55. Ney, A., *Synthese* 192 (10): 3105–3124 (2015).
56. Ney, A., *Synthese* 1–23 (2017).
57. Ney, A., *Finding the World in the Wavefunction* (Oxford University Press, forthcoming).
58. North, J., in: D. Albert and A. Ney (eds.) *The Wave Function: Essays on the Metaphysics of Quantum Mechanics* (Oxford University Press, 2013).
59. Norsen, T., *Foundations of Physics* 40 (12): 1858–1884 (2010).
60. Prior, E. C., *Dispositions* (Aberdeen University Press, 1985).

61. Schrödinger, E., *Die Naturwissenschaften* 23: 807–812, 823–828, 844–849 (1935).
62. Sober, E., *Ockham's Razors: A User's Manual* (Cambridge University Press, 2015).
63. Solè, A., *Studies in History and Philosophy of Modern Physics* 44: 365–378 (2015).
64. Suàrez, M., *Synthese* 192 (10): 3203–3228 (2015).
65. Teller, P., *The British Journal for the Philosophy of Science* 37: 71–81 (1986).

ON THE OBJECTIVITY OF QUANTA

OTÁVIO BUENO
Department of Philosophy, University of Miami, Coral Gables, FL 33124, USA
E-mail: otaviobueno@mac.com

Whether quanta exist or not is a complex question that engages with a variety of issues in the realism/anti-realism debate and concerning the interpretation of quantum mechanics, ranging from the kinds of commitments and requirements needed to determine the existence of quantum particles to the sorts of empirical and instrumental control of the relevant phenomena. In this paper, I argue that one need not settle the issue of the existence of quanta to determine their objectivity, that is, to settle whether they are what they are independently of what one takes them to be. The objectivity of quanta is a separate issue from those concerning their existence and metaphysical specification of the kind of thing quanta are. Along the way, I discuss the similarities and differences between mathematical objectivity and the objectivity of quanta, and consider the role that the framework in which quanta are formulated plays in addressing the issue of their objectivity. In the end, the objectivity of quanta and their existence are importantly different. One can hold on to the former while being agnostic about the latter.

Keywords: Objectivity; quanta; quasi-set theory; realism; anti-realism; existence.

1. Introduction

Questions about the existence of quanta cut across different interpretations of quantum mechanics, with different kinds of realism providing considerations for the existence of quanta, primarily based on unprecedented empirical support, and various forms of anti-realism advancing arguments that support the difficulties associated with the commitment to their existence. Several such arguments are mainly concerned with the challenges to specify the nature of quanta and the unclarity of what it takes to settle the issue.

In this paper, I argue that independently of whether quanta exist or not, they are objective: they are what they are independently of what one takes them to be. (This does not mean, of course, that it is determined what they are: this is an ongoing issue.) I examine the similarity between mathematical objectivity and the objectivity of quanta, while highlighting important differences between them. I also appraise the role played by the underlying framework in which quanta are formulated to address the issue of their objectivity. In the end, I argue that the objectivity of quanta does not require the existence of quanta: accepting the former does not require commitment to the latter.

2. Existence, non-existence, representational inequivalence

Questions of existence are notoriously difficult to settle in general, particularly for objects whose nature is contentious. After all, it is unclear how we can determine that something exists without also specifying what it is that exists. But in the case of quanta such specifications are elusive and dependent on particular frameworks.

In QFT (quantum field theory), strictly speaking, quanta do not exist, since it is a theory of fields rather than particles and the latter have no independent standing in it (see [8]). However, one could insist, it is not in the context of a theory but in the outcomes of experiments that empirical evidence for quanta is to be found. After all, it is in experimental results that particle-like behavior is detected, and to make sense of the results seems to require a suitable conceptualization, an interpretation, of quanta. As Robert Wald notes:

> [...] quantum field theory is the quantum theory of a field, not a theory of 'particles'. However, when we consider the manner in which a quantum field interacts with other systems to which it is coupled, an interpretation of states in [Fock space] in terms of 'particles' naturally arises. It is, of course, essential that this be the case if quantum field theory is to describe observed phenomena, since 'particle-like' behavior is commonly observed [13, pp. 46-47]; cited in [7].

Considerations of this kind suggest the need for quanta in the proper understanding of experimental practices in quantum mechanics. Note, however, that experiments alone will not settle the matter of the nature of quanta, since experimental results are compatible with different characterizations of the particles' nature. What quanta ultimately are is not determined by experiments alone. Any such answer will also depend on the interpretation of the experimental results together with particular interpretations of quantum mechanics.

This immediately raises a puzzle. After all, there is no place for particles in relativistic quantum theories. As David Malament [11] initially established, and as Rob Clifton and Hans Halvorson [8] improved further, given a localization system over Minkowski spacetime satisfying localizability, translation covariance, energy bounded below, and microcausality, there is no chance for the detection of a particle in any local region of space. As Malament notes:

> [...] in the attempt to reconcile quantum mechanics with relativity theory [...] one is driven to a field theory; all talk about 'particles' has to be understood, at least in principle, as talk about the properties of, and interactions among, quantized fields [11, p. 1].

As a result, talk of particles within RQFT (relativistic quantum field theory) is simply a *façon de parler* [8]. But if there are no particles, how can one reconcile the theory with the observable features of experiments?

An additional puzzle emerges concerning representational inequivalence. In fact, attempts to specify quanta in QFT lead to inequivalent particle representations: Rindler quanta and Minkowski quanta [7]. Clifton and Halvorson take these inequivalent quanta to be *complementary*. On their view, we should:

> thoroughly abandon the idea that Minkowski and Rindler observers moving through the same field are both trying to detect the presence of particles *simpliciter*. Their motions cause their detectors to couple to *different* incompatible particle observables of the field, making their perspectives on the field necessarily complementary. Furthermore, taking this complementarity seriously means saying that neither the Minkowski nor Rindler perspective yields the uniquely 'correct' story about the particle content of the field, and that both are necessary to provide a complete picture [7, pp. 318-319].

The problem, however, is that it is unclear how the combination of both Minkowski and Rindler quanta can provide a *complete* picture: they offer incompatible accounts of *different* phenomena (in fact, phenomena of different kinds) rather than complementary accounts of the same events. It is the coverage of different kinds of phenomena that complicates the completeness claim. One would need to establish that the two descriptions, taken together, offer an exhaustive account of the particle content of fields. However, it is not clear how that can be achieved, given the different demands on what is required to discern the presence of these objects. To be detected, Rindler quanta demand a detector that is sufficiently accelerated in a Minkowski vacuum, whereas Minkowski quanta typically demand detectors in particle physics experiments performed, to a good extent, in Minkowskian spacetime. This suggests a difference in kind between these quanta, given the nature of the idealization involved in Rindler quanta. As a result, it is unclear how the exhaustiveness condition, required for completeness, can be satisfied.

In fact, Matthias Egg [9] questions whether the Rindler and the Minkowski representations are even on a par. As part of his defense of causal realism, he insists that we have reason to favor the Minkowski representation, which is, on his view, causally warranted, in contrast with the Rindler representation, which, he insists, is not. We lack, according to Egg, a proper theoretical conception of the latter. As he points out:

> Setting aside astroparticle physics, the spacetime regions in which we perform particle physics experiments are, to a very good approximation, Minkowskian, so we have overwhelming causal warrant for the countability of Minkowski quanta. And we do not have the same kind of warrant for Rindler quanta. In fact, it is not even clear how a Rindler detector (i.e., a detector sufficiently accelerated to detect Rindler quanta in a Minkowski vacuum) is to be modelled theoretically [...]. In other words, we do not have a clear notion of what it would mean to actually count Rindler quanta, and this implies [...] that the countability of Rindler quanta is not a *material property*. Consequently, the Rindler quanta hypothesis violates the criterion of material inference and it therefore lacks causal warrant [9, p. 168].

As opposed to Rindler quanta, however, Minkowski quanta are not subject to these concerns on Egg's view:

> [...] the theoretical warrant for Rindler quanta does not threaten the causal warrant for Minkowski quanta, because the two kinds of quanta do not pertain to *different explanations of the same phenomena*, but to explanation of *different phenomena* (measurements of a Minkowski detector vs. measurements of a Rindler detector) [9, p. 168].

Granting that we are indeed dealing with two different phenomena, the issue that has animated the discussion up to this point still remains. One would need to specify what the detection of a Minkowski quanta amounts to and what the nature of these objects ultimately is. However, addressing these issues depends on particular conceptions of the phenomena — particular accounts of the nature of Minkowski quanta — and this, in turn, depends on particular interpretations of the theories under consideration. As a result, it is unclear that the issues can be resolved in a straightforward empirical way, as just a matter of devising suitable detection procedures for these objects. After all, from an ontology point of view, even if something is thereby detected, what exactly has been detected still needs to be specified. What results is a form of ontological indeterminacy.

3. Quanta and quasi-sets: Some troubles

Underlying the disagreement regarding the proper characterization of quanta, and the resulting indeterminacy, there is an additional concern. On some interpretations of quantum mechanics, quanta are objects for which identity is not well defined. After all, the permutation of two indistinguishable quantum particles does not change the state the system is in. In fact, nothing in the quantum-mechanical description of such state distinguishes the particles, let alone individuates them. As a result, in contrast with particles from classical physics, quantum particles seem to lack identity. After all, on the classical account of identity, numerically distinct objects are not the same, but the quantum-mechanical formalism is unable to distinguish quantum particles of the same kind. In classical physics, the permutation of indistinguishable particles would have changed the resulting system's state, given that such particles, even if indistinguishable, are numerically distinct; in fact, they are distinct individuals [10].

To address this concern, and in light of the ontological indeterminacy involved in the characterization of quanta, these objects are better conceptualized in a framework that does not require that all objects exhibit identity conditions. Quasi-set theory provides the most thoroughly developed framework to represent and characterize this phenomenon [10]. On this approach, quanta are thought of as non-individuals, that is, as objects for which identity does not apply. Aggregates of quanta, in turn, can be formulated via quasi-sets, that is, as collections of objects for which is not require that they have well-specified identity conditions. They are only supposed to be indistinguishable from one another.

We saw above some of the difficulties associated with specifying the nature of quanta. Related challenges emerge in the particle interpretation of quantum field theory, and Steven French and Décio Krause note that, when considering the theory:

> [...] even if one were to insist that the 'essential reality' is a set of fields, yet still their 'particle grin' cannot be dismissed. Granted that quanta are not well defined in situations where we have interaction, it is the number of quanta in an 'aggregate' (that is, the cardinality of the appropriate quasi-set) that is typically measured in a scattering experiment, for example [10, p. 383].

Making sense of this 'particle grin' is a crucial component of the interpretation of quanta. It involves determining what such particles are like and finding ways of reconciling their nature with the outcome of experimental results. Without this, it is unclear how to accommodate properly the empirical data. Interestingly, within the quasi-set-theoretic framework, the number of quanta in a given aggregate is characterized in terms of the cardinality of the corresponding quasi-set. In this way, quasi-set-theoretical foundations for quantum mechanics offer a sophisticated framework to interpret the outcome of measurements in scattering experiments. But what can be said about the nature of quanta?

Crucial to this task is the distinction between individuals and non-individuals: the former satisfy both identity and individuation conditions, whereas the latter lack at least one, and possibly both, of them [1]. Identity conditions specify what it takes for an object to be (or to remain) the same and what it takes for it not to. Individuation conditions, in addition, specify the requirements for an object to be individuated, to be uniquely singled out. Separated here are objects (broadly understood), as things that have identity conditions, and individuals (also broadly understood), as things that, in addition, can be individuated, singled out as one. This is a crucial distinction, as French and Krause highlight (using 'entity' where I use 'object'):

> [...] just as the Scholastics separated distinguishability from individuality, we insist on a further conceptual distinction between entities and individuals. Indeed, it is hard to see what sense can be made of the notion of 'quanta' without such a distinction [10, p. 384].

On this view, quanta, lacking both identity and individuation conditions, are non-individuals. They are the kind of thing to which identity does not apply (since they lack well-defined identity conditions) and that cannot be singled out (for they lack individuation conditions too). Despite that, suitably interpretated, quanta are things — understood here as something that does not require identity — that manifest in experiments and, in this respect, as French and Krause note, are objective:

> In our terms, 'entity' is a wider notion than 'individual' and field quanta are precisely examples of non-individual entities which are objective in the sense of being partially manifested in experiments [10, p. 384].

Although I do not follow French and Krause's terminological use of 'entity' to refer to something that may not have identity conditions (as noted, I reserve a neutral term like 'object' for that), the distinction between objects and non-individuals that they have underlined is significant: non-individuals are objects that lack identity and individuation conditions. These are the *typical* non-individuals. *Transitory* non-individuals, in turn, lack individuation conditions but have identity conditions. Given the lack of individuation conditions, it is not possible to single out uniquely a transitory non-individual. It has identity conditions (identity apply to such object), but without individuation conditions, the sameness of such an object cannot be determined across time. Hence, its transitory status. From one moment to the next, it is not determined whether a transitory non-individual remains the same or not, since some other non-individual of the same type would be indistinguishable from it, and it is not possible to single them out. Finally, objects that have individuation conditions, and hence that can be singled out over time, will also have identity conditions: individuation conditions provide (partial) identity conditions for such objects at least by allowing one to distinguish such objects from others. Objects that have both identity and individuation conditions are *individuals*. One can single out each zebra by its unique stripe pattern. This individuation condition will also distinguish one zebra from another (and from other objects), thus offering also (partial) identity conditions for it, which indicate which objects *differ* from it.

Also important is the point that non-individuals are objective since they are "partially manifested in experiments" [10, p. 384]. The partiality emerges from the indeterminacy regarding the nature of such non-individuals, since what is manifested experimentally does not fully specify the nature of quanta. But given what quanta are, it is unclear that there is any other way for them to be manifested but via experiments, in which quanta can appear and be made publicly available (after suitable instrumental manipulation and proper interpretation). Figuring in suitably interpreted mathematical equations is not a form of even *partial* manifestation for quanta, since they do not literally appear in such equations: they may be referred to in them, but this is not a form of appearance, given that quanta themselves are not displayed in such contexts.

Having said that, the concern raised in the previous section still emerges here: even if one shifts the focus from the reality of quanta to their objectivity, as French and Krause [10, p. 383] propose, it is radically underdetermined precisely what one is being objective about. So, from an ontological perspective, the concern ultimately remains. Indeed:

> Questions as to the 'reality' of such quanta may be sidestepped in favour of the claim that they are, at least, 'objective'. But objective what? Substances? Trope bundles? Again each option comes with a price attached but at least the quanta package is less expensive than might be thought [10, p. 383].

Despite the significant benefits of quasi-set theory as a formal framework for the representation of quanta as non-individuals, there is still a price to be paid: the presuppositions made by the quasi-set-theoretic framework itself. It identifies quanta with particular quasi-sets, and thus it turns the latter into abstract entities: after all, neither sets nor quasi-sets are located in spacetime. So, by identifying quanta with certain quasi-sets, we cannot take them to be concrete objects. But quanta need to be concrete if we are to make sense of the role they play in experiments, such as in empirical detections, given that we cannot empirically detect abstract entities.

It may be argued that there is no need to identify quanta with quasi-sets: all that is needed is to use quasi-sets to *represent* quanta without identifying the latter with the former. This is certainly a possibility. The problem with this response, however, is that it fails to address the issue that prompted the discussion in the first place: what exactly *are* quanta? To the extent that the issue is left open by the quasi-set-theoretic approach — quanta are only represented by invoking quasi-sets without, however, settling the issue of what they ultimately are — one still faces the ontological indeterminacy that has guided the discussion so far.

More generally, the trouble is that the conceptualization of quanta is here implemented in terms of a particular mathematical, quasi-set-theoretic framework, but what is represented is meant to be something physical.

What is needed then is a sharp distinction between the mathematical and the physical: the representational and the objectual, as it were. Mathematical, including (quasi-)set-theoretic, frameworks provide representational devices, which can be used to account for objective features of the phenomena, such as what follows (or what does not follow) from empirical bits of information about them. Clearly, it is not up to us (or the relevant researchers) what are the properties that quanta have (or fail to have). These are properties of the quanta themselves (whether we can know them or not).

In mathematics, there is objectivity without presupposing the existence of mathematical objects. Once certain mathematical principles are introduced (such as the axioms for groups, sets, or quasi-sets) and once a logic is specified (whether classical or non-classical), what follows from such principles is not up to us: it is an objective fact about the structures in question. The existence of any of the objects that are quantified over need not be assumed, since one is simply considering the relation between a logic (a particular consequence relation in a given language) and the relevant principles: what *follows* from such principles given the logic.

The objectivity of quanta can also be articulated along these lines. Given a particular theoretical characterization of quanta (and a logic), what follows from such characterization is not up to us, quite independently of the existence of quanta. However, since quanta are supposed to be physical (concrete constituents of the world), an additional requirement is in place as well: they are the kinds of thing to which one should have empirical access to, at least via suitable experimental settings. This additional requirement is not addressed, at least not directly, when quanta are represented within quasi-set theory.

4. An alternative approach

Is there an alternative way of addressing these difficulties? In what follows, I will sketch a proposal that begins to consider them. The account is quite deflationary in that it is not committed to the existence of quanta; in fact, it is agnostic on this issue. Despite that, the proposal preserves quanta's objectivity while providing grounds as to why the specification of their nature, however it is implemented, is not required. The key difference with previous approaches is an argument to the effect that such characterization, if understood as providing the ultimate nature of quanta, is headed to failure. A significant role for understanding is then identified.

4.1. *Ontologically neutral quantification and empirical detection*

An important feature of the proposal advanced here is a particular conception of quantification. Quantifiers need not be taken to be ontologically committing. Instead, ontologically neutral quantification can be adopted ([2] and [3]). Consider, for instance:

(1) There are sets that do not exist (they are too big for that).
(2) If QM is true, there are vectors in Hilbert spaces.

None of these statements require the *existence* of the objects one quantifies over. Rather, quantification over certain objects need not settle the issue as to whether such objects exist or not. To do that, an existence predicate, E, is introduced. Satisfaction conditions for this predicate depend on the particular metaphysical conception that one adopts: ontological independence, detection, observation, spatiotemporal location, verificability, etc. It is unclear, however, that one can characterize necessary conditions for the satisfaction of that predicate without begging the question against some metaphysical views. Nominalists who insist that only concrete objects, which are located in spacetime, exist clearly beg the question against platonists. The same goes for those who deny the existence of numbers since they are ontologically dependent on our linguistic practices and psychological processes [2]; a point that platonists, again, reject.

With the existence predicate E in place, it is straightforward to formulate the two statements above clearly separating quantification from ontological commitment:

(1) $\exists x(Sx \wedge \neg Ex)$, in which '$S$' stands for 'is a set'.
(2) $T(QM) \rightarrow \exists x(Hx \wedge \neg Ex)$, in which '$T$' and '$H$' stand for, respectively, a truth predicate and 'is a vector in a Hilbert Space'.

The objectivity of mathematics is preserved even when ontologically neutral quantifiers are used. Consider, for example, the statement to the effect that *there are infinitely many prime numbers.* This statement follows from Peano Arithmetic: once formulated in this theory, it is not up to us whether it holds or not. As a result, the statement is *objective*, whether numbers exist or not, since it is a matter of what follows or not from the principles of the theory. Moreover, note that the statement does *not* determine the *nature* of prime numbers: whether these objects are abstract entities, Fregean objects, nodes in a structure, mereological atoms, concrete entities, or something else altogether is left entirely open. There is no need to specify the metaphysical nature of numbers in order to understand their properties and their behavior as they are characterized in mathematical practice.

In the case of quanta, a similar objectivity is also found. Relative to each formulation of quanta (whether they are Rindler quanta, Minkowski quanta, or certain quasi-sets), quanta are what they are *independently* of what one takes them to be. But, similarly to what goes on in mathematics, none of these formulations determines the *nature* of quanta: the issue is left entirely open, given that there are different, non-equivalent, interpretations of these objects. Moreover, despite quantification over quanta, one need not be ontologically committed to their existence. Given ontologically neutral quantifiers, the existence of quanta is an *additional* issue that cannot simply be settled from the fact that they are quantified over in our best theories.

There still a difference, however, between objectivity in mathematics and in physics: in the case of mathematical frameworks, there is, of course, no expectation that one should be able to detect empirically the mathematical objects that are posited in the framework. Taken as part of a purely mathematical setting, mathematical objects provide no constraints on the physical world (except perhaps for cardinality considerations). In the case of physics, simply positing certain objects in a framework does *not* secure commitment to the existence of these objects: one needs to secure their empirical *detection*, even if their nature cannot be fully characterized.

This is clearly illustrated by the Dirac equation, which is compatible with three radically different interpretations: one in which the equation's negative energy solutions are simply disregarded (as is commonly done in classical mechanics); one interpretation in which negative energy solutions stand for "holes" in spacetime (but the resulting account, which assigns the same mass to electrons and protons, is empirically inadequate); and, finally, one interpretation in which negative energy solutions stand for a new particle that has the same mass as the electron but the opposite charge, namely, something that, with additional empirical work, can be interpreted as the positron (for details and references to the relevant literature, see [3]). Since the same equation is compatible with empirically distinct interpretations, the equation itself does not specify what is going on in the world. Rather, different interpretations are needed to connect the equation to the empirical data (an important task in theoretical physics). One then needs to supply independent empirical detection of the suitably interpreted physical objects that are posited by the interpretation (a crucial task in experimental physics).

These considerations suggest a significant distinction in scientific practice between theoretical prediction and detection ([2] and [5]). From a suitably interpreted theory, one can make theoretical predictions. For instance, the prediction of the existence of Neptune from Newtonian equations and Bode law provides a clear case: after making the prediction of Neptune's likely orbit, astronomers still needed to detect empirically the planet before its existence could be confirmed. A similar situation emerges in the case of the positron that was just mentioned: it was not enough to offer an empirical prediction of the particle as Dirac did; instrumental access via proper detection still needed to be secured. In other words, theoretical predictions in general are *not sufficient* for the commitment to the existence of the predicted objects, since such objects can be thought of as useful fictions, or artifacts of the theoretical framework, or results from a mistaken interpretation of the formalism, among other possibilities — all of which conflict with the objects' existence. Until the predicted objects are empirically *detected*, their existence is typically left open. After their detection, including the checks that what was detected was indeed what is thought to have been detected (something that not always can be easily settled), the objects have their existence usually resolved within scientific practice.

In the case of quanta, however, the situation involves an *additional constraint*: the specification of what exactly has been detected still needs to be determined. However, as noted above, it is unclear what the nature of these objects ultimately is.

4.2. *Aggregates without sets or quasi-sets*

To address this issue, it is important to provide a formulation of quanta that does not depend on sets, whether classical sets or quasi-sets. After all, classical sets presuppose extensionality: $x = y \leftrightarrow \forall z(z \in x \leftrightarrow z \in y)$. And this requires quanta to be *individuals*, at least in the minimal sense of being entities for which identity is defined.

As we saw, non-classical sets, such as quasi-sets, do not presuppose extensionality. Something weaker, however, is required, namely, weak extensionality, which can be formulated in terms of an indistinguishability relation: "[quasi-]sets which have the be *same* quantity of elements of the be *same* sort are indistinguishable" ([10, p. 290], italics added). (Note the use of be *identity*, in the form of sameness, in this particular formulation of weak extensionality.) However, in this formulation, quanta would end up being *non-individuals*, that is, entities for which identity is not defined. What is needed, then, is a framework to formulate quanta that is entirely independent from classical or non-classical sets, since either settles an aspect of the nature of quanta: if quanta are identified with sets, they become individuals; if they are identified with quasi-sets, they become non-individuals.

As an alternative, aggregates of quanta can be thought of in terms of an *equivalence relation*, which is not understood as being formulated in either set theory or in quasi-set theory. Rather, it is just a relation that is reflexive, symmetric, and transitive. Indistinguishability is one such relation. It groups quanta together as those items that are indistinguishable from one another. Note that since electrons and protons are *distinguishable* from one another (they have different charges), they are not grouped together in an equivalence relation. In this way, distinct *kinds* of equivalence relations are formed: even though *within* a given kind (such as, *being indistinguishable from electrons*), quanta are *not* distinguishable from one another, they be *are* distinguishable across distinct kinds (given that, say, electrons and protons are distinguishable).

If one starts with objects (things for which identity may or may not apply), one can then form equivalence relations among such objects, grouping them together with objects that are indistinguishable from them. To the resulting *kinds* of indistinguishable objects, one can apply unsharp functions, which should not be understood as sets (or quasi-sets), but as particular relations, characterized in terms of their arguments and values (each are aggregates), so that each object in the argument is associated with a single object as its value. The function is unsharp since it does not require the identity of the objects in an aggregate, but only their indistinguishability [4]. By studying the properties of these kinds, within the constraints provided by quantum mechanics, one can focus on the relevant aggregates without assuming that they are either sets or quasi-sets. This is as it should be. After all, we are interested in the *quanta* rather than on their sets or quasi-sets. In this way, one can then consider the *objectivity* of quanta as what they are independently of how they are taken to be.

On this approach, the nature of quanta is ultimately left open: any attempt to settle this issue simply takes one back to particular approaches in ontology, such as whether quanta are individuals or non-individuals, whether they are universals, particulars, or tropes. It is unclear, however, that one could resolve such an issue. These ontological approaches are all empirically equivalent: with suitable interpretative adjustments as needed, they yield precisely the same empirical results one would obtain from quantum mechanics alone. And to make sense of experimental practices with quanta, there is no need to resolve the issue of which of these approaches (if any) offers the best account of the phenomena, given that the practice itself is silent about the nature of the objects it engages with. Metaphysical characterizations of quanta, given their removal from empirical information, provide at best a way of indicating how such objects could be. In this respect, by exploring these characterizations, we gain some understanding of how the world could be if quantum mechanics plus each metaphysical characterization were true [12]. Although we need not, and could not, settle the issue of which of these characterizations is ultimately true, by exploring their implications, we gain valuable modal understanding of the range of what is possible regarding these objects. And this is something worth finding out more about.

5. Conclusion

The question of the existence of quanta is plagued with difficulties regarding their nature, formulation, and interpretation. The issue of the objectivity of quanta is perhaps a bit more tractable, but even there one needs to be careful to avoid some pitfalls that emerge from the formulation of quanta in terms of sets or quasi-sets, given the presuppositions that these representations make regarding them, by taking quanta to be, respectively, individuals or non-individuals. An alternative way forward consists in considering quanta in terms of aggregates, which are understood here as (non-set-theoretic or quasi-set-theoretic) equivalence relations of indistinguishable objects. In this way, at least in principle, quanta's objectivity can be preserved without commitment to their existence or with any particular attempt to determine their nature.

References

1. Arenhart, J., Krause, D., and Bueno, O. (2019). Making Sense of Nonindividuals in Quantum Mechanics. In: O. Lombardi, S. Fortin, C. López, and F. Holik (eds.), *Quantum Worlds: Perspectives on the Ontology of Quantum Mechanics.* Cambridge: Cambridge University Press, 185–204.
2. Azzouni, J. (2004). *Deflating Existential Consequence: A Case for Nominalism.* New York: Oxford University Press.
3. Bueno, O. (2005). Dirac and the Dispensability of Mathematics, *Studies in History and Philosophy of Modern Physics* **36**, 465–490.
4. Bueno, O. (2019). Weyl, Identity, Indiscernibility, Realism. In: A. Cordero (ed.), *Philosophers Look at Quantum Mechanics.* Cham: Springer, 199–214.
5. Chakravartty, A. (2007). *A Metaphysics for Scientific Realism: Knowing the Unobservable.* Cambridge: Cambridge University Press.

6. Clifton, R. (2004). *Quantum Entanglement: Selected Papers.* (Edited by Jeremy Butterfield and Hans Halvorson.) Oxford: Oxford University Press.
7. Clifton, R., and Halvorson, H. (2000). Are Rindler Quanta Real? Inequivalent Particle Concepts in Quantum Field Theory, *Journal of Mathematical Physics* **41**, 1711–1717. (Reprinted in [6]; page references are to the reprinted version.)
8. Clifton, R., and Halvorson, H. (2002). No Place for Particles in Relativistic Quantum Theories, *Philosophy of Science* **69**, 1–28. (Reprinted in [6]; page references are to the reprinted version.)
9. Egg, M. (2014). *Scientific Realism in Particle Physics: A Causal Approach.* Berlin: De Gruyter.
10. French, S., and Krause, D. (2006). *Identity in Physics: A Historical, Philosophical, and Formal Analysis.* Oxford: Oxford University Press.
11. Malament, D. (1996). In Defense of Dogma: Why There Cannot Be a Relativistic Quantum Mechanics of (Localized) Particles. In: R. Clifton (ed.), *Perspectives on Quantum Reality.* Dordrecht: Kluwer, 1–10.
12. van Fraassen, B.C. (1991). *Quantum Mechanics: An Empiricist View.* Oxford: Clarendon Press.
13. Wald, R.M. (1994). *Quantum Field Theory in Curved Spacetime and Black Hole Thermodynamics.* Chicago: University of Chicago Press.

WHY ARE WE OBSESSED WITH "UNDERSTANDING" QUANTUM MECHANICS?

STEPHEN BOUGHN

Departments of Physics and Astronomy, Haverford College,
Haverford, PA, USA
E-mail: sboughn@haverford.edu

For the last hundred years, philosophers and physicists alike have been obsessed with the quest of comprehending the mysteries of quantum mechanics. I'll argue that those who still pursue an understanding of quantum riddles are burdened with a classical view of reality and fail to truly embrace the fundamental quantum aspects of nature.

Keywords: Quantum foundations; the measurement problem.

1. Introduction

Richard Feynman famously declared [1], "I think I can safely say that nobody really understands quantum mechanics."[a] Sean Carroll lamented the persistence of this sentiment in a recent opinion piece ([2] entitled "Even Physicists Don't Understand Quantum Mechanics: Worse, they don't seem to want to understand it."[b] Quantum mechanics is arguably the greatest achievement of modern science and by and large we absolutely understand quantum theory. Rather, the "understanding" to which these two statements evidently refer concerns the ontological status of theoretical constructs.[c] For example, "Do quantum wave functions accurately depict physical reality?" The quantum measurement problem represents a collection of such queries and the conundrums to which they lead. Carroll offers two unsolved quantum riddles: 1) How is it that a quantum object exists as a superposition of different possibilities when we're not looking at it and

[a]The 1964 Messenger Lectures at Cornell University, Lecture 6.
[b]NY Times, Sunday Review, Sept. 7, 2019.
[c]Unlike Carroll, I doubt that Feynman was bothered by this lack of "understanding". See Section 4, Final Remarks.

then snaps into just a single location when we do look at it? and 2) Is the quantum mechanical wave function a complete and comprehensive representation of the world or does it have no direct connection with reality at all? According to Carroll,

> Until physicists definitively answer these questions, they can't really be said to understand quantum mechanics thus Feynman's lament. Which is bad, because quantum mechanics is the most fundamental theory we have, sitting squarely at the center of every serious attempt to formulate deep laws of nature. If nobody understands quantum mechanics, nobody understands the universe.

On the other hand, these questions have been pondered since the dawn of quantum mechanics nearly a century ago and if they haven't been definitively answered by now, perhaps the source of the quandary is not with the answers but with the questions themselves. Inarguably, during that time numerous physicists and even more philosophers endeavored to "understand" quantum mechanics in this way. Some have claimed success; however, I doubt that many would agree. Most physicists are content with foregoing such metaphysical issues, falling back on Bohr's pragmatic Copenhagen interpretation, and then get on with the business of doing physics.

I confess that during my student days, and even thereafter, I was mightily bothered by these quantum mysteries and enjoyed spending time and effort worrying about them. Of course, as Carroll also laments, I avoided letting this avocation interfere with my regular physics research, otherwise, my academic career undoubtedly would have suffered.[d] As I approached the twilight of my career (I'm now retired), I happily resumed my ambition to "understand quantum mechanics" and have ended up writing several papers on the subject.[e] Furthermore, as others before me, I now proudly profess that I finally understand quantum mechanics! Even so, I'm somewhat chagrinned that my understanding is essentially the same as that expressed by Niels Bohr in the 1930s, minus some of Bohr's more philosophical trappings.[f] Some have criticized my epiphany with remarks to the effect that

---

[d] I'm reminded of the adage on the wall of the pool room in my undergraduate dining hall. "To play a good game of billiards is the mark of a gentleman. To play too good a game of billiards is the mark of an ill-spent youth."

[e] I'll spare you a listing and will reference only one, my overall view of the philosophical foundations of physics [3].

[f] Heisenberg once remarked, in all due reverence, that Bohr was "primarily a philosopher, not a physicist" [4].

I am too dismissive of the wonderful mysteries of quantum mechanics by relegating its role to that of an algorithm for making predictions while at the same time too reverent of insights provided by classical mechanics. I've come to believe that, quite to the contrary, those who still pursue an understanding of Carroll's quantum riddles are burdened with a classical view of reality and fail to truly embrace the fundamental quantum aspects of nature. In the remainder of this essay I'll illustrate this point of view with four famous quantum conundrums and leave it to the reader to conjure more examples.

2. The Copenhagen interpretation

I'm sure that most of you are familiar with the "Copenhagen interpretation" of quantum mechanics and so I will not elaborate on its details. However, I need to address some of the many misstatements attributed to it.[g] My understanding of Bohr's interpretation comes primarily from a 1972 paper by Henry P. Stapp[h]. and the reader is referred to that paper as well as another by Don Howard [6] for an account of the mythology of the Copenhagen interpretation. Contrary to the claims of some, Bohr did not assert that quantum mechanics provides a complete description of an individual system but rather that (in Stapp's words) "...quantum theory provides a complete account of atomic phenomena...", that is, "...no theoretical construction can yield experimentally verifiable predictions about atomic phenomena that cannot be extracted from a quantum theoretical description." Neither did Bohr subscribe to the notion of the collapse of the wave function (that came from von Neumann and Dirac) nor did he believe in a privileged role for the observer. Bohr did not maintain that classical mechanics was a necessary component of the foundations of quantum mechanics, on the contrary, he claimed that all experimental apparatus are, in principle, subject to the laws of quantum mechanics. The quantum-classical divide to which he referred was simply the realization that the description of experiments must be in common language that can be comprehended by scientists and technicians alike. A point he made time and time again: it is the quantum of action that necessitates the probabilistic character of quantum mechanics. In fact, the statistical nature of quantum mechanics

[g] "The Copenhagen Interpretation" was not a label approved by Bohr. Heisenberg introduced it in 1955.

[h] In fact this paper [5], which I read as a graduate student, has had a profound effect on my general view of the philosophical foundations of physics [3].

and the resulting conflicts with a classical notion of reality is in large part responsible for the quantum measurement problem and Carroll's quantum riddles. I'll refrain from commenting further on Bohr's interpretation except to the extent necessary in the following discussions of four quantum conundrums.

3. Four quantum conundrums

The following four related questions are associated with the quantum measurement problem and all are of the sort for which Carroll insists we have definitive answers if we are ever "to formulate deep laws of nature". The first is the famous Einstein-Podolsky-Rosen "paradox", which many consider resolved; however, it presents a clear-cut example of how it is that the burden of a classical view of reality can seemingly confound our understanding of quantum mechanics.

3.1. *The Einstein-Podolsky-Rosen paradox*

The 1935 paper [7] by Einstein, Podolsky, and Rosen (EPR) expressed those authors dissatisfaction with the then current formulation of quantum mechanics. The conclusion was neither that quantum mechanics yields incorrect predictions nor that quantum mechanics is nonlocal, i.e., violates special relativity. Rather it was that quantum mechanics "...does not provide a complete description of the physical reality." The EPR manuscript was written in Einstein's absence by Podolsky and Einstein was not happy with it. He chose to couch his objection on the apparent action at a distance implied by quantum mechanics as will be discussed in the next section. Nevertheless, Einstein did agree with the conclusion that quantum mechanics does not provide a complete description of a single system.[i] The argument in EPR was a rather byzantine analysis involving a definition of elements of reality; however, in the end the conclusion was that both the position and the momentum of a particle were elements of physics reality. The hidden assumption running through the paper is that a single quantum system, in this case a single particle, has a well-defined location and momentum at each instant of time whether or not these quantities are observed. This is the essence of a classical world view. Without that assumption, the argument collapses.

---

[i]I agree that quantum mechanics does not provide a complete description of the physical world but for an entirely different reason. (See Section 3.4 below.)

A year later, Einstein [8] offered a resolution to the EPR paradox, the ensemble interpretation of quantum mechanics. As Einstein described it:

> The Ψ function does not in any way describe a condition which could be that of a single system; it relates rather to many systems, to 'an ensemble of systems' in the sense of statistical mechanics... Such an interpretation eliminates also the [EPR] paradox recently demonstrated by myself and two collaborators...

The ensemble interpretation, while minimalist, is considered by some to be the most reasonable interpretation of quantum mechanics (e.g., see [9]. While Einstein admitted that "such an interpretation eliminates also the [EPR] paradox", he added "To believe this is logically possible without contradiction; but, it is so very contrary to my scientific instinct that I cannot forego the search for a more complete conception". In other words, he would still strongly suspect that quantum mechanics is incomplete. The ensemble interpretation seems to be immune to the EPR conclusion that quantum mechanics cannot provide a complete description of a single system by requiring that quantum mechanics only apply to ensembles of systems. However, the specter of classical reality is still present for what does the ensemble represent if not a collection of single particles each with a definite position and momentum. Ballentine's presentation [9] of the ensemble interpretation reinforces this notion. If so, it's hard to imagine how Einstein or Ballentine could accept quantum mechanics as a complete description of reality and thereby resolve the EPR paradox.[j]

3.2. *Bell's theorem, entanglement, and quantum non-locality*

Many credit the EPR paper with bringing the notion of entangled quantum states and action at a distance to the attention of the physics community; however, the major figures in physics at the time were already arguing about entanglement and Einstein had been bothered by action at a distance since the 1927 Solvay conference [6]. As I mentioned above, Einstein was not happy with the EPR paper. In a letter to Schrödinger written a month

[j]As I pointed out in the Introduction, Bohr did not assert that quantum mechanics provides a complete description of an individual system; however, why couldn't one apply a statistical interpretation to a single system. For example, I would be happy to place a bet on the outcome of a quantum transition in a single system if the odds were heavily in my favor (using a Bayesian notion of probability).

after the appearance of EPR he wrote [6], "...the little essay...has not come out as well as I really wanted; on the contrary, the main point was, so to speak, buried by the erudition." In that same letter Einstein chose to base his argument for incompleteness on what he termed the "separation principle" according to which the real state of affairs in one part of space cannot be affected instantaneously or superluminally by events in a distant part of space. Suppose AB is the joint state of two systems, A and B, that interact and subsequently move away from each other to different locations. (Schrödinger [10] coined the term entangled to describe such a joint state.) In his letter to Schrödinger, Einstein explained [6]

> After the collision, the real state of (AB) consists precisely of the real state A and the real state of B, which two states have nothing to do with one another. The real state of B thus cannot depend upon the kind of measurement I carry out on A [separation principle]. But then for the same state of B there are two (in general arbitrarily many) equally justified [wave functions] Ψ_B, which contradicts the hypothesis of a one-to-one or complete description of the real states.

His conclusion was again that quantum mechanics cannot provide a complete description of reality. In this argument Einstein moves from the classical notion that the position and momentum of a particle are elements of physics reality to the presumption that the quantum wave function of a particle must be unique if it is to provide a complete description of the *real physical state* of the particle. Quantum mechanics does not satisfy this requirement thus it cannot provide a complete description of the system. This also involves a classical view of reality; it simply replaces the words "the position and momentum of a particle are elements of physical reality" with "quantum wave functions are elements of physical reality". Quantum mechanics doesn't allow this characterization.

Einstein's argument invokes action at a distance, "the real state of B thus cannot depend upon the kind of measurement I carry out on A", which had bothered him since 1927. This brings us to John Bell's 1964 paper [11], "On the Einstein Podolsky Rosen Paradox". Bell was interested in "hidden variable" theories, classical theories with unknown parameters that could faithfully reproduce the predictions of quantum mechanics. His conclusion was that the only way this could be accomplished was with superluminal interactions, the "spooky action at a distance" that vexed Einstein. Bell chose a spin singlet system as an example of an entangled state. In his notation, let $A(\hat{a}, \lambda)$ and $B(\hat{b}, \lambda)$ represent the results of the spin measurements of particles 1 and 2 in the $\hat{a}$ and $\hat{b}$ directions respectively. Bell's condition of locality is that the correlation of any two measurements, $P(\hat{a}, \hat{b})$ be given by $P(\hat{a}, \hat{b}) = \int d\lambda \rho(\lambda) A(\hat{a}, \lambda) B(\hat{b}, \lambda)$ where λ represents the hidden variables with statistical distribution $\rho(\lambda)$. A moment's thought should convince one that the product of measurement outcomes in this expression precludes the types of correlations (quantum interference) that quantum mechanics demands and so it's not surprising that the subsequent analysis of a sequence of such measurements is in conflict with the predictions of standard quantum mechanics. The expression for $P(\hat{a}, \hat{b})$ is clearly motivated by a classical view of reality, but to be fair the purpose of Bell's paper was to investigate the possibility of a classical, hidden variable theory. His conclusion was that any such theory could not be Lorentz invariant, i.e., it must be non-local. However, in a subsequent paper [12], he formalized this "notion of local causality" and, from the violation of a set of derived inequalities, concluded that not only is quantum mechanics nonlocal[k] but also there exits a "gross nonlocality of nature". Again, here is a quantum riddle that rests wholly on the persistence of a classical view of reality.

3.3. *Wave function collapse and the paradox of Schrödinger's cat*

Recall Carroll's first quantum riddle: *How is it that a quantum object exists as a superposition of different possibilities when we're not looking at it and then snaps into just a single location when we do look at it?* This snapping into a single location is an example of what's referred to as collapse of the wave function and Carroll wants to know why this happens. There have been attempts to modify quantum mechanics so that such a collapse is a theoretical prediction; however, there are no experimental observations that suggest such a modification is necessary. Standard quantum mechanics is self-consistent and its predictions are in agreement with all current observations. Then why even ask the question? As far as I can tell, the only reason is a desire to attach a classical reality to the quantum wave function as did Einstein in his argument for the incompleteness of quantum mechanics. If one forgoes this notion, there is absolutely no reason to pursue the matter further.

A related issue is the paradox of Schrödinger's cat [10]. In Schrödinger's

[k] I've always found it curious that the fact that standard quantum mechanics forbids superluminal signaling hasn't disabused those who assert quantum nonlocality of their claims.

1935 gedanken experiment, a cat is penned up in a chamber with a radioactive substance that is monitored by a Geiger counter. The half-life of the substance is such that after one hour there is a 50% chance that a single radioactive atom will have decayed in which case the counter discharges. If so, then a relay releases a hammer that shatters a flask of hydrocyanic acid and the cat dies. If the counter does not discharge, the flask is not broken and the cat lives. After one hour, the wave function of the entire system expresses this situation by having equal parts of an alive cat/undecayed atom and a dead cat/decayed atom and thus the “indeterminacy originally restricted to the atomic domain becomes transformed into macroscopic indeterminacy” [10]. When the box is opened and the cat plus atom system observed, the wave function collapses into one of these two states. This is absurd. Surely, the cat is either alive or dead before the box is opened even if we don’t know which is the case. A way out of this conundrum is to claim that the wave function collapses as soon as the (macroscopic) Geiger counter clicks. So perhaps the collapse mechanism does make sense after all. On the other hand, the predictions of standard quantum mechanics are completely consistent with what we observe and until we have experimental evidence to the contrary, why purse the issue further. The only reason I can think of is to satisfy our notion of classical reality.

3.4. *The quantum-classical divide*

As a final example, let’s consider the question of what constitutes an observation that supposedly causes the superposition of positions of a particle to “suddenly snap into just a single location.” As Carroll puts it, “Why are observations special? What counts as an ‘observation’, anyway?” According to the Copenhagen interpretation, or so the standard argument goes, the act of measurement or observation must be described in terms of classical physics. Because classical physics is not (generally) couched in probabilistic terms, the classical measurement of a quantum system yields the specific result required by the probabilistic interpretation of the wave function. While possibly resolving the measurement dilemma, this explanation immediately raises two related questions: 1) When and where does the classical measurement occur, i.e., where is the quantum-classical divide? and 2) How does one describe the physical interaction across this divide? In posing these two questions, the specter of the notion of classical reality has again made its unwelcome appearance. The conviction that every interaction involves some physical process that is located at a specific time and place unquestionably arises from a classical view of reality.

A related conundrum is why is it necessary to revert to classical physics in order to understand quantum mechanics. Why didn't Bohr seem to worry about the unseemly merger of classical and quantum formalisms? I suspect the answer is because he didn't actually consider that any such merger is required. While Bohr endeavored to be extremely careful in expressing his ideas, his prose is often obscure. However, consider his following brief description of a measurement [13]:

> The decisive point is to recognize that the description of the experimental arrangement and the recordings of observations must be given in plain language, suitably refined by the usual terminology. This is a simple logical demand, since by the word 'experiment' we can only mean a procedure regarding which we are able to communicate to others what we have done and what we have learnt.

Nowhere in this description does he refer to classical physics. To be sure, Bohr often used the term "classical" in describing measurements but as Camilleri and Schlosshauer [14] point out, "Bohr's doctrine of classical concepts is not primarily an interpretation of quantum mechanics...but rather is an attempt by Bohr to elaborate an epistemology of experiment." That the prescription for how to perform an experiment must be subsumed within the theoretical formalism of quantum mechanics follows only by analogy with classical physics. Again, it seems that a classical worldview lurks behind this quantum conundrum.[l] While I just characterized the requirement that experimental prescriptions be included in the theoretical formalism as emerging from a classical worldview; in fact, classical physics already fails to satisfy this criterion. I've argued elsewhere [3] that measurement prescriptions also lie wholly outside the formalism of classical mechanics.

It is because the descriptions of measurements are wholly removed from the formalism of quantum mechanics that I agree with EPR's conclusion that quantum mechanics "does not provide a complete description of the physical reality."[m] Dyson [16] made the incompleteness more explicit through four sensible gedanken experiments couched in the ordinary

[l]Howard [15] has endeavored, successfully I believe, to formalize Bohr's notion of "classical" descriptions of measuring instruments in terms of density matrices and thereby insures consistency with the formalism of quantum mechanics. However, from my pragmatic, experimentalist's point of view, I'm satisfied with Bohr's epistemology of experiment [3].

[m]On the other hand, I freely admit that I'm not entirely sure what it means for any theory to provide a complete description of physical reality. (See [3].)

language of measurements and experimental physics that defy quantum mechanical explanation. Quantum mechanics quite simply cannot be applied to all conceivable situations. For Dyson, the dividing line between classical and quantum physics is the same as the dividing line between the past and the future. The wave function constitutes a statistical prediction of future events. After the event occurs, the wave function doesn't collapse rather it becomes irrelevant. The results of past observations, facts, are the domain of classical physics; the probabilities of future events, wave functions, are the domain of quantum physics. Ergo, quantum mechanics does not provide a complete description of nature.

4. Final remarks

I began my essay, as did Carroll, with Feynman's declaration that "nobody really understands quantum mechanics". Carroll refers to it as "Feynman's lament"; however, I seriously doubt that Feynman was bothered by Carroll's quantum riddles. Rather, I suspect that Feynman was simply referring to the wonderful mysteries of quantum mechanics and, perhaps, admonishing those who enter the house of quantum mechanics to leave their notions of classical reality at the door. In fact, immediately following his Messenger Lecture "lament", Feynman [1] councils us

> So do not take the lecture too seriously, thinking that you really have to understand in terms of some model what I am going to describe, but just relax and enjoy it. I am going to tell you what nature behaves like. If you will simply admit that maybe she does behave like this, you will find her a delightful, entrancing thing.

The seemingly intractable riddles of quantum mechanics weren't the first physicists have encountered. The notion of luminiferous aether confounded physicists beginning in the 18th century. In this case, the "classical" view of reality was that all waves, including light, require physical media in which wave disturbances propagate. After Maxwell, our notion of reality expanded to include dynamic electromagnetic fields and the aether medium was discarded. Another example is provided by the nature of time in the context of the special theory of relativity. In this case, the classical notion of absolute time leads to such unanswerable questions as whether or not two events occur simultaneously. Ultimately, we had to abandon the notion of absolute time and accept a new reality as prescribed by relativity theory. This latter conundrum was resolved much more timely, within a few years of Einstein's seminal 1905 paper. The point is that new phenomena often require new points of view that end up changing our notions of reality. Embracing these changes leads to a better understanding of the natural world. So my advice to Carroll is, embrace the wonderful quantum world and the quantum reality it implies. Do not try to shoehorn the quantum world into a classical reality.

Finally, I absolutely support the pursuit of deep understanding of our physical theories, a tried and true endeavor that has been crucial for our quest of understanding the physical world around us. The appearance of new phenomena, the unification of theories, and even the pursuit of elegance and simplicity are all legitimate reasons for such an avocation and the successes have been spectacular: spin, superconductivity, neutrinos, anti-matter, black holes, gravitational waves, and many, many more. The purpose of my essay is simply to point out that attempts to understand quantum mechanics by imposing a classical reality on what are essentially quantum phenomena inevitably leads one down a rabbit hole. Why not "relax and enjoy it"?

Acknowledgements

As always I thank my two muses, Freeman Dyson and Marcel Reginatto, for encouraging my pursuit of understanding the foundations of quantum mechanics.

References

1. R. Feynman, *The Character of Physical Law*, The 1964 Messenger Lectures at Cornell University (The Modern Library, New York 1994).
2. S. Carroll, "Even Physicists Don't Understand Quantum Mechanics: Worse, they don't seem to want to understand it.", *New York Times, Sunday Review*, Sept. 7, 2019.
3. S. Boughn, "On the Philosopical Foundations of Physics: An Experimentalist's Perspective", arXiv:1905.07359 [physics.hist-ph; quant-ph] 2019.
4. A. Pais, *Niels Bohr's Times* (Clarendon Press, Oxford 1991).
5. H. Stapp, "The Copenhagen interpretation", *American Journal of Physics* **40**, 1098–1116, 1972.
6. D. Howard, "Revisiting the Einstein-Bohr dialogue", *Iyyun: The Jerusalem Philosophical Quarterly* **56**, 57–90, 2007.
7. A. Einstein, B. Podolsky, N.Rosen, "Can quantum-mechanical description of reality be considered complete?", *Physical Review* **47**, 777–780, 1935.
8. A. Einstein, "Physics and Reality", *Journal of the Franklin Institute* **221**, 349–382, 1936.
9. L. Ballentine, "The Statistical Interpretation of Quantum Mechanics", *Reviews of Modern Physics* **42**, 358–381, 1970.

10. E. Schrödinger, "Die gegenwärtige Situation in der Quantenmechanik", *Naturwissenschaften* **23**, 807–812, 1935. [English translation in: J. Trimmer, *Proceedings of the American Philosophical Society* **124**, 323–338, 1980.]
11. J. Bell, "On the Einstein Podolsky Rosen Paradox", *Physics* **1**, 195–200, 1964.
12. J. Bell "The theory of local beables", presented at the sixth GIFT seminar, appears in *Speakable and Unspeakable in Quantum Mechanics* (Cambridge Univ. Press, Cambridge 2004).
13. N. Bohr, *Essays 1958/1962 on Atomic Physics and Human Knowledge* (Ox Bow Press, Woodbridge, CT 1963).
14. K. Camilleri, M. Schlosshauer, "Niels Bohr as philosopher of experiment: Does decoherence theory challenge Bohr's doctrine of classical concepts?", *Studies in History and Philosophy of Modern Physics* **49**, 73–83, 2015.
15. D. Howard, "What Makes a Classical Concept Classical? Toward a Reconstruction of Niels Bohr's Philosophy of Physics", in *Niels Bohr and Contemporary Philosophy*, eds. J. Faye and H. Forse (Kluwer Acad. Pub. 1994).
16. F. Dyson, "Thought experiments—In Honor of John Wheeler", Contribution to Wheeler Symposium, Princeton, 2002, reprinted in *Bird and Frogs* by F. Dyson, pp. 304–321 (World Scientific, Singapore 2015).

WHEN BERTLMANN WEARS NO SOCKS. COMMON CAUSES INDUCED BY MEASUREMENTS AS AN EXPLANATION FOR QUANTUM CORRELATIONS

DIEDERIK AERTS

Center Leo Apostel for Interdisciplinary Studies and Department of Mathematics, Vrije Universiteit Brussel, 1050 Brussels, Belgium
E-mail: diraerts@vub.ac.be

MASSIMILIANO SASSOLI DE BIANCHI

Center Leo Apostel for Interdisciplinary Studies, Vrije Universiteit Brussel, 1050 Brussels, Belgium and Laboratorio di Autoricerca di Base, 6917 Barbengo, Switzerland
E-mail: msassoli@vub.ac.be

It is well known that correlations produced by common causes in the past cannot violate Bell's inequalities. This was emphasized by Bell in his celebrated example of Bertlmann's socks. However, if common causes are induced by the very measurement process i.e., actualized at each run of a joint measurement, in a way that depends on the type of joint measurement that is being executed (hence, the common causes are contextually actualized), the resulting correlations are able to violate Bell's inequalities, thus providing a simple and general explanation for the origin of quantum correlations. We illustrate this mechanism by revisiting Bertlmann's socks example. In doing so, we also emphasize that Bell's inequalities, in their essence, are about demarcating 'non-induced by measurements' (non-contextual) from 'induced by measurements' (contextual) common causes, where the latter would operate at a non-spatial level of our physical reality, when the jointly measured entangled entities are microscopic in nature.

Keywords: Bell's inequalities; entanglement; no-signaling; contextuality; common causes; extended Bloch representation; hidden-measurement interactions.

1. Introduction

In a groundbreaking paper, inspired by the analysis of the EPR paradox situation in quantum theory [1], John Bell derived in 1964 mathematical inequalities that are able to test our common sense conception of physical reality, when we are confronted with the phenomenon of quantum entanglement [2]. For quite some time, his results remained largely ignored by the physics community, but things changed when in the Seventies of last century Alain Aspect proposed a specific experimental scheme using entangled photons and polarizers having their orientations randomly changing in time, credibly testing the violation predicted by quantum mechanics [3], and that in the Eighties he performed experiments (with the collaboration of Philippe Grangier, Gérard Roger and Jean Dalibard) [4–7] showing for the first time, in a convincing way, the actual violation of Bell's inequalities, later confirmed by numerous experimental groups in the decades to come (see also [8] for a detailed description of these historical experiments and [9] for a recent test).

Also in the Eighties, in an article entitled "Bertlmann's socks and the nature of reality," Bell famously mentioned the habit of his colleague Reinhold Bertlmann of always wearing socks of different colors, to point out the difference between quantum and non-quantum correlations, emphasizing that the color-correlations one can observe when meeting Dr. Bertlmann have nothing to do with those produced by quantum entanglement [10].[a] In this article, we consider a variation of Bell's story about Bertlmann's socks, to explain that the fundamental demarcation his inequalities provide is between common causes that are induced by the measurement processes (and because of that they are contextual) and common causes that are not induced by the measurement processes (and because of that they are not contextual).

Our fictional Dr. Bertlmann (whose habits are slightly different from those of the real Dr. Bertlmann, as described by Bell) is known to always wear socks of different colors. If one of his two socks is pink, with certainty the other will be observed to be non-pink, hence, a (anti)correlation will be invariably present between the colors of his two socks. The Dr. Bertlmann of our story also loves pink very much, and always makes sure to start by wearing a pink sock, on the foot he initially pays attention to, then immediately after he wears a non-pink sock, on the other foot. He always takes care to also have a handkerchief in each one of the two front pockets of his trousers, but in this case he likes to always have them of the same color,

[a]The episode was reported by Bertlmann as follows [11]: "One day I was sitting in our computer room with my computer cards, when my colleague Gerhard Ecker rushed in, waving a preprint in his hands. He shouted, "Reinhold, look, now you're famous!" I could hardly believe my eyes as I read and reread the title of a paper by John, "Bertlmann's socks and the nature of reality." I was totally stunned. As I read the first page, my heart stood still."

which fifty percent of the times will be pink and the other fifty percent of the times non-pink.

From time to time, Bertlmann meets his two friends Alice and Bob, who always go around together. When this happens, Alice and Bob are very interested in knowing about Bertlmann's socks and handkerchiefs. More precisely, when Alice meets Bertlmann, depending on her mood, she decides to perform one of the following two measurements, relative to Bertlmann's left side of his body. Measurement A is about the pinkness property of Bertlmann's left handkerchief. To perform it, she simply asks Bertlmann to tell her the color of his left handkerchief. If the answer is "pink," the observation is considered to be successful and the outcome is denoted A_1. Otherwise, the outcome is denoted A_2. The second measurement that Alice can decide to perform, denoted A', is about the color-correlation between Bertlmann's left handkerchief and left sock. To perform it, she asks Bertlmann to tell her if the handkerchief in his left pocket and his left sock are both pink or both non-pink. If his response is affirmative, the observation is considered to be successful and the outcome is denoted A'_1. Otherwise, the outcome is denoted A'_2. Bob operates exactly in the same way as Alice, but with respect to Bertlmann's right side of his body. His two measurements are denoted B and B', with outcomes B_1 and B_2, and B'_1 and B'_2, respectively.

So, when Bertlmann meets Alice and Bob, they will jointly ask him the questions relative to the measurements they decide to perform, according to their mood of the moment. This defines four possible joint measurements, which we will denote AB, AB', $A'B$ and $A'B'$, where AB is the joint measurement where Alice performs A and Bob performs B, AB' is the joint measurement where Alice performs A and Bob performs B', and so on. Assuming for simplicity that Bertlmann has the same propensity to put the pink sock on the left foot or on the right one, it is easy to convince oneself that we have the probabilities:

$$\begin{aligned}
&\mathcal{P}(A_1,B_1)=\frac{1}{2}, \quad \mathcal{P}(A_1,B_2)=0, \quad \mathcal{P}(A_2,B_1)=0, \quad \mathcal{P}(A_2,B_2)=\frac{1}{2}\\
&\mathcal{P}(A_1,B'_1)=\frac{1}{4}, \quad \mathcal{P}(A_1,B'_2)=\frac{1}{4}, \quad \mathcal{P}(A_2,B'_1)=\frac{1}{4}, \quad \mathcal{P}(A_2,B'_2)=\frac{1}{4}\\
&\mathcal{P}(A'_1,B_1)=\frac{1}{4}, \quad \mathcal{P}(A'_1,B_2)=\frac{1}{4}, \quad \mathcal{P}(A'_2,B_1)=\frac{1}{4}, \quad \mathcal{P}(A'_2,B_2)=\frac{1}{4}\\
&\mathcal{P}(A'_1,B'_1)=0, \quad \mathcal{P}(A'_1,B'_2)=\frac{1}{2}, \quad \mathcal{P}(A'_2,B'_1)=\frac{1}{2}, \quad \mathcal{P}(A'_2,B'_2)=0
\end{aligned} \tag{1}$$

which we can use to calculate the expectation values:

$$\begin{aligned}
E(A,B) &= \mathcal{P}(A_1,B_1)+\mathcal{P}(A_2,B_2)-\mathcal{P}(A_1,B_2)-\mathcal{P}(A_2,B_1)=1\\
E(A,B') &= \mathcal{P}(A_1,B_1')+\mathcal{P}(A_2,B_2')-\mathcal{P}(A_1,B_2')-\mathcal{P}(A_2,B_1')=0\\
E(A',B) &= \mathcal{P}(A_1',B_1)+\mathcal{P}(A_2',B_2)-\mathcal{P}(A_1',B_2)-\mathcal{P}(A_2',B_1)=0\\
E(A',B') &= \mathcal{P}(A_1',B_1')+\mathcal{P}(A_2',B_2')-\mathcal{P}(A_1',B_2')-\mathcal{P}(A_2',B_1')=-1
\end{aligned} \tag{2}$$

Clearly, the Bell-CHSH inequality [12], $|\text{CHSH}| \leq 2$, with CHSH given by the combination $E(A,B)+E(A,B')+E(A',B)-E(A',B')$, or by any other combination obtained by interchanging the roles of A and A' and/or those of B and B', is not violated by the above averages. This because, even though correlations exist between Alice's and Bob's outcomes, they are all the consequence of common causes in the past, which were actualized before Alice and Bob decided to select and perform their measurements. We shall call these past common causes 'non-induced by measurements common causes', because they are not actualized during and by the experimental contexts created by Alice and Bob. Note that a straightforward consequence of a common cause not being induced by the measurement process is that Alice's outcome probabilities cannot depend on which measurement Bob decides to jointly perform with her, and vice versa, which means that the so-called no-signaling conditions (also called marginal laws) are automatically obeyed.

2. When Bertlmann wears no socks

What if the presence of the common causes, responsible for the observed correlations, would be instead induced by the measurements set-ups themselves? To explore this possibility, we now consider a different situation.[b] We assume that during the summer period, because of the heat, our fictional Dr. Bertlmannn decides in the morning not to wear any socks. However, not being sure if he will need them later on during the day, he always keeps a pair of them in his briefcase (a pink one and a non-pink one). On the other hand, he continues to put handkerchiefs of same color in the two pockets of his trousers, as usual.

This time, when he meets his friends, and Alice asks him to tell if his left handkerchief and his left sock are color-correlated (assuming Alice is performing A' and Bob is performing B), since he wears no socks, but is

[b]Our example is inspired by the macroscopic non-local box presented in [13].

eager not to disappoint his friend and respond to her question, he quickly turns around and puts a sock on his left foot, following his habit of always wearing a pink sock on the foot he initially pays attention to, which in the present situation is the left one, because of Alice's question. Following this action, he can respond to Alice's interrogation, and of course when he puts the pink sock on the left foot, he will also put the remaining non-pink sock on the right foot. The situation where Alice performs A and Bob performs B' is similar, with this time the pink sock ending up on the right foot. On the other hand, when Alice and Bob jointly perform measurements A' and B', Bertlmann has to break the symmetry of the situation by deciding himself on which foot to pay attention first, which will be the one receiving the pink sock.

Assuming as before that Bertlmann has the same propensity to put the pink sock on the left foot or on the right foot, it is easy to convince oneself that we now have the probabilities:

$$\begin{aligned}
&\mathcal{P}(A_1,B_1)=\frac{1}{2}, \quad \mathcal{P}(A_1,B_2)=0, \quad \mathcal{P}(A_2,B_1)=0, \quad \mathcal{P}(A_2,B_2)=\frac{1}{2}\\
&\mathcal{P}(A_1,B_1')=\frac{1}{2}, \quad \mathcal{P}(A_1,B_2')=0, \quad \mathcal{P}(A_2,B_1')=0, \quad \mathcal{P}(A_2,B_2')=\frac{1}{2}\\
&\mathcal{P}(A_1',B_1)=\frac{1}{2}, \quad \mathcal{P}(A_1',B_2)=0, \quad \mathcal{P}(A_2',B_1)=0, \quad \mathcal{P}(A_2',B_2)=\frac{1}{2}\\
&\mathcal{P}(A_1',B_1')=0, \quad \mathcal{P}(A_1',B_2')=\frac{1}{2}, \quad \mathcal{P}(A_2',B_1')=\frac{1}{2}, \quad \mathcal{P}(A_2',B_2')=0
\end{aligned} \tag{3}$$

Note that (3) differs from (1) only for what concerns the probabilities relative to the situations where Alice and Bob perform different typologies of measurements. To understand why, consider that there are only two possibilities prior to the execution of the joint measurements: Bertlmann has pink handkerchiefs in his two pockets, or non-pink ones. Since he wears no socks, in the joint measurements AB' the pink sock will go to his right foot and the non-pink one to his left foot. Hence, we have two possible outcomes: one such that the handkerchiefs are pink and the right sock is also pink, which corresponds to outcome (A_1, B_1'), the other such that the handkerchiefs are non-pink and the right sock is pink, i.e., not of the same color of the handkerchiefs, which corresponds to outcome (A_2, B_2'). Hence, $\mathcal{P}(A_1, B_1') = \frac{1}{2}$ and $\mathcal{P}(A_2, B_2') = \frac{1}{2}$, and we can reason in the same way for the other probabilities in (3).

The expectation values are now:

$$\begin{aligned}
E(A,B) &= \mathcal{P}(A_1,B_1)+\mathcal{P}(A_2,B_2)-\mathcal{P}(A_1,B_2)-\mathcal{P}(A_2,B_1)=1\\
E(A,B') &= \mathcal{P}(A_1,B'_1)+\mathcal{P}(A_2,B'_2)-\mathcal{P}(A_1,B'_2)-\mathcal{P}(A_2,B'_1)=1\\
E(A',B) &= \mathcal{P}(A'_1,B_1)+\mathcal{P}(A'_2,B_2)-\mathcal{P}(A'_1,B_2)-\mathcal{P}(A'_2,B_1)=1\\
E(A',B') &= \mathcal{P}(A'_1,B'_1)+\mathcal{P}(A'_2,B'_2)-\mathcal{P}(A'_1,B'_2)-\mathcal{P}(A'_2,B'_1)=-1
\end{aligned} \tag{4}$$

which means that the Bell-CHSH inequality is violated up to its algebraic maximum, as is clear that $E(A,B)+E(A,B')+E(A',B)-E(A',B')=4$.

Note that the no-signaling conditions are also satisfied. Indeed, $\mathcal{P}_B(A_1)\equiv\mathcal{P}(A_1,B_1)+\mathcal{P}(A_1,B_2)=\frac{1}{2}+0=\frac{1}{2}$ and $\mathcal{P}_{B'}(A_1)\equiv\mathcal{P}(A_1,B'_1)+\mathcal{P}(A_1,B'_2)=\frac{1}{2}+0=\frac{1}{2}$. Also, $\mathcal{P}_B(A_2)\equiv\mathcal{P}(A_2,B_1)+\mathcal{P}(A_2,B_2)=0+\frac{1}{2}=\frac{1}{2}$ and $\mathcal{P}_{B'}(A_2)\equiv\mathcal{P}(A_2,B'_1)+\mathcal{P}(A_2,B'_2)=0+\frac{1}{2}=\frac{1}{2}$. In other words, $\mathcal{P}_B(A_i)=\mathcal{P}_{B'}(A_i)$, $i=1,2$, and it is easy to check that we also have, with obvious notation, $\mathcal{P}_B(A'_i)=\mathcal{P}_{B'}(A'_i)$, $\mathcal{P}_A(B_i)=\mathcal{P}_{A'}(B_i)$ and $\mathcal{P}_A(B'_i)=\mathcal{P}_{A'}(B'_i)$, $i=1,2$. In other words, Alice and Bob, with their joint measurements, can create correlations but cannot use them to communicate.

3. Induced by measurements common causes

The above example of 'Bertlmann wearing no socks' shows that the idea that quantum correlations would result from 'induced by measurements' common causes' is a perfectly licit one. However, despite its simplicity and generality, it has surprisingly not attracted significant interest in the scientific community so far (although the idea was first presented, albeit with a slightly different terminology, since the eighties of the last century [14]). Note that in this article our focus is uniquely on the 'Bertlmann wearing no socks' situation, because of its affinity with Bell's historical example, but it is possible and easy to replace Dr. Bertlmann by a passive physical entity, like an elastic band, with the outcomes being for instance created by Alice and Bob when they jointly pull its two ends, causing it to break at some unpredictable point, so that the lengths of the obtained fragments will necessarily be correlated; see [13] for the details.

Note also that the very quantum formalism already suggests that when a system is not in an eigenstate of the measurement being executed, the outcome is literally created (i.e., actualized) by the latter, and when a joint measurement is executed, joint outcomes will consequently be actualized; and if they are correlated, it is natural to assume that the joint measurement also actualized the common causes that are at the origin of the correlations. Also, if the common causes are actualized only when the joint measurements are executed, there will be no 'unique set of common causes' characterizing all the correlations of the four joint measurements AB, AB', $A'B$ and $A'B'$, which is the reason why the Bell-CHSH inequality can be violated (as we will better explain in the next section).

Quantum micro-physical entities in entangled states can be described similarly to the situation of Bertlmann wearing no socks. For example, a system formed by two electrons in a singlet state can be conceptualized as a situation where the two electrons "wear no spin," i.e., a situation where the two-entities haven't yet actualized specific spin directions, which therefore remain potential until the meeting with Alice's and Bob's Stern-Gerlach apparatuses, forcing them to acquire one, in the same way Alice and Bob, when they meet Bertlmann, also "force" him to wear socks of specific colors. This can be more clearly seen by also observing that a singlet state $|s\rangle \propto |+\rangle_{\hat{u}}|-\rangle_{\hat{u}} - |-\rangle_{\hat{u}}|+\rangle_{\hat{u}}$, where $|+\rangle_{\hat{u}}$ and $|-\rangle_{\hat{u}}$ represent the "up" and "down" eigenstates of the one-entity spin observables along the $\hat{u}$-direction, is a rotationally invariant vector. This means that there is nothing particular about the direction $\hat{u}$ that is used to explicitly write it, in terms of "up" and "down" components: one can equivalently choose a different direction $\hat{w} \neq \hat{u}$ and still find, after some algebraic calculation, that $|s\rangle \propto |+\rangle_{\hat{w}}|-\rangle_{\hat{w}} - |-\rangle_{\hat{w}}|+\rangle_{\hat{w}}$. In other words, the zero-spin state $|s\rangle$ resides in a one-dimensional subspace of the two-entity Hilbert space (obtained from the antisymmetric tensor product of the two one-entity state spaces) which is invariant under the action of the 3D rotation group. Hence, it is to be interpreted as a "no-spin state," where spins are genuinely potential, thus non-spatial. In our Bertlmann's example (which could receive an explicit quantum mathematical representation in complex Hilbert space, following the method presented in [15]), we can easily see how the the choice of the joint measurement to be performed (the experimental context) affects the outcomes' correlations. Measurement $A'B'$ is an indeterministic context as regards the foot receiving the pink sock. Indeed, because of the questions jointly addressed by Alice and Bob, Bertlmann has to actualize a thought, or sensation, about one of his two feet, on which he will bring his attention. Such thought, or sensation, in combination with Bertlmann's habit of wearing first the pink sock, is an example of a 'induced by measurements (contextual) common cause', i.e., of a common cause which will not be the same when a different joint measurement is executed. In fact, measurements $A'B$ and AB' are deterministic contexts as regards the foot that will receive the pink sock: $A'B$ always produces a left pink sock and AB' always produces a right pink sock. So, the common cause is different from that of measurement $A'B'$ — hence contextual — as it does not involve anymore a decision-making from Bertlmann, whose initial attention towards one of his feet is now externally triggered by Alice, in measurement $A'B$, or by Bob, in measurement AB'.

The fundamental distinction between 'non-induced by measurements' and 'induced by measurements' common causes was emphasized by one of us already in 1990, however not directly referring to the causes of the correlations, but to the correlations themselves. More precisely, correlations resulting from 'non-induced by measurements' (non-contextual) common causes were referred to as 'correlations of the first kind', whereas correlations originating from 'induced by measurements' (contextual) common causes, actualized during the measurement processes, were referred to as 'correlations of the second kind' [16]. Among the reasons that led us to write this article, there is also that of proposing this new designation, which we hope will be able to more efficiently capture the attention of physicists on this simple and natural explanation for quantum correlations.

4. Bell's locality assumption

When two outcomes are correlated, their joint probability will not factorize into a product of probabilities for the individual outcomes, and this independently of the fact that the correlations are of the first or of the second kind, i.e., resulting from 'non-induced by measurements' or 'induced by measurements' common causes, respectively.

Let us consider first the "standard" situation of Bertlmann leaving home with socks already on his feet. According to (1), we have $\mathcal{P}(A_1) = \mathcal{P}(A_1, B_1) + \mathcal{P}(A_1, B_2) = \mathcal{P}(A_1, B_1') + \mathcal{P}(A_1, B_2') = \frac{1}{2}$ and $\mathcal{P}(B_1) = \mathcal{P}(A_1, B_1) + \mathcal{P}(A_2, B_1) = \mathcal{P}(A_1', B_1) + \mathcal{P}(A_2', B_1) = \frac{1}{2}$. On the other hand, $\mathcal{P}(A_1, B_1) = \frac{1}{2}$, hence $\mathcal{P}(A_1, B_1) \neq \mathcal{P}(A_1)\mathcal{P}(B_1)$.

As emphasized by Bell, for example in [10], it is however reasonable to assume that if the common causes of the correlations are known and can be held fixed, the probabilities associated with possible residual fluctuations will factorize. This is what is usually called 'Bell's locality assumption'. Let us denote by λ_i, $i = 1, 2, 3, 4$, the common causes at the origin of the correlations of the first kind (1). These can be identified with the four possible states of mind of Bertlmann when wearing his socks and handkerchiefs. More precisely, λ_1 describes the state of mind such that Bertlmann decides to wear pink handkerchiefs and a pink left sock (and therefore a non-pink right sock); λ_2 describes the state of mind such that Bertlmann decides to wear pink handkerchiefs and a non-pink left sock; λ_3 describes the state of mind such that Bertlmann decides to wear non-pink handkerchiefs and a non-pink left sock; λ_4 describes the state of mind such that Bertlmann decides to wear non-pink handkerchiefs and a pink left sock.

Consider, as an example, the situation where we know that the state of mind of Bertlmann is λ_1. Then the probabilities become:

$$\begin{aligned}
&\mathcal{P}(A_1, B_1; \lambda_1) = 1, \quad \mathcal{P}(A_1, B_2; \lambda_1) = 0, \quad \mathcal{P}(A_2, B_1; \lambda_1) = 0 \\
&\mathcal{P}(A_2, B_2; \lambda_1) = 0, \quad \mathcal{P}(A_1, B_1'; \lambda_1) = 0, \quad \mathcal{P}(A_1, B_2'; \lambda_1) = 1 \\
&\mathcal{P}(A_2, B_1'; \lambda_1) = 0, \quad \mathcal{P}(A_2, B_2'; \lambda_1) = 0, \quad \mathcal{P}(A_1', B_1; \lambda_1) = 1 \\
&\mathcal{P}(A_1', B_2; \lambda_1) = 0, \quad \mathcal{P}(A_2', B_1; \lambda_1) = 0, \quad \mathcal{P}(A_2', B_2; \lambda_1) = 0 \\
&\mathcal{P}(A_1', B_1'; \lambda_1) = 0, \quad \mathcal{P}(A_1', B_2'; \lambda_1) = 1, \quad \mathcal{P}(A_2', B_1'; \lambda_1) = 0 \\
&\mathcal{P}(A_2', B_2'; \lambda_1) = 0
\end{aligned} \tag{5}$$

and we also have for the marginals:

$$\begin{aligned}
\mathcal{P}(A_1; \lambda_1) = \mathcal{P}(B_1; \lambda_1) = \mathcal{P}(B_2'; \lambda_1) = \mathcal{P}(A_1'; \lambda_1) = 1 \\
\mathcal{P}(A_2; \lambda_1) = \mathcal{P}(B_2; \lambda_1) = \mathcal{P}(B_1'; \lambda_1) = \mathcal{P}(A_2'; \lambda_1) = 0
\end{aligned} \tag{6}$$

which means that, say for joint measurement AB', we have the factorization:

$$\mathcal{P}(A_i, B_j'; \lambda_1) = \mathcal{P}(A_i; \lambda_1)\mathcal{P}(B_j'; \lambda_1), \quad i, j = 1, 2, \tag{7}$$

and the same holds true for the other joint measurements, when the states of mind λ_2, λ_3 and λ_4 are fixed. Also, with the assumption (here just for simplicity) that Bertlmann's different states of mind are equiprobable, it is easy to check that we have the equalities:

$$\mathcal{P}(A_i, B_j') = \frac{1}{4}\sum_{k=1}^{4} \mathcal{P}(A_i, B_j'; \lambda_k) = \frac{1}{4}\sum_{k=1}^{4} \mathcal{P}(A_i; \lambda_k)\mathcal{P}(B_j'; \lambda_k) \tag{8}$$

for all $i, j = 1, 2$, and the same holds for the other joint measurements.

It immediately follows from (8) that the Bell-CHSH inequality will be satisfied. This is a standard derivation, but we reproduce it below for facilitating our subsequent discussion. Using (8), we can write:

$$
\begin{aligned}
E(A,B') &= \mathcal{P}(A_1,B'_1)+\mathcal{P}(A_2,B'_2)-\mathcal{P}(A_1,B'_2)-\mathcal{P}(A_2,B'_1)\\
&= \frac{1}{4}\sum_{k=1}^{4}[\mathcal{P}(A_1,B'_1;\lambda_k)+\mathcal{P}(A_2,B'_2;\lambda_k)-\mathcal{P}(A_1,B'_2;\lambda_k)-\mathcal{P}(A_2,B'_1;\lambda_k)]\\
&= \frac{1}{4}\sum_{k=1}^{4}[\mathcal{P}(A_1;\lambda_k)-\mathcal{P}(A_2;\lambda_k)][\mathcal{P}(B'_1;\lambda_k)-\mathcal{P}(B'_2;\lambda_k)]\\
&= \frac{1}{4}\sum_{k=1}^{4}E(A;\lambda_k)E(B';\lambda_k) \qquad (9)
\end{aligned}
$$

and same for the averages of the other joint measurements. Observing that we can write:

$$
\begin{aligned}
E(A,B)\pm E(A,B') &= \frac{1}{4}\sum_{k=1}^{4}E(A;\lambda_k)[E(B;\lambda_k)\pm E(B';\lambda_k)]\\
E(A',B)\mp E(A',B') &= \frac{1}{4}\sum_{k=1}^{4}E(A';\lambda_k)[E(B;\lambda_k)\mp E(B';\lambda_k)] \qquad (10)
\end{aligned}
$$

and that by definition $|E(A;\lambda_k)|, |E(A';\lambda_k)|, |E(B;\lambda_k)|, |E(B';\lambda_k)| \leq 1$, we have:

$$
\begin{aligned}
|E(A,B)\pm E(A,B')| &\leq \frac{1}{4}\sum_{k=1}^{4}|E(B;\lambda_k)\pm E(B';\lambda_k)|\\
|E(A',B)\mp E(A',B')| &\leq \frac{1}{4}\sum_{k=1}^{4}|E(B;\lambda_k)\mp E(B';\lambda_k)| \qquad (11)
\end{aligned}
$$

Finally, observing that we also have $|E(B;\lambda_k)\pm E(B';\lambda_k)|+|E(B;\lambda_k)\mp E(B';\lambda_k)|\leq 2$, we obtain:

$$
\begin{aligned}
&|E(A,B)+E(A,B')+E(A',B)-E(A',B')|\\
&\quad\leq |E(A,B)+E(A,B')|+|E(A',B)-E(A',B')|\\
&\quad\leq \frac{1}{4}\sum_{k=1}^{4}[|E(B;\lambda_k)\pm E(B';\lambda_k)|+|E(B;\lambda_k)\mp E(B';\lambda_k)|]\leq 2 \qquad (12)
\end{aligned}
$$

In other words, although common causes that are not induced by measurements do not imply that the latter, when jointly performed by Alice and Bob, would be separate, in the sense that, say for joint measurement AB', we would have $\mathcal{P}(A_i,B'_j)=\mathcal{P}(A_i)\mathcal{P}(B'_j)$, since we have still the validity of the average (8), they nevertheless imply that the Bell-CHSH inequality is satisfied.

Let us now consider the situation of Bertlmann wearing no socks, when he meets Alice and Bob, i.e., the situation where the common causes are induced by the measurements. Why in this case the Bell-CHSH inequality can be violated? This is so because we cannot anymore write the expectation values of the joint measurements as in (9), i.e., as sums of products of the expectation values for the individual measurements, with the sums running over the same common causes. The reason we cannot do this is that now the common causes are not the same for the different joint measurements, i.e., they are contextual. Let us show this in some detail.

For joint measurement AB, where Alice and Bob only ask about the color of Bertlmann's handkerchiefs, there are only two possible causes at the origin of the observed correlations. One corresponds to the state of mind of Bertlmann when he decides to wear pink handkerchiefs, let us denote it μ_1, the other corresponds to the state of mind of Bertlmann when he decides to wear non-pink handkerchiefs, let us denote it μ_2.

Consider joint measurement $A'B$. There is now a combination of the causes μ_1 and μ_2 with the event of meeting Alice asking about the color-correlation between the left sock and the left handkerchief, and also jointly meeting Bob only asking about the color of the right handkerchief. This triggers a deterministic action from Bertlmann, wearing a pink sock on his left foot (and consequently a non-pink sock on the right foot). So, we have again two possible common causes at the origin of the observed outcomes, let us call them ν_1 and ν_2, which are clearly different from those of joint measurement AB, with ν_1 giving rise to outcome (A'_1, B_1) and ν_2 giving rise to outcome (A'_2, B_2).

For joint measurement AB', reasoning in a similar way, we find that we have again two possible common causes at the origin of the observed outcomes, which we will denote σ_1 and σ_2, with the former giving rise to outcome (A_1, B'_1) and the latter to outcome (A_2, B'_2).

Finally, for joint measurement $A'B'$, the situation is similar to that described in the previous section, with the four common causes λ_1, λ_2, λ_3 and λ_4. So, we can now write the following four uniform averages of products of probabilities $(i, j = 1, 2)$:

$$\mathcal{P}(A_i, B_j) = \frac{1}{2}\sum_{k=1}^{2} \mathcal{P}(A_i; \mu_k)\mathcal{P}(B_j; \mu_k)$$

$$\mathcal{P}(A'_i, B_j) = \frac{1}{2}\sum_{k=1}^{2} \mathcal{P}(A'_i; \nu_k)\mathcal{P}(B_j; \nu_k)$$

$$\mathcal{P}(A_i, B'_j) = \frac{1}{2}\sum_{k=1}^{2} \mathcal{P}(A_i; \sigma_k)\mathcal{P}(B'_j; \sigma_k)$$

$$\mathcal{P}(A'_i, B'_j) = \frac{1}{4}\sum_{k=1}^{4} \mathcal{P}(A'_i; \lambda_k)\mathcal{P}(B'_j; \lambda_k) \tag{13}$$

The above allows us to still write for the expectation values:

$$E(A, B) = \frac{1}{2}\sum_{k=1}^{2} E(A; \mu_k)E(B; \mu_k)$$

$$E(A', B) = \frac{1}{2}\sum_{k=1}^{2} E(A'; \nu_k)E(B; \nu_k)$$

$$E(A, B') = \frac{1}{2}\sum_{k=1}^{2} E(A; \sigma_k)E(B'; \sigma_k)$$

$$E(A', B') = \frac{1}{4}\sum_{k=1}^{4} E(A'; \lambda_k)E(B'; \lambda_k) \tag{14}$$

However, when considering the sums and differences in (10), we are now unable to factorize the different expressions, as we did in the right hand sides of (10), as is clear that the averages are not anymore functions of the same variables for the common causes at the origin of the correlated outcomes. Hence, the Bell-CHSH inequality cannot be proven anymore and can in principle be violated, even up to its maximum value, as it is the case in our example.

5. Hidden-measurement interactions

Generally speaking, a measurement is a process involving the interaction of an entity (which is the entity to be measured) with an experimental context, usually described as another entity, called the measuring apparatus. One also needs to specify the operations to be performed in order to correctly obtain such interaction and the rule to be used to interpret its effects, in terms of possible outcomes of the measurement.

Quantum measurements are typically associated with indeterministic experimental contexts, corresponding to situations such that even when there is a full knowledge of the pre-measurement state, there is a maximum and irreducible lack of knowledge about the interaction with the measurement apparatus. To put it another way, in a quantum measurement the causes producing the effects of the different possible outcomes are induced by the very measurement process (and in that sense are contextual), i.e., actualized in an unpredictable way, as in a (weighted) symmetry breaking process, at each run of the measurement [17].

This mechanism, based on "contextual causes," which is likely at the origin of quantum indeterminism, becomes particularly evident when the standard formalism is extended in what was called the 'extended Bloch representation (EBR)' of quantum mechanics [18,19]. It is not the scope of this article to enter into the details of the EBR, which provides a natural completion of the quantum formalism. Let us simply mention here that it uses a generalized Bloch sphere representation in which it is possible to describe not only the states of a measured entity, but also the causes that can contextually produce the different outcomes, so much so that in this approach the Born rule can be derived in a non-circular way, instead of being just postulated.

In the EBR formalism, the 'non-induced by measurements' causes at the origin of the different possible outcomes are called 'hidden-measurement interactions'. However, the term "interaction" should not be understood here in the sense of a 'fundamental force' described by a potential term in the system's Lagrangian, but as the cause of a symmetry breaking actualizing a potential property of the system, resulting from bringing the measured system in contact with the measuring apparatus (in our example, resulting from Bertlmann meeting with Alice and Bob).[c]

Quantum joint measurements are just a specific type of measurements performed on bipartite entities, where the operations to be carried out on the two components forming the entity, traditionally described as Alice's and Bob's measurements, can be distinguished and are performed in a coincident way, with the resulting effects being described as couples of outcomes. When the entity is in an entangled state, these couples of outcomes will be correlated in such a way that Bell's inequalities can be violated. From the perspective of the EBR, this is what one would expect to happen. Indeed, the couples of outcomes produced by Alice's and Bob's joint measurements are precisely described in this approach as the effects of common causes actualized at each run of the experiment, in a way that depends on which measurements Alice and Bob have decided to jointly execute, i.e., in a contextual way. In other words, the situation described in our above simplified

[c]The term "hidden," in "hidden-measurement interactions," is to be understood in the sense of "inaccessible," i.e., in the sense of an aspect of the measurement process that the experimenter cannot take control of, without altering the experiment in such a way that it would not correspond anymore to the observation of the same physical property.

example of Dr. Bertlmann (when he does not wear socks) does capture the essence of a mechanism which can be described in all generality within the mathematical formalism of the EBR of quantum mechanics.

6. Final remarks

Our analysis shows that the fundamental ingredient for obtaining a violation of the Bell-CHSH inequality is that the common causes at the origin of the observed correlations when, say, AB is performed, are different from those at the origin of the observed correlation when, say, $A'B$ is performed, as evidenced in (13). In the quantum formalism, this manifests in the fact that the observables associated with AB and $A'B$ do not commute, and when these observables are written as tensor products, this of course reduces to the non-commutability of A and A'.

However, as the 'Dr. Bertlmann wearing no socks' experimental situation demonstrates, A and A' (and similarly B and B') can very well be compatible (i.e., jointly executable) measurements and one can still have a violation of the Bell-CHSH inequality, also a maximal one for that matter. So, the incompatibility responsible for the violation of the Bell-CHSH inequality truly lies at the non-local level of the overall joint measurements, which are characterized by different sets of hidden-measurement interactions, i.e., by different common causes, actualized in a contextual way.

Bell's inequalities are in that sense to be understood as providing a fundamental demarcation between the following two situations. When they are violated, it means that the observed correlations are the result of common causes that are contextually induced by the measurements themselves; when they are obeyed, it means that the common causes are not induced by the measurements, or at least not in a way that is different for each measurement. Also, this mechanism of the measurements being at the origin of the observed correlations, is not a prerogative of the micro-physical systems only, as the many examples of macroscopic entities violating Bell's inequalities analyzed during the years have clearly shown (see [23] and the references cited therein).

The important difference, when Bell's inequalities are violated by classical macro-physical entities, like Dr. Bertlmann's socks and handkerchiefs, compared to when they are violated by quantum micro-physical ones, like the entangled spins of two electrons, is that the connections out of which the common causes are created for the macro-entities are spatial elements of reality, whereas for the micro-entities they are non-spatial. Note that our fictional Dr. Bertlmann character can also be interpreted as a bipartite entity, as we can easily distinguish his left foot and left pocket from his right foot and right pocket. These left and right physical aspects, although spatially separated, remain intimately connected at a more abstract level, through Bertlmann's habits of always wearing socks of different colors and handkerchiefs of the same color.[d]

The same is true when considering micro-entities, like two entangled electrons in a singlet state. Even if the two electrons, once they have traveled far away from the source, can only be detected in widely separated spatial regions, they nevertheless remain connected at a more abstract level, thus manifesting a genuine non-spatial nature, allowing them to still behave as a whole entity. And when Alice and Bob jointly act on such an interconnected bipartite entity, correlations will be contextually created out of its whole structure. In other words, the common causes at the origin of the correlations have to be understood not only as being induced by measurements, and therefore genuinely contextual, but also genuinely non-spatial.

In our 'Dr. Bertlmann not wearing socks' experimental situation, the so-called no-signaling conditions (also called marginal laws) are satisfied, in accordance with what is predicted by the quantum formalism, when the joint measurements are assumed to be correctly described by tensor product observables. In other words, the statistics of outcomes that Alice obtains does not depend on the measurements Bob is jointly performing, and vice versa, so that Alice and Bob cannot use their joint measurements to communicate [20,21]. However, it is easy to define measurements such that the marginals laws will also be violated. This is the case for instance for the historical 'vessels of water model' [22], where joint measurements on a system formed by two vessels connected through a tube and containing a certain amount of transparent water were considered, and for similar models that have been analyzed during the years [23]. In the 'Dr. Bertlmann not wearing socks' situation, if instead of A' we consider a measurement A'', consisting in asking Bertlmann to tell the color of his left sock, with A''_1 corresponding to the "pink" answer and A''_2 to the "non-pink" answer, and similarly for B'', for the right sock, then it is easy to check that the four joint measurements AB, AB'', $A''B$ and $A''B''$ will also maximally violate the Bell-CHSH inequality, but this time will also violate the no-signaling conditions.

[d]In that respect, our example is somehow in between spatiality and non-spatiality. Indeed, where are Bertlmann's habits located? Are they in space? These are challenging questions that will receive very different answers depending on the person to whom they are asked.

The reason we have presented in this article a situation not violating the marginal laws, is that we wanted to emphasize that the 'induced by measurements' mechanism is fully compatible with the no-signaling requirement, but it is important to also emphasize that it is general enough to also allow for the description of situations where the marginal laws are not necessarily obeyed. Now, in many experiments the latter have been observed to be violated [24–29], although it is still not clear what the correct interpretation of these violations should be. Note however that when an entangled bipartite system is understood as forming an undivided non-spatial whole, and joint measurements are seen as processes contextually actualizing common causes producing the correlations, in principle the latter can violate not only the Bell-CHSH inequality but also the marginal laws. This does not mean, as is usually believed, that a superluminal communication would for this be possible. Indeed, if space-like separated correlations originating from a common cause in the past are not in violation of relativity theory (as 'Bertlmann's socks' example illustrates), then the same will also be true if the common cause is induced by the very measurement. This means that experimental situations where 'induced by measurements common causes' also give rise to a violation of the marginal laws cannot be used to achieve any (statistical) faster-than-light communication, as no genuine faster-than-light signal traveling in space can be initiated and controlled by either Alice or Bob. For more details on this specific aspect, we refer the interested reader to [23].

We observe that it is quite well accepted among physicists that quantum mechanics cannot be embedded in a 'locally causal theory', i.e., a theory such that, to quote Bell [30]: "the direct causes (and effects) of events are near by, and even the indirect causes (and effects) are no further away than permitted by the velocity of light." In this article, we have emphasized that the reason for the irreducible non-locality of the theory would not be the existence of some unknown signals propagating "quantum information" in space at some superluminal speeds (which should be of at least seven orders of magnitude larger than the speed of light, according to [31]), but a mechanism of actualization of potential common causes, during the execution of joint measurements, explaining why space-like separated correlations can be easily created violating Bell's inequalities.

We also mention that when we introduce 'induced by measurements common causes', hence contextual common causes, as an explanation for the violation of Bell's inequalities, we do not want to merely highlight a well-known result, namely that certain quantum effects are incompatible with non-contextual hidden variable models. The new insight we want to put forward here is that the 'induced by measurements common causes' are physically real and not just the result of an *ad hoc* mathematical construction. Indeed, in our 'Bertlmann wears no socks' example, the 'induced by measurements common causes' are easy to point out, hence are not some mysterious "hidden" variables. The same is true in many other experiments with macroscopic objects, like an elastic band that is pulled from its two ends until it breaks into two fragments of correlated lengths [13,32], or a volume of water that is extracted using two siphons, thus producing two correlated quantities of water, in two separate reference containers [22], or two dice connected via a rigid rod that are jointly rolled and thus produce correlated pairs of "upper faces" [33], or even in experiments with abstract entities, like specific conceptual combinations that are brought into more concrete states in well-designed psychological experiments (see [23] and the references cited therein).

This is just to emphasize that experimental contexts designed in such a way that the joint measurements will induce the common causes actualizing the correlations, will naturally violate Bell's inequalities, regardless of whether these experiments are performed on entities that are quantum or classical, microscopic or macroscopic, abstract or concrete. Of course, when performing experiments on micro-physical entities, we cannot access the level where the 'induced by measurements common causes' are actualized, as we can do instead with the 'Bertlmann wears no socks' example and the other above-mentioned examples. However, considering that two micro-physical entities in entangled states do form a whole that gets separated by joint measurements, similarly to how a whole elastic gets separated into two fragments when pulled from its ends, an explanation of the origin of quantum correlations in terms of 'common causes induced by measurements' is a very natural and general one, applicable to all sorts of systems.

When a rock explodes in two pieces flying apart, many properties of the two pieces of rock will be correlated. Similarly, when two pieces of a same playing card are hidden in two envelopes and sent to distant places, where they are opened, the outcomes will be correlated (for instance, their color will be the same, either red or black). Such examples are typically given as situations where Bell's inequalities are not violated, in contrast to the more mysterious character of a quantum violation. But as we explained in this article, the reason for this difference is that the measurements considered on the 'flying apart pieces of rocks', or the 'remote opening of the envelopes', are not as such 'inducing' the common causes at the origin of the correlations, as is the case for the 'breaking of the elastic'. Instead, they only 'observe already existing correlations and register them in a passive way'. In other words, in our highlighting that Bell's inequalities are naturally violated by 'common causes induced by measurements', in relation to both microscopic and macroscopic entities, the fact that the measurements produce the correlations, instead of merely observing them, is absolutely crucial.

References

1. Einstein, A., Podolski, B. & Rosen, N., Can quantum-mechanical description of physical reality be considered complete?, *Physical Review* **47**, 777–780.
2. Bell, J. (1964). On the Einstein Podolsky Rosen Paradox. *Physics* **1**, pp. 195–200.
3. Aspect, A. (1976). Proposed experiment to test the nonseparability of quantum mechanics. *Physical Review D* **14**, 1944–1951.
4. Aspect, A., Grangier, P. & Roger, G. (1981). Experimental Tests of Realistic Local Theories via Bell's Theorem. *Physical Review Letters* **47**, 460.
5. Aspect, A., Grangier, P. & Roger, G. (1982a). Experimental realization of Einstein-Podolsky-Rosen-Bohm Gedankenexperiment: A new violation of Bell's Inequalities. *Physical Review Letters* **49**, 91–94.
6. Aspect, A., Dalibard, J. & Roger, G. (1982b). Experimental test of Bell's Inequalities using time-varying analyzers. *Physical Review Letters* **49**, 1804–1807.
7. Aspect, A. & Grangier, P. (1985). About resonant scattering and other hypothetical effects in the orsay atomic-cascade experiment tests of Bell inequalities: A discussion and some new experimental data. *Lettere al Nuovo Cimento* **43**, 345–348.
8. Bertlmann, R. A. (1990). *Foundations of Physics* **20**, 1191–1212.
9. Hensen, B., Bernien, H., Dréau, A. E., Reiserer, A., Kalb, N., Blok, M. S., Ruitenberg, J., Vermeulen, R. F. L., Schouten, R. N., Abellán, C., Amaya, W., Pruneri, V., Mitchell, M. W., Markham, M., Twitchen, D. J., Elkouss, D., Wehner, S., Taminiau T. H. & Hanson R. (2016). Loophole-free Bell inequality violation using electron spins separated by 1.3 kilometres. *Nature* **526**, 682–686.
10. Bell, J.S. (1981). Bertlmann's socks and the nature of reality. *Journal de Physique Colloques* **42** (C2), C2-41-C2-62. Doi: 10.1051/jphyscol:1981202.
11. Bertlmann, R. A. (2015). Magic moments with John Bell. *Physics Today* **68**, 40–45, doi: 10.1063/PT.3.2847.
12. Clauser, J. F., Horne, M. A., Shimony, A. & Holt, R.A. (1969). Proposed experiment to test local hidden-variable theories. *Physical Review Letters* **23**, 880–884.
13. Aerts, S. (2005). A realistic device that simulates the non-local PR box without communication. http://arxiv.org/abs/quant-ph/0504171.
14. Aerts, D. (1984). How do we have to change quantum mechanics in order to

describe separated systems. In: S. Diner, D. Fargue, G. Lochak and F. Selerri (Eds.), *The Wave-Particle Dualism* (419–431). Dordrecht: Springer.
15. Aerts, D. & Sozzo, S. (2014a). Quantum entanglement in conceptual combinations. *International Journal of Theoretical Physics* **53**, 3587–3603.
16. Aerts, D. (1990). An attempt to imagine parts of the reality of the microworld. In: J. Mizerski, et al. (Eds.), *Problems in Quantum Physics II; Gdansk '89*, World Scientific Publishing Company, Singapore, 3–25.
17. Aerts, D. & Sassoli de Bianchi, M. (2017). Quantum measurements as weighted symmetry breaking processes: The hidden measurement perspective. *International Journal of Quantum Foundations* **3**, 1–16.
18. Aerts, D. & Sassoli de Bianchi M. (2014). The Extended Bloch Representation of Quantum Mechanics and the Hidden-Measurement Solution to the Measurement Problem, *Annals of Physics* **351**, 975–1025.
19. Aerts, D. & Sassoli de Bianchi, M. (2016). The Extended Bloch Representation of Quantum Mechanics. Explaining Superposition, Interference and Entanglement, *Journal of Mathematical Physics* **57**, 122110.
20. Ghirardi, G. C., Rimini, A. & Weber, T. (1980). A General Argument against Superluminal Transmission through the Quantum Mechanical Measurement Process. *Lett. Nuovo Cimento* **27**, 293–298.
21. Ballentine, L. E. & Jarrett, J. P. (1987). Bell's theorem: Does quantum mechanics contradict relativity? *American Journal of Physics* **55**, 696–701.
22. Aerts, D. (1982). Example of a macroscopical situation that violates Bell inequalities. *Lettere al Nuovo Cimento* **34**, 107–111.
23. Aerts, D., Aerts Arguëlles, J., Beltran, L., Geriente, S., Sassoli de Bianchi, M., Sozzo, S. & Veloz, T. (2019). Quantum entanglement in physical and cognitive systems: A conceptual analysis and a general representation. *European Physical Journal Plus* **134**: 493.
24. Adenier, G. & Khrennikov, A. (2007). Is the fair sampling assumption supported by EPR experiments?, *J. Phys. B: Atomic, Molecular and Optical Physics* **40**, 131–141.
25. De Raedt, H., Michielsen, K. & Jin, F. (2012). Einstein-Podolsky-Rosen-Bohm laboratory experiments: Data analysis and simulation, *AIP Conf. Proc. 1424*, 55–66.
26. De Raedt H., Jin, F. & Michielsen, K. (2013). Data analysis of Einstein-Podolsky-Rosen-Bohm laboratory experiments. *Proc. of SPIE 8832, The Nature of Light: What are Photons? V*, 88321N.
27. Adenier, G. & Khrennikov, A. (2017). Test of the no-signaling principle in the Hensen loophole-free CHSH experiment, *Fortschritte der Physik (Progress in Physics)* **65**, 1600096.
28. Bednorz A. (2017). Analysis of assumptions of recent tests of local realism, *Phys. Rev. A* **95**, 042118.
29. Kupczynski, M. (2017). Is Einsteinian no-signalling violated in Bell tests?, *Open Phys.* **15**, 739–753.
30. Bell, J. (2004). La nouvelle cuisine. Respectfully dedicated to the great chef. In: *Speakable and Unspeakable in Quantum Mechanics*. Cambridge University Press, 232–248.

31. Zbinden, H., Brendel, J., Gisin, N. & Tittel, W. (2001). Experimental test of non-local quantum correlation in relativistic configurations. *Phys Rev A* **63**: 022111.
32. Sassoli de Bianchi, M. (2013). Using simple elastic bands to explain quantum mechanics: A conceptual review of two of Aerts' machine-models. *Central European Journal of Physics* **11**, 147–161.
33. Sassoli de Bianchi, M. (2013). Quantum dice. *Annals of Physics* **336**, 56–75.

ARE BELL-TESTS ONLY ABOUT LOCAL INCOMPATIBILITY?

DIEDERIK AERTS

Center Leo Apostel for Interdisciplinary Studies and Department of Mathematics, Vrije Universiteit Brussel, 1050 Brussels, Belgium
E-mail: diraerts@vub.ac.be

MASSIMILIANO SASSOLI DE BIANCHI

Center Leo Apostel for Interdisciplinary Studies, Vrije Universiteit Brussel, 1050 Brussels, Belgium and Laboratorio di Autoricerca di Base, 6917 Barbengo, Switzerland
E-mail: msassoli@vub.ac.be

The view exists that Bell-tests would only be about *local* incompatibility of quantum observables and that quantum *non-locality* would be an unnecessary concept in physics. In this note, we emphasize that it is not incompatibility at the *local* level that is important for the violation of Bell-CHSH inequality, but incompatibility at the *non-local* level of the joint measurements. Hence, non-locality remains a necessary concept to properly interpret the outcomes of certain joint quantum measurements.

Keywords: Bell-CHSH inequality; Bell-test, contextuality; compatibility; non-locality; non-spatiality; marginal laws, no-signaling conditions.

Some authors have argued that since Bell-CHSH inequality is only violated under the condition of local incompatibility of Alice's and Bob's observables, Bell-tests should only be considered as special tests of incompatibility of said local observables, hence Bell's introduction of the very notion of non-locality would be misleading and the term "non-locality" should be dismissed altogether [1–5]. More precisely, addressing here the more specific point raised in [2], and following Khalfin and Tsirelson's algebraic method [6], we observe that by taking the square of the CHSH operator

$$C = A \otimes (B + B') + A' \otimes (B - B'), \tag{1}$$

then using the fact that the observables A, A', B and B' have ± 1 eigenvalues, one can write:

$$C^2 = 4\mathbb{I} + [A, A'] \otimes [B, B']. \quad (2)$$

Since the inequality $C^2 \leq 4\mathbb{I}$ (which implies $|\langle\psi|C|\psi\rangle| \leq 2$, for all ψ, which is the usual statement of the Bell-CHSH inequality; see for instance [7] for the details) can only be violated if $[A, A'] \neq 0$ and/or $[B, B'] \neq 0$, one finds that the Bell-CHSH inequality cannot be violated if Alice's and/or Bob's observables are compatible, i.e., commute. Based on this observation, one might be tempted to conclude that Bell-tests cannot truly highlight the presence of non-locality, but only of local incompatibility (local non-commutability) of Alice's and Bob's measurements.

The above reasoning is however incomplete, as it does not take into account the reason why local non-commutativity is necessary in the first place. To show this, let us start considering the situation where Alice's and Bob's measurements are compatible, so that we have the commutation relations $[A, A'] = 0$ and $[B, B'] = 0$. If so, one can in principle define a single measurement scheme for Alice, consisting in jointly measuring the two observables A and A', as well as a single measurement scheme for Bob, consisting in jointly measuring the two observables B and B'. Let us denote $\mathcal{A}$ and $\mathcal{B}$ the observables associated with these two bigger local measurements, performed by Alice and Bob, respectively. If A and A' are 2-outcome observables, this means that $\mathcal{A}$ is associated with the 4 outcomes (A_1, A'_1), (A_2, A'_1), (A_1, A'_2), (A_2, A'_2), so that the outcome-probabilities for the two sub-measurements A and A' can be deduced as marginals of the outcome-probabilities of such bigger local measurement; and the same holds true for Bob's observable $\mathcal{B}$, associated with the 4 outcomes (B_1, B'_1), (B_2, B'_1), (B_1, B'_2) and (B_2, B'_2).

If we additionally assume that the measurement defined by jointly executing Alice's and Bob's measurements is properly described in terms of a tensor product observable $\mathcal{A} \otimes \mathcal{B}$, as is usually done in standard quantum mechanics, i.e., by the product of the two commuting observables $\mathcal{A} \otimes \mathbb{I}$ and $\mathbb{I} \otimes \mathcal{B}$, it is clear that the overall experimental situation can be described in terms of a single measurement, defined by the action of Alice and Bob jointly performing $\mathcal{A}$ and $\mathcal{B}$. Such single measurement would produce the following 16 possible outcomes:

$$((A_1, A'_1), (B_1, B'_1)),\ ((A_1, A'_1), (B_2, B'_1)),\ ((A_1, A'_1), (B_1, B'_2)),$$

$$((A_1, A'_1), (B_2, B'_2)),\ ((A_2, A'_1), (B_1, B'_1)),\ ((A_2, A'_1), (B_2, B'_1)),$$
$$((A_2, A'_1), (B_1, B'_2)),\ ((A_2, A'_1), (B_2, B'_2)),\ ((A_1, A'_2), (B_1, B'_1)),$$
$$((A_1, A'_2), (B_2, B'_1)),\ ((A_1, A'_2), (B_1, B'_2)),\ ((A_1, A'_2), (B_2, B'_2)),$$
$$((A_2, A'_2), (B_1, B'_1)),\ ((A_2, A'_2), (B_2, B'_1)),\ ((A_2, A'_2), (B_1, B'_2)),$$
$$((A_2, A'_2), (B_2, B'_2)). \quad (3)$$

From their probabilities, one can easily deduce those of the 4 different possible joint sub-measurements, for instance the one obtained by considering sub-measurement A in association with sub-measurement B. More precisely, the outcome-probability $\mathcal{P}(A_1, B_1)$, of obtaining outcome A_1 for A and outcome B_1 for B, would be given by the sum:

$$\begin{aligned}\mathcal{P}(A_1, B_1) &= \mathcal{P}((A_1, A'_1), (B_1, B'_1)) + \mathcal{P}((A_1, A'_1), (B_1, B'_2)) \\ &\quad + \mathcal{P}((A_1, A'_2), (B_1, B'_1)) + \mathcal{P}((A_1, A'_2), (B_1, B'_2)),\end{aligned} \quad (4)$$

and similarly for the other outcome-probabilities.

Now, the probabilities deduced from a single measurement situation can always fit into a single Kolmogorovian probability space, and therefore be represented in terms of deterministic, non-contextual hidden-variables; see for instance the representation theorem in [8]. In other words, if all measurements performed by Alice and Bob are compatible, no probabilistic structure extending beyond the classical one can be revealed. This means that in order to highlight the existence of elements of reality that cannot be described by classical probability models, and therefore by classical hidden-variables theories, one needs to consider situations where an entity is not always subjected to the same experimental context, i.e., to the same measurement, however big such measurement is. Bell's work was precisely about identifying and analyzing experimental situations able to produce joint probabilities that cannot be modeled using a single Kolmogorovian probability space, in order to test whether the type of correlations identified by EPR [9] did truly exist in reality.

In the case of the Bell-CHSH inequality, 4 different measurements are required and represented by the tensor product observables $A \otimes B$, $A' \otimes B$, $A \otimes B'$ and $A' \otimes B'$. Using $A^2 = A'^2 = B^2 = B'^2 = \mathbb{I}$, one can then deduce the the 6 commutation relations:

$$
\begin{aligned}
&[A \otimes B, A' \otimes B] = [A, A'] \otimes \mathbb{I}, \\
&[A \otimes B, A \otimes B'] = \mathbb{I} \otimes [B, B'], \\
&[A' \otimes B, A' \otimes B'] = \mathbb{I} \otimes [B, B'], \\
&[A \otimes B', A' \otimes B'] = [A, A'] \otimes \mathbb{I}, \\
&[A \otimes B, A' \otimes B'] = [A, A'] \otimes BB' + A'A \otimes [B, B'], \\
&[A' \otimes B, A \otimes B'] = [A', A] \otimes BB' + AA' \otimes [B, B']. \qquad (5)
\end{aligned}
$$

It is clear from the above that for the 4 observables $A \otimes B$, $A' \otimes B$, $A \otimes B'$ and $A' \otimes B'$, to describe measurements that cannot be incorporated into a single measurement scheme, we must have $[A, A'] \neq 0$ and/or $[B, B'] \neq 0$, i.e., Alice's and Bob's local observables must not all commute. However, this transfer of the incompatibility requirement from the non-local to the local level of the observables, is only the consequence of the fact that a specific representational choice has been a priori adopted: that of describing all joint measurements between Alice and Bob as product measurements relative to a *unique* tensor product representation of the state space. But this is a very special situation, which will not necessarily apply to all experimental situations. For instance, it will certainly be invalid if in addition to the Bell-CHSH inequality also the marginal laws (also called no-signaling conditions) are violated, as observed in many experiments [10–15].[a]

But even when the marginal laws are obeyed, a single tensor product representation for all the observables will not work if the Bell-CHSH inequality is violated beyond Tsirelson's bound, as it is the case for, say, the "Bertlmann wears no socks" experiment described in [18]. This is an experimental situation where the Bell-CHSH inequality is maximally violated even though Alice's measurements A and A', and Bob's measurements B and B', are perfectly compatible at the local level. In other words, this is a situation that cannot be described by (2), as the violation originates from an incompatibility which manifests at the non-local level of the joint measurements, the reason being that the correlations are created by the joint action of Alice and Bob, in a purely contextual way, i.e., the common causes at the origin of the correlations are contextually actualized.

So, while being true that incompatibility does play a central role in the construction of a Bell-test experiment, and in the understanding of its

[a]Note that the marginal laws can be easily violated when joint measurements are performed on spatially interconnected macroscopic entities, as well as on conceptual entities that are connected through meaning; see [17] and the references cited therein.

rationale, it is incompatibility at the global level of the joint measurements that is fundamental to have, which only reduces to local incompatibility when all the entanglement can be "pushed" into the state of the system. This is only possible, within the standard Hilbertian formulation of quantum mechanics, if the Bell-CHSH inequality is violated below Tsirelson's bound and all marginal laws are satisfied. In more general situations, entanglement needs to be allocated also at the level of measurements, as they cannot anymore be all described as product observables relative to a same tensor product representation. This is of course a manifestation of contextuality: the possibility of using a tensor product representation for the observables describing joint measurements becomes contextual, in the sense that one needs to adopt a different isomorphism for each joint measurement, in order to allocate the entanglement resource only in the state; see [16,17] for the details.

To put the above differently, the incompatibility of the different joint measurements means that one cannot find a non-contextual hidden-variables representation for the observed outcome-probabilities, i.e., one cannot find common causes in the past explaining *all* the correlations that are revealed by the experimental data. So, one is forced to recognize that these correlations were not all pre-existing the measurements, that some of them (or all of them) were contextually created by the latter. And this means that the common causes at the origin of the correlations are in turn genuinely contextual, where contextual means here that they are actualized at each run of a joint measurement in a way that depends on the type of joint measurement that is being executed [18]. Thus, if it is correct to say that Bell-test experiments are about evidencing the presence of incompatibility, i.e., the fact that not all measurements can be jointly performed, this does not mean that no conclusion can be drawn about the underlying reality producing such incompatibility. Indeed, we know that Alice's and Bob's laboratories can be located at arbitrary distance in space and that despite that, their joint action can still create correlations in a way that depends on the operations they jointly and simultaneously perform. But if their remotely performed joint actions are able to create correlations, or better, the common causes that are at their origin, how can a discussion about non-locality be avoided? We do not refer here to a notion of non-locality in the naif sense of "something spooky traveling in space at superluminal speed," but in the sense of something that can operate from a (non-spatial) layer of our physical reality, not being affected by spatial distances.

One can of course dislike the idea that our world, at its core, would be non-local, i.e., non-spatial, and certainly saying what something "is not" is just a first step in an investigation. The next step is about explaining what the nature of a non-spatial entity would be, and how it would relate to our spatial domain, in which it can leave traces, for instance in the form of impacts in our measuring apparatuses. To tentatively take that second step, our group in Brussels worked out in the past years a challenging hypothesis, according to which the micro-physical entities would be endowed of a conceptual nature, similar to that of the human concepts [19,20]. Non-spatiality would then be an expression of the fact that the micro-physical entities, being essentially conceptual in nature,[b] can be in more or less abstract states, with the less abstract ones (i.e., the more concrete ones) being precisely those associated with the condition of "being in space."

A few additional remarks are in order, to better understand what a Bell-test can tell us about the reality of a composite micro-system. We mentioned that a single tensor product representation can only be used when the Bell-CHSH inequality is violated below Tsirelson's bound and all marginal laws are satisfied. This should lead one to reflect on the practice of mathematically representing joint measurements by product operators, and then have superpositions of product states to describe the presence of entanglement.

If we accept the idea that density operators can also represent genuine states, then the situation is not in conflict with the general physical principle saying that a composite system exists, and therefore is in a well-defined state, only if its sub-systems also exist, and therefore are also in well-defined states [21]. On the other hand, if we believe that only vector-states (i.e., pure states) can represent genuine states, then the choice of representing the state of a composite system as a superposition of product states, not allowing then to attach vector-states to the sub-systems, becomes questionable, as physically unintelligible (how could the sub-systems exist if they are not in well-defined states?).

A different possibility would be to drop the requirement of describing entanglement as a property of the state, i.e., to use a product state to describe the state of the system and then non-product operators to describe the joint measurements. In other words, when confronted with a violation of the Bell-CHSH inequality, there is no a priori requisite to mathematically

[b] Just as electromagnetic waves and acoustic waves share the same undulatory nature, but remain completely distinct entities, the same would apply to human conceptual entities and the microscopic entities: they would share a same conceptual nature, while remaining completely distinct entities.

model the experimental situation in terms of entangled states: entangled (non-product) measurements can also be used, and in fact must be used if the marginal laws are also disobeyed.

But even this situation is not full satisfactory, as is clear that when a system evolves, a product state will generally transform into a non-product state, hence the interpretational problem remains and the use of density operators to describe genuine states seems to be inescapable [21].

The above issue can be better understood if one observes, as one of us did many decades ago, that in the Hilbert space formalism separated entities cannot be consistently modeled. The reason for this is that the Hilbertian formalism is, structurally speaking, too specific, as it satisfies an axiom called the 'covering law' [22–24]. This means that non-locality, which in ultimate analysis means non-separability, is intrinsically part of the quantum formalism, so much so that locality/separability cannot be even properly expressed within it (not for a lack of states, but for a lack of properties). In other words, the very decision of using a Hilbert space to model a physical system already and unavoidably introduces non-locality/non-separability in its description.

Note that the Bell-CHSH inequality, being expressed only in terms of probabilities, is independent of the mathematical formalism used to model an experimental situation. Also, Bell was primarily interested in separability, and not whether the probability structure of a composite system would be Kolmogorovian or non-Kolmogorovian, and in particular if the probabilities of the different joint measurements could be described as the marginals of a unique joint probability distribution. In particular, Bell wasn't focused on the marginal laws being satisfied or not. This question only came about later, with the analysis of Fine [26], Pitowsky [27] and more recently Dzhafarov and Kujala [28].

Note also that the special attention placed on the marginal laws only resulted from their interpretation as 'no-signaling conditions' [29,30]. Indeed, it has been argued that if violated they could be used to achieve faster than light communication. However, a more attentive analysis of the situation shows that this is not necessarily the case, for at least two reasons: it is not clear what are the times involved, in order to handle a large enough statistical ensemble of identically prepared systems, and if, when they are properly accounted for, they would still allow for an effective supraluminal communication. Also, and more importantly, the existence of correlations separated by space-like intervals does not per se imply that they result from underlying phenomena propagating in space faster than the speed of light, as the numerous models investigated by our group have clearly shown; see for instance the discussions in [7,17,18,31].

Our digression on marginal laws allows us to "close the circle" of our analysis. We mentioned that the description of all the observables associated with the different joint measurements in a Bell-test experiment, as product observables relative to a single tensor product representation, is what creates the illusion that Bell-tests would only be about local incompatibility. There is however no a priori physical justification to believe that such a peculiar representation would correspond to the general case. Certainly, it is not forced upon us by the available empirical data, which in fact tell us a rather different story, considering that the marginal laws are typically violated.

The proof of the marginal laws also relies on the existence of a single tensor product representation, which however, again, is not imposed by the quantum formalism (see in particular the discussion in [32]), hence should be justified by the data.

To conclude, Bell-tests are not just about local incompatibility: taking into account the available data, and until proven to the contrary, they also are about non-local incompatibility, hence about non-locality. Furthermore, the marginal laws are not genuine no-signaling conditions, as their violation does not necessarily imply a faster than light propagation of signals in space, and they also result of the peculiar choice (which requires a physical justification) of representing all joint measurements in a given experimental situation as product measurements relative to a unique tensor product decomposition.

References

1. Khrennikov, A. (2017). Bohr against Bell: Complementarity versus nonlocality, *Open Phys.* **15**, 734–738.
2. Khrennikov, A. (2019). Violation of the Bell's type inequalities as a local expression of incompatibility, *J. Phys.: Conf. Ser.* **1275**, 012018.
3. Khrennikov, A. (2019). Get Rid of Nonlocality from Quantum Physics, *Entropy* **21**, 806.
4. Khrennikov, A. (2020). Two faced Janus of quantum nonlocality, *Entropy* **22**, 303.
5. Griffiths, R. B. (2020). Nonlocality Claims are Inconsistent with Hilbert Space Quantum Mechanics, *Phys. Rev. A* **101**, 022117.
6. Khalfin, L. A. & Tsirelson, B. S. (1985). Quantum and quasi-classical analogs of Bell inequalities. In: *Symposium on the Foundations of Modern Physics 1985* (ed. Lahti et al.; World Scientific Publishing), 441–460.

7. Sassoli de Bianchi, M. (2020). Violation of CHSH inequality and marginal laws in mixed sequential measurements with order effects. *Soft Computing* **24**, 10231–10238.
8. Aerts, D. & Sozzo, S. (2012a). Entanglement of conceptual entities in Quantum Model Theory (QMod). *Quantum Interaction. Lecture Notes in Computer Science 7620*, 114–125.
9. Einstein, A. Podolsky, B. & Rosen, N. (1935). Can Quantum-Mechanical Description of Physical Reality Be Considered Complete?, *Phys. Rev.* **47**, 777–780.
10. Adenier, G. & Khrennikov, A. (2007). Is the fair sampling assumption supported by EPR experiments?, *J. Phys. B: Atomic, Molecular and Optical Physics* **40**, 131–141.
11. De Raedt, H., Michielsen, K. & Jin, F. (2012). Einstein-Podolsky-Rosen-Bohm laboratory experiments: Data analysis and simulation, *AIP Conf. Proc. 1424*, 55–66.
12. De Raedt H., Jin, F. & Michielsen, K. (2013). Data analysis of Einstein-Podolsky-Rosen-Bohm laboratory experiments, *Proc. of SPIE 8832*, The Nature of Light: What are Photons? V, 88321N.
13. Adenier, G. & Khrennikov, A. (2017). Test of the no-signaling principle in the Hensen loophole-free CHSH experiment, *Fortschritte der Physik* (Progress in Physics) **65**, 1600096.
14. Bednorz A. (2017). Analysis of assumptions of recent tests of local realism, *Phys. Rev. A* **95**, 042118.
15. Kupczynski, M. (2017). Is Einsteinian no-signalling violated in Bell tests?, *Open Phys.* **15**, 739–753.
16. Aerts, D. & Sozzo, S. (2014). Quantum Entanglement in Concept Combinations. Int. *J. Theor. Phys.* **53**, 3587–3603.
17. Aerts, D., Aerts Arguëlles, J., Beltran, L., Geriente, S., Sassoli de Bianchi, M., Sozzo, S. & Veloz, T. (2019). Quantum entanglement in physical and cognitive systems: A conceptual analysis and a general representation. *European Physical Journal Plus* **134**: 493.
18. Aerts, D. & Sassoli de Bianchi, M. (2019). When Bertlmann wears no socks. Common causes induced by measurements as an explanation for quantum correlations. arXiv:1912.07596 [quant-ph]. Published in this volume.
19. Aerts, D., Sassoli de Bianchi, M., Sozzo, S. & Veloz, M. (2020). On the Conceptuality interpretation of Quantum and Relativity Theories. *Foundations of Science* **25**, 5–54.
20. Aerts, D. & Beltran, L. (2019). Quantum Structure in Cognition: Human Language as a Boson Gas of Entangled Words. *Foundations of Science* **25**, 755–802. https://doi.org/10.1007/s10699-019-09633-4.
21. Aerts, D. & Sassoli de Bianchi, M. (2016). The extended Bloch representation of quantum mechanics: Explaining superposition, interference, and entanglement, *Journal of Mathematical Physics* **57**, 122110.
22. Aerts, D. (1980). Why is it impossible in quantum mechanics to describe two or more separated entities. *Bulletin de l'Academie royale de Belgique, Classes des Sciences* **66**, 705–714.

23. Aerts, D. (1982). Description of many physical entities without the paradoxes encountered in quantum mechanics, *Found. Phys.* **12**, 1131–1170.
24. Aerts, D. (1984). The missing elements of reality in the description of quantum mechanics of the EPR paradox situation, *Helv. Phys. Acta* **57**, 421–428.
25. Sassoli de Bianchi, M. (2019). On Aerts' overlooked solution to the EPR paradox. In: *Probing the Meaning of Quantum Mechanics. Information, Contextuality, Relationalism and Entanglement.* D. Aerts, M.L. Dalla Chiara, C. de Ronde & D. Krause (eds.) World Scientific, pp. 185–201.
26. Fine, A. (1982). Joint distributions, quantum correlations, and commuting observables. *Journal of Mathematical Physics* **23**, 1306–1310.
27. Pitowsky, I. (1989). *Quantum Probability, Quantum Logic.* Lecture Notes in Physics 321. Berlin: Springer.
28. Dzhafarov, E. N. & Kujala, J. V. (2017). *Fortschr. Phys.* **65**, 1600040.
29. Ballentine, L. E. & Jarrett, J. P. (1987). Bell's theorem: Does quantum mechanics contradict relativity? *American Journal of Physics* **55**, 696–701; doi: 10.1119/1.15059.
30. Ghirardi, G. C., Rimini, A. & Weber, T. (1980). A General Argument against Superluminal Transmission through the Quantum Mechanical Measurement Process. *Lett. Nuovo Cimento* **27**, 293–298.
31. Aerts, D., Aerts Arguëlles, J., Beltran, L., Geriente, S., Sassoli de Bianchi, M., Sozzo, S. & Veloz, T. (2018). Spin and wind directions II: A Bell State quantum model. *Foundations of Science* **23**, 337–365.
32. Kennedy, J. B. (1995). On the Empirical Foundations of the Quantum No-Signalling Proofs, *Philosophy of Science* **62**, 543–560.

A PARTICLE ONTOLOGICAL INTERPRETATION OF THE WAVE FUNCTION

SHAN GAO

Research Center for Philosophy of Science and Technology, Shanxi University, Taiyuan 030006, P. R. China
E-mail: gaoshan2017@sxu.edu.cn

The ontological meaning of the wave function is an important problem in the metaphysics of quantum mechanics. The conventional view is wave function realism, according to which the wave function represents a real physical field in a fundamental high-dimensional space. In this chapter, I present a new analysis of the meaning of the wave function, and the analysis suggests a particle ontological interpretation of the wave function. According to this interpretation, the wave function of a quantum system describes the state of random discontinuous motion of particles in three-dimensional space, and in particular, the modulus squared of the wave function gives the probability density that the particles appear in every possible group of positions in space. Moreover, I argue that the difference between particle ontology and field ontology may result in different predictions under certain reasonable assumption, and it is the former, not the latter, that is consistent with the predictions of quantum mechanics. Finally, I discuss similar pictures of motion of particles suggested by others. It is argued that Bell's Everett (?) theory also implies the picture of random discontinuous motion of particles.

Keywords: Wave function; particle ontology; field ontology; random discontinuous motion of particles.

1. Introduction

The meaning of the wave function has been a hot topic of debate since the early days of quantum mechanics. Recent years have witnessed a growing interest in this long-standing question [17-20, 33, 37]. If the wave function is indeed ontic, then exactly what physical state does it represent? The conventional view is wave function realism, according to which the wave function represents a real, physical field in a fundamental high-dimensional configuration space [3]. This is arguably the simplest and most straightforward way of thinking about the wave function realistically. However, this interpretation is plagued by the problem of how to explain our three-dimensional impressions, and it is still debatable whether there is a satisfying solution to this problem [26, 27, 29, 30, 38]. This motivates a few authors to suggest that the wave function represents a property of discrete particles in our ordinary three-dimensional space [26, 27, 29, 30], although they do not give a concrete ontological picture of these particles in space and time and specify what property the property is. In this chapter, I will present some arguments supporting this view, and introduce a concrete interpretation of the wave function in terms of particle ontology in three-dimensional space.[a]

The plan of this chapter is as follows. In Section 2, I give a critical review of the arguments for and against wave function realism. In Section 3, I present a new ontological analysis of the wave function. It is realized that a better way to explore the relationship between the wave function and the physical entity it describes is not only analyzing the structure of the wave function itself, but also analyzing the whole Schrödinger equation, which governs the evolution of the wave function over time. I argue that what the wave function of an N-body system describes is not one physical entity, either a continuous field or a discrete particle, in an $3N$-dimensional space, but N physical entities in our ordinary three-dimensional space. Moreover, by an analysis of the entangled states of an N-body quantum system, I further argue that these physical entities are not continuous fields but discrete particles, and the motion of these particles is discontinuous and random. In Section 4, I introduce an ontological interpretation of the wave function in terms of random discontinuous motion of particles. According to this interpretation, the wave function of an N-body quantum system describes the state of random discontinuous motion of N particles, and in particular, the modulus squared of the wave function gives the probability density that the particles appear in every possible group of positions in space. At a deeper level, the wave function may represent the propensity property of the particles that determines their random discontinuous motion. In Section 5, the suggested particle ontological interpretation of the wave function and wave function realism are compared. It is argued that the difference between particle ontology and field ontology may result in different predictions under certain reasonable assumption, and it is the former, not the latter, that is consistent with the predictions of quantum mechanics. In Section 6, I briefly review similar pictures of motion of particles suggested by other people. It

[a]Some contents of this chapter are presented in Ref. [20].

is argued that Bell's Everett (?) theory also implies the picture of random discontinuous motion of particles. Conclusions are given in the last section.

2. Wave function realism

The wave function of a physical system is in general a mathematical object defined in a high-dimensional configuration space. For an N-body system, the configuration space in which its wave function is defined is $3N$-dimensional. Before presenting my analysis of the nature of configuration space and the meaning of the wave function, I will first examine a widely-discussed view, wave function realism,[b] which regards the wave function as a description of a real, physical field in a fundamental high-dimensional space [3-5].

In recent years, wave function realism seems to become an increasingly popular position among philosophers of physics and metaphysicians [34]. This view is composed of two parts. The first part says that configuration space is a real, fundamental space. Albert [3] writes clearly,

> The space *we* live in, the space in which any realistic interpretation of quantum mechanics is necessarily going to depict the history of the world as *playing itself out* ... is *configuration*-space. And whatever impression we have to the contrary (whatever impression we have, say, of living in a three-dimensional space, or in a four-dimensional space-time) is somehow flatly illusory. [3, p. 277]

The second part of this view states what kind of entity the wave function is in the configuration space. Again, according to Albert [3],

> The sorts of physical objects that wave functions *are*, on this way of thinking, are (plainly) *fields* - which is to say that they are the sorts of objects whose states one specifies by specifying the values of some set of numbers at every point in the space where they live, the sorts of objects whose states one specifies (in *this* case) by specifying the values of two numbers (one of which is usually referred to as an *amplitude*, and the other as a *phase* at every point in the universe's so-called *configuration* space.

[b]Since a realist interpretation of the wave function does not necessarily imply that the wave function describes a real, physical field in configuration space, the appellation "wave function realism" seems misleading. But for the sake of convenience I will still use this commonly used appellation in the following discussion.

> The values of the amplitude and the phase are thought of (as with all fields) as intrinsic properties of the points in the configuration space with which they are associated. [3, p. 278]

Note that configuration space conventionally refers to an abstract space that is used to represent possible configurations of particles in three-dimensional space, and thus when assuming wave function realism it is not accurate to call the high-dimensional space in which the wave function exists "configuration space". For wave function realism, there are no particles and their configurations, and the high-dimensional space is also fundamental, whose dimensionality is defined in terms of the number of degrees of freedom needed to capture the wave function of the system. But I will still use the appellation "configuration space" in my discussion of wave function realism for the sake of convenience.

There are two main motivations for adopting wave function realism. The first, broader motivation is that it seems to be the simplest, most straightforward, and most flat-footed way of thinking about the wave function realistically [3]. In Lewis's [26] words,

> The wavefunction figures in quantum mechanics in much the same way that particle configurations figure in classical mechanics; its evolution over time successfully explains our observations. So absent some compelling argument to the contrary, the prima facie conclusion is that the wavefunction should be accorded the same status that we used to accord to particle configurations. Realists, then, should regard the wavefunction as part of the basic furniture of the world... This conclusion is independent of the theoretical choices one might make in response to the measurement problem; whether one supplements the wavefunction with hidden variables (Bohm 1952), supplements the dynamics with a collapse mechanism (Ghirardi, Rimini and Weber 1986), or neither (Everett 1957), it is the wavefunction that plays the central explanatory and predictive role [26].

The end result is then the assumption of a physical space with a geometrical structure isomorphic to the configuration space and a set of physical properties isomorphic to the amplitude and phase of the wave function.

The second, more specific motivation for adopting wave function realism is entanglement or nonseparability of quantum mechanical states. In classical mechanics, the state of a system of N particles can be represented as a point in a $3N$-dimensional configuration space, and the configuration space representation is simply a convenient summary of the positions of all these particles. In quantum mechanics, however, the wave function of an entangled N-body system, which is defined in a $3N$-dimensional configuration space, cannot be broken down into individual three-dimensional wave functions of its subsystems. Thus the configuration space representation cannot be regarded as a convenient summary of the individual subsystem states in three-dimensional space; there are physical properties of an entangled N-body system that cannot be represented in terms of the sum of the properties of N subsystems moving in three-dimensional space [26, 27]. Therefore, wave function realism seems to be an inescapable consequence, according to which there exist a configuration space entity, the wave function, as a basic physical ingredient of the world.

In the following, I will first analyze the above motivations for wave function realism and then analyze its potential problems. To begin with, the first broader motivation to adopt wave function realism is debatable, since the approach of reading off the nature of the physical entity represented directly from the structure of the mathematical representation is problematic [28]. The problem has two aspects. On the one hand, a physical theory may have different mathematical formulations. If one reads off the physical ontology of the theory directly from the mathematical structure used to formulate the theory, then one will in general obtain different, conflicting ontologies of the same theory. This means that one at least needs to consider the reasons to choose one ontology rather than another.

Take standard quantum mechanics as an example [28]. When the wave function is multiplied by a constant phase, the new wave function yields the same empirical predictions. Thus standard quantum mechanics has two empirically equivalent mathematical formulations: one in terms of vectors in Hilbert space, and the other in terms of elements of projective Hilbert space. Obviously, taking the mathematics of these two formulations at face value will lead to different physical ontologies. Moreover, choosing the former, namely assuming wave function realism, is obviously inconsistent with the widely accepted view that gauge degrees of freedom are not physical, and thus one also needs to explain why the overall phase of the wave function, which is in principle unobservable, represents a real physical degree of freedom. In contrast, choosing the latter will avoid this thorny issue. As a result, even if it is reasonable to read off the physical ontology directly from the mathematical structure, it seems that one should not choose the Hilbert space formulation and assume wave function realism.[c]

[c]For a more detailed analysis of this problem see [28].

On the other hand, a mathematical representation may represent different physical ontologies, and one also needs to explain why choose one ontology rather than another. Although wave function realism is the straightforward way of thinking about the wave function realistically, there are also other possible realistic interpretations of the wave function. For example, as argued by Monton [26, 29, 30], the wave function of an N-body quantum system may represent the property of N particles in three-dimensional space. Moreover, according to their analysis, this interpretation is devoid of several potential problems of wave function realism (see later discussion). Thus, it seems that even if one chooses the Hilbert space formulation one should not assume wave function realism either.

To sum up, as Maudlin [28] concluded, studying only the mathematics in which a physical theory is formulated is not the royal road to grasp its ontology. In my view, we also need to study the connection between theory and experience in order to grasp the ontology of a physical theory.

Next, I will analyze the second concrete motivation to adopt wave function realism, the existence of quantum entanglement. Indeed, quantum mechanical states are distinct from classical mechanical states in their nonseparability or entanglement. The wave function of an entangled N-body system, which is defined in a $3N$-dimensional space, cannot be decomposed of individual three-dimensional wave functions of its subsystems. Schrödinger took this as the defining feature of quantum mechanics. However, contrary to the claims of Ney [33], this feature does not necessarily imply that the $3N$-dimensional space is fundamental and real, and the wave function represents a real, physical field in this space. As pointed out by French [13], there is at least an alternative understanding of entanglement in terms of the notion of "nonsupervenient" relations holding between individual physical entities existing in three-dimensional space [14, 41]. In addition, as noted above, one can also interpret the wave function of an N-body quantum system as the property of N particles in three-dimensional space [26, 29, 30].[d] The existence of these alternatives undoubtedly reduces the force of the second motivation to adopt wave function realism.

In the following, I will analyze the potential problems of wave function realism. The most obvious problem is how to explain our three-dimensional

[d]It is worth noting that there is also another possibility, namely the wave function is a multiple-field, a configuration of which assigns properties to sets of N points in three-dimensional space [9, 12, chap. 5].

impressions, which has been called the "problem of perception" [38]. Albert clearly realized this problem:

> The particularly urgent question (again) is where, in this picture, all the tables, and chairs, and buildings, and people are. The particularly urgent question is how it can possibly have come to pass, on a picture like this one, that there appear to us to be *multiple* particles moving around in a *three-dimensional* space [4, p. 54].

The first possible solution to this problem is the so-called instantaneous solution, which attempts to extract an image of a three-dimensional world from the instantaneous $3N$-dimensional wave function [26]. However, as argued by Monton [29], this solution faces a serious objection. It is that a point in configuration space alone does not pick out a unique arrangement of objects in three-dimensional space, as this solution obviously requires. A point in configuration space is given by specifying the values of $3N$ parameters, but nothing intrinsic to the space specifies which parameters correspond to which objects in three-dimensional space. Therefore, the $3N$-dimensional wave function at a particular instant cannot underpin our experiences of a three-dimensional world at that instant (except there is a preferred coordinatization of the $3N$-dimensional space that wave function realism seems to lack).

The second possible solution is the so-called dynamical solution, which attempts to show that the dynamical evolution of the $3N$-dimensional wave function over time produce the illusion of N particles moving around in a three-dimensional space [3, 26]. The key is to notice that the Hamiltonian governing the evolution of the wave function takes a uniquely simple form for a particular grouping of the coordinates of the wave function into ordered triples. In a classical mechanical world, the same form of Hamiltonian provides a notion of three-dimensional inter-particle distance, which can play a natural physical role since it "reliably measures the degree to which the particles in question can dynamically affect one another" [3]. Thus inhabitants living in such a world will perceive the system as containing N particles in a three-dimensional space. How about the quantum mechanical world then? According to Albert [3], if inhabitants in such a world don't look too closely, the appearances they encounter will also correspond with those of their classical counterparts. The reason is roughly that for a everyday object its true representation in terms of the evolution of a wave function can be approximated by the corresponding evolution of a point in a classical configuration space [26].[e] Therefore, even though wave function realism says that the world is $3N$-dimensional, we perceive it as having only three dimensions.

However, as argued by Lewis [26, 27], this dynamical solution to the "problem of perception" is also problematic. The reason is that the dynamical laws are not invariant under the coordinate transformations of the $3N$-dimensional space, but invariant under the coordinate transformations of a three-dimensional space. This indicates that the configuration space has a three-dimensional structure. In other words, although the wave function is a function of $3N$ independent parameters, but the transformational properties of the Hamiltonian require that these parameters refer to only three different spatial directions. Then, according to Lewis [26], Albert's dynamical solution is not only impossible but also unnecessary. It is impossible because the Hamiltonian takes exactly the same form under every choice of coordinates in three-dimensional space, and thus no choice makes it particularly simple. It is unnecessary because the outcome that the coordinates are naturally grouped into threes is built into the structure of configuration space, and thus does not need to be generated as a mere appearance based on the simplicity of the dynamics.

It is worth emphasizing that Lewis's argument based on invariance of dynamical laws also poses a serious threat to wave function realism. It strongly suggests that the quantum mechanical configuration space is not a real, fundamental space, but a space of configurations of particles existing in a three-dimensional space, quite like the classical mechanical configuration space. Correspondingly, the wave function is not a physical field in the high-dimensional configuration space either. This conclusion is also supported by a recent analysis of Myrvold [31]. He argued that since quantum mechanics arises from a relativistic quantum field theory, and in particular, the wave function of quantum mechanics and the configuration space in which it is defined are constructed from field operators defined on ordinary spacetime, the configuration space is not fundamental, but rather is derivative of structures defined in three-dimensional space, and the wave function is not like a physical field either.

[e]However, as admited by Albert [3, p. 56], "The business of actually filling in the details of these accounts is not an altogether trivial matter and needs to be approached separately, and anew, for each particular way of solving the measurement problem, and requires that we attend carefully to exactly how it is that the things we call particles actually manifest themselves in our empirical experience of the world." For a further analysis of this issue see Ref. [5].

However, as Lewis [26] also admitted, the above analysis does not provide a direct interpretation of the wave function. In his view, the wave function is a distribution over three-dimensional particle configurations, while a distribution over particle configurations is not itself a particle configuration. Then, what is the state of distribution the wave function represents? Are there really particles existing in three-dimensional space? I will try to answer these intriguing questions in the next section.

3. A new ontological analysis of the wave function

A better way to explore the relationship between the wave function and the physical entity it describes is not only analyzing the structure of the wave function itself, but also analyzing the whole Schrödinger equation, which governs the evolution of the wave function over time. The Schrödinger equation contains more information about a quantum system than the wave function of the system, an important piece of which is the mass and charge properties of the system that are responsible for the gravitational and electromagnetic interactions between it and other systems.[f] Mainly based on an analysis of these properties, I will argue that what the wave function of an N-body system describes is not one physical entity, either a continuous field or a discrete particle, in a $3N$-dimensional space, but N physical entities in our ordinary three-dimensional space. Moreover, by a new ontological analysis of the entangled states of an N-body quantum system, I will further argue that these physical entities are not continuous fields but discrete particles, and the motion of these particles is discontinuous and random.

3.1. *Understanding configuration space*

In order to know the meaning of the wave function, we need to first know the meaning of the coordinates on the configuration space of a quantum system, in which the wave function of the system is defined. As we have seen above, there are already some analyses of this issue. Here I will give a new analysis.

[f]In this sense, the wave function is not a complete description of a physical system, since it contains no information about the mass and charge of the system. Note that other authors have already analyzed the Hamiltonian that governs the evolution of the wave function in analyzing the meaning of the wave function, but what they have analyzed is only the coordinates of the Hamiltonian, not the other important parameters of the Hamiltonian such as mass and charge and their relations with the coordinates.

One way to understand configuration space is to see how the wave function transforms under a Galilean transformation between two inertial frames. Consider the wave function of an N-body quantum system in an inertial frame S with coordinates (x, y, z, t), $\psi(x_1, y_1, z_1, \ldots, x_N, y_N, z_N, t)$. The coordinates $(x_1, y_1, z_1, \ldots, x_N, y_N, z_N)$ are coordinates on the $3N$-dimensioanl configuration space of the system, and the wave function of the system is defined in this space. Now, in another inertial frame S' with coordinates (x', y', z', t'), where $(x', y', z', t') = G(x, y, z, t)$ is the Galilean transformation, the wave function becomes $\psi^{'}(x_1^{'}, y_1^{'}, z_1^{'}, \ldots, x_N^{'}, y_N^{'}, z_N^{'}, t')$, where $(x_i^{'}, y_i^{'}, z_i^{'}, t') = G(x_i, y_i, z_i, t)$ for $i = 1, \ldots, N$. Then the transformation of the arguments of the wave function already tells us the meaning of these arguments or the meaning of the coordinates on the configuration space of the system. It is that the $3N$ coordinates of a point in the configuration space of an N-body quantum system are N groups of three position coordinates in three-dimensional space. Under the Galilean transformation between two inertial coordinate systems S and S', each group of three coordinates (x_i, y_i, z_i) of the $3N$ coordinates on configuration space also transforms according to the Galilean transformation.

In addition, the interaction Hamiltonian of an N-body quantum system says the same thing. For example, the interaction Hamiltonian of an N-body quantum system under an external potential $V(x, y, z)$ is $\sum_{i=1}^{N} V(x_i, y_i, z_i)$, and the corresponding term in the Schrödinger equation is $\sum_{i=1}^{N} V(x_i, y_i, z_i)\psi(x_1, y_1, z_1, \ldots, x_N, y_N, z_N, t)$. Obviously the arguments of the potential function are the three position coordinates in our ordinary three-dimensional space, so do the corresponding group of three arguments of the wave function.

Another way to understanding configuration space is to resort to experience. The Born rule (for projective measurements) also tells us the meaning of the coordinates on the configuration space of a quantum system. According to the Born rule, the modulus squared of the wave function of an N-body quantum system, $|\psi(x_1, y_1, z_1, \ldots, x_N, y_N, z_N, t)|^2$, represents the probability density that the first subsystem is detected in position (x_1, y_1, z_1) in our three-dimensional space and the second subsystem is detected in position (x_2, y_2, z_2) in our three-dimensional space and so on. Thus, each group of three coordinates (x_i, y_i, z_i) of the $3N$ coordinates on configuration space are the three position coordinates in our ordinary three-dimensional space.

In summary, I have argued that the $3N$ coordinates of a point in the configuration space of an N-body quantum system are N groups of three position coordinates in our ordinary three-dimensional space.

3.2. *Understanding subsystems*

In order to know the ontological meaning of the wave function of an N-body quantum system, we also need to understand the characteristics of the subsystems which constitute the system.

First of all, the Schrödinger equation tells us something about subsystems. In the Schrödinger equation for an N-body quantum system, there are N mass parameters $m_1, m_2, \ldots, m_N$ (as well as N charge parameters etc). These parameters are not natural constants, but properties of the system; they may be different for different systems. Moreover, each mass parameter describes the same mass property, and it may assume different values for different subsystems. Therefore, it is arguable that these N mass parameters describe the same mass property of N subsystems. In other words, an N-body quantum system contains N subsystems or N physical entities, each of which has its respective mass and charge properties, and the wave function of the system describes the state of these physical entities. This conclusion is obvious when the wave function of an N-body quantum system is a product state of N wave functions.

Moreover, these N physical entities exist in our three-dimensional space, not in the $3N$-dimensional configuration space. The reason is that in the Schrödinger equation for an N-body quantum system, each mass parameter m_i is *only* correlated with each group of three coordinates (x_i, y_i, z_i) of the $3N$ coordinates on configuration space, while these three coordinates (x_i, y_i, z_i), according to the above analysis, are the three position coordinates in our three-dimensional space. Here it is also worth noting that the configuration space (as a fundamental space) cannot accomodate mass and charge distributions. For example, consider a product state of a two-body quantum system, in which there are point masses m_1 and m_2 in two positions in our three-dimensional space. This requires that in the corresponding *single* position in the six-dimensional configuration space of the system there should exist *two* distinguishable point masses m_1 and m_2. But in every position in the configuration space there can only exist one total point mass (if the space is real and fundamental).

Secondly, the connection of quantum mechanics with experience also tells us the same thing. Recall the Born rule says that the modulus squared of the wave function of an N-body quantum system, $|\psi(x_1, y_1, z_1, \ldots, x_N, y_N, z_N, t)|^2$, represents the probability density that the first subsystem with mass m_1 is detected in position (x_1, y_1, z_1) and the second subsystem with mass m_2 is detected in position (x_2, y_2, z_2) and so on. Thus the Born rule also says that each subsystem with its respective mass m_i is correlated with each group of three coordinates (x_i, y_i, z_i) of the $3N$ coordinates on configuration space. Moreover, the results of the measurements of the masses and charges of an N-body quantum system also indicate that the system is composed of N subsystems with their respective masses and charges.

To sum up, it is arguable that for an N-body quantum system, there are N subsystems or N physical entities with respective masses and charges in our three-dimensional space.

3.3. *Understanding entangled states*

In the following, I will present a new ontological analysis of the entangled states of an N-body quantum system. The analysis may provide an important clue to the ontological meaning of the wave function.

Consider a two-body system whose wave function is defined in a six-dimensional configuration space. First of all, suppose the wave function of the system is localized in one position $(x_1, y_1, z_1, x_2, y_2, z_2)$ in the space at a given instant. This wave function can be decomposed into a product of two wave functions which are localized in positions (x_1, y_1, z_1) and (x_2, y_2, z_2) in the same three-dimensional space, our ordinary three-dimensional space, respectively. It is uncontroversial that this wave function describes two independent physical entities, which are localized in positions (x_1, y_1, z_1) and (x_2, y_2, z_2) in our three-dimensional space, respectively. Moreover, as I have argued previously, the Schrödinger equation that governs the evolution of the system further indicates that these two physical entities have respective masses such as m_1 and m_2 (as well as respective charges such as Q_1 and Q_2 etc).

Next, suppose the wave function of the two-body system is localized in two positions $(x_1, y_1, z_1, x_2, y_2, z_2)$ and $(x_3, y_3, z_3, x_4, y_4, z_4)$ in the six-dimensional configuration space at a given instant. This is an entangled state, which can be generated from a product state by the Schrödinger evolution of the system. In this case, there are still two physical entities with the original masses and charges in three-dimensional space, since the Schrödinger evolution does not create or annihilate physical entities,[g] and the mass and charge properties of the system do not change during its evolution either. According to the above analysis, the wave function of the

[g]In other words, when the state of the two physical entities evolves from a product state to an entangled state, the interaction between them does not annihilate any of them from the three-dimensional space.

two-body system being localized in position $(x_1, y_1, z_1, x_2, y_2, z_2)$ means that physical entity 1 with mass m_1 and charge Q_1 exists in position (x_1, y_1, z_1) in three-dimensional space, and physical entity 2 with mass m_2 and charge Q_2 exists in position (x_2, y_2, z_2) in three-dimensional space. Similarly, the wave function of the two-body system being localized in position $(x_3, y_3, z_3, x_4, y_4, z_4)$ means that physical entity 1 exists in position (x_3, y_3, z_3) in three-dimensional space, and physical entity 2 exists in position (x_4, y_4, z_4) in three-dimensional space. These are two ordinary physical situations. Then, when the wave function of these two physical entities is an entangled state, being localized in both positions $(x_1, y_1, z_1, x_2, y_2, z_2)$ and $(x_3, y_3, z_3, x_4, y_4, z_4)$, how do they exist in three-dimensional space?

Since the state of the physical entities described by the wave function is defined either at a precise instant or during an infinitesimal time interval around a given instant as the limit of a time-averaged state, there are two possible existent forms. One is that the above two physical situations exist at the same time at the precise given instant in three-dimensional space. This means that physical entity 1 exists in positions (x_1, y_1, z_1) and (x_3, y_3, z_3), and physical entity 2 exists in positions (x_2, y_2, z_2) and (x_4, y_4, z_4). Since there is no correlation between the positions of the two physical entities, the wave function that describes this existent form is not an entangled state but a product state, which is localized in four positions $(x_1, y_1, z_1, x_2, y_2, z_2)$, $(x_3, y_3, z_3, x_4, y_4, z_4)$, $(x_1, y_1, z_1, x_4, y_4, z_4)$, and $(x_3, y_3, z_3, x_2, y_2, z_2)$ in the six-dimensional configuration space. Thus this possiblity is excluded.

The other possible existent form, which is thus the actual existent form, is that the above two physical situations exist "at the same time" during an arbitrarily short time interval or an infinitesimal time interval around the given instant in three-dimensional space. Concretely speaking, the situation in which physical entity 1 is in position (x_1, y_1, z_1) and physical entity 2 is in position (x_2, y_2, z_2) exists in one part of the continuous time flow, and the situation in which physical entity 1 is in position (x_3, y_3, z_3) and physical entity 2 is in position (x_4, y_4, z_4) exists in the other part. The restriction is that the temporal part in which each situation exists cannot be a continuous time interval during an arbitrarily short time interval; otherwise the wave function describing the state in the time interval will be not the original superposition of two branches, but one of the branches. This means that the set of the instants at which each situation exists is not a continuous instant set but a discontinuous, dense instant set. At some discontinuous instants, physical entity 1 with mass m_1 and charge Q_1 exists in position (x_1, y_1, z_1) and physical entity 2 with mass m_2 and charge Q_2 exists in position (x_2, y_2, z_2), while at other discontinuous instants, physical entity 1 exists in position (x_3, y_3, z_3) and physical entity 2 exists in position (x_4, y_4, z_4). By this way of time division, the above two physical situations exist "at the same time" during an arbitrarily short time interval or during an infinitesimal time interval around the given instant.

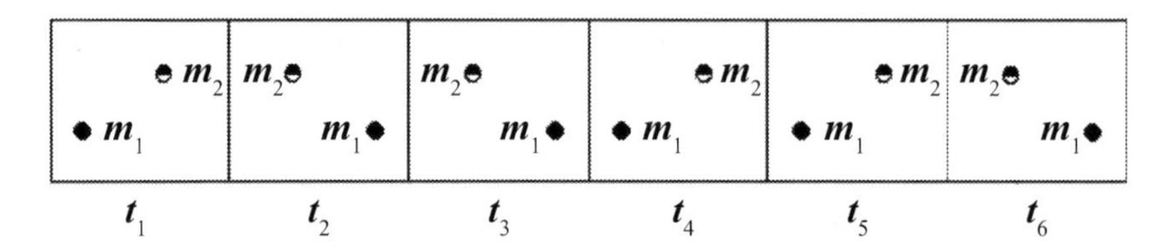

Fig. 1. Two entangled physical entities in space at six neighboring instants.

This way of time division implies a picture of discontinuous motion for the involved physical entities, which is as follows. Physical entity 1 with mass m_1 and charge Q_1 jumps discontinuously between positions (x_1, y_1, z_1) and (x_3, y_3, z_3), and physical entity 2 with mass m_2 and charge Q_2 jumps discontinuously between positions (x_2, y_2, z_2) and (x_4, y_4, z_4). Moreover, they jump in a precisely simultaneous way. When physical entity 1 jumps from position (x_1, y_1, z_1) to position (x_3, y_3, z_3), physical entity 2 always jumps from position (x_2, y_2, z_2) to position (x_4, y_4, z_4), and vice versa. In the limit case where position (x_2, y_2, z_2) is the same as position (x_4, y_4, z_4), physical entities 1 and 2 are no longer entangled, while physical entity 1 with mass m_1 and charge Q_1 still jumps discontinuously between positions (x_1, y_1, z_1) and (x_3, y_3, z_3). This means that the picture of discontinuous motion also exists for one-body systems. As argued before, since quantum mechanics does not provide further information about the positions of the physical entities at each instant, the discontinuous motion described by the theory is essentially random too.

The above analysis may also tell us what these physical entities are. A physical entity in our three-dimensional space may be a continuous field or a discrete particle. For the above entangled state of a two-body system, since each physical entity is only in one position in space at each instant (when there are two positions it may occupy), it is not a continuous field but a localized particle. In fact, there is a more general reason why these physical entities are not continuous fields in three-dimensional space. It is that for an entangled state of an N-body system we cannot even define

N continuous fields in three-dimensional space which contain the whole information of the entangled state.

Since a general position entangled state of a many-body system can be decomposed into a superposition of the product states of the position eigenstates of its subsystems, the above analysis applies to all entangled states. Therefore, it is arguable that an N-body quantum system is composed not of N continuous fields but of N discrete particles in our three-dimensional space. Moreover, the motion of these particles is not continuous but discontinuous and random in nature, and especially, the motion of entangled particles is precisely simultaneous.[h]

4. The wave function as a description of random discontinuous motion of particles

In classical mechanics, we have a clear physical picture of motion. It is well understood that the trajectory function $x(t)$ in the theory describes continuous motion of a particle. In quantum mechanics, the trajectory function $x(t)$ is replaced by a wave function $\psi(x,t)$. If the particle ontology is still viable in the quantum world, then it seems natural that the wave function should describe some sort of more fundamental motion of particles, of which continuous motion is only an approximation in the classical domain, as quantum mechanics is a more fundamental theory of the physical world, of which classical mechanics is an approximation. The previous analysis provides a strong support for this conjecture. It says that a quantum system is a system of particles that undergo random discontinuous motion (RDM in brief).[i] Here the concept of particle is used in its usual sense. A particle is a small localized object with mass and charge, and it is only in one position in space at each instant. As a result, the wave function in quantum mechanics can be regarded as a description of the more fundamental motion of particles, which is essentially discontinuous and random. In this section, I will give a more detailed analysis of RDM of particles and the ontological meaning of the wave function.

[h]Note that the analysis and its results given in this section also hold true for other instantaneous properties. I will discuss this point later.

[i]We may say that an electron is a quantum particle in the sense that its motion is not continuous motion described by classical mechanics, but random discontinuous motion described by quantum mechanics.

4.1. *A mathematical viewpoint*

Compared with continuous motion of particles, the picture of RDM of particles seems strange and unnatural for most people. This is not beyond expectations, since we are most familiar with the apparent continuous motion of objects in our everyday world. However, it can be argued that random discontinuous motion is more natural and logical than continuous motion from a mathematical point of view. Let us see why this is the case.

The motion of a particle can be described by a functional relation between each instant and its position at this instant in mathematics. In this way, continuous motion is described by continuous functions, while discontinuous motion is described by discontinuous functions. The question is: Which sort of functions universally exist in the mathematical world? This question can also be put in another more appropriate way. Since motion does not exist at an instant, the state of motion of a particle at a given instant is defined not by its position at the precise instant, but by its positions during an infinitesimal time interval around the instant. This means that the state of motion of a particle at a given instant will be described by a set of points in space and time, in which each point represents the position of the particle at each instant during an infinitesimal time interval around the given instant. Then the question is: What is the general form of such a set of points? Is it a continuous line? Or is it a discontinuous set of points? The former corresponds to continuous motion, while the latter corresponds to discontinuous motion.

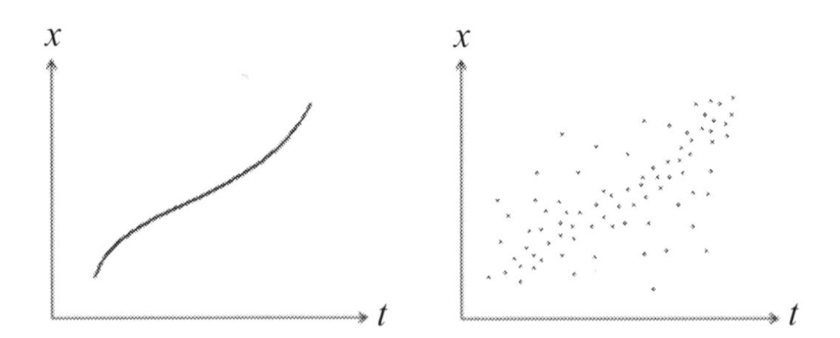

Fig. 2. Continuous motion vs. discontinuous motion.

The right answers to these questions had not been found until the late 19th and early 20th centuries. Before then only continuous functions were studied in mathematics. The existence of continuous functions accords with everyday experience. In particular, the motion of macroscopic objects is apparently continuous, and thus can be directly described by continuous functions. However, mathematics became more and more dependent on logic rather than on experience in the second half of the 19th century. Many discontinuous functions were invented during that period. For example, a famous discontinuous function is that its value is zero at each rational point and is one at each irrational point.

At first, most elder mathematicians were very hostile to discontinuous functions. They called them pathological functions. As Poincaré remarked in 1899,

> Logic sometimes makes monsters. For half a century we have seen a mass of bizarre functions which appear to be forced to resemble as little as possible honest functions which serve some purpose. More of continuity, or less of continuity, more derivatives, and so forth... In former times when one invented a new function it was for a practical purpose; today one invents them purposely to show up defects in the reasoning of our fathers. (Quoted in [24, p. 973])

However, several young mathematicians, notably Borel and Lebesgue, took discontinuous functions seriously. They discovered that these functions could also be strictly analyzed with the help of the set theory. As a result, they led a revolution in mathematical analysis, which transformed classical analysis into modern analysis.[j] A core notion introduced by them is measure. A measure of a set is a generalization of the concepts of the length of a line, the area of a plane figure, and the volume of a solid, and it can be intuitively understood as the size of the set. Length can only be used to describe continuous lines, which are very special sets of points, while measure can be used to describe more general sets of points. For example, the set of all rational points between 0 and 1 in a real line is a dense set of points, and its measure is zero. Certainly, the measure of a line still equals to its length.

The above questions can now be answered by using the measure theory. All functions form a set, whose measure can be set to one. Then according to the measure theory, the measure of the subset of all continuous functions is zero, while the measure of the subset of all discontinuous functions is one.

---

[j] It is well known that in the same period there also happened a revolution in physics, which transformed classical mechanics into quantum mechanics. I will argue below that there is a connection between modern analysis in mathematics and quantum mechanics in physics.

This means that continuous functions are extremely special functions, and nearly all functions are discontinuous functions. As Poincaré also admitted,

> Indeed, from the point of view of logic, these strange functions are the most general... If logic were the sole guide of the teacher, it would be necessary to begin with the most general functions. (Quoted in [24, p. 973])

Similarly, a general set of points in an infinitesimal region of space-time, which represents a general local state of motion of a particle, is a discontinuous, dense set of points. And a continuous line in the region is an extremely special set of points, and the measure of the subset of all continuous lines is zero.

Therefore, from a mathematical point of view, random discontinuous motion, which is described by general discontinuous functions and discontinuous sets of points, is more natural and logical than continuous motion, which is described by extremely special continuous functions and continuous lines. Moreover, if the great book of nature is indeed written in the language of mathematics, as Galileo once put it, then it seems that it is discontinuous motion of particles, not continuous motion of particles, that universally exists in the physical world.

4.2. *Describing random discontinuous motion of particles*

In the following, I will give a strict description of RDM of particles based on the measure theory. For the sake of simplicity, I will mainly analyze one-dimensional motion. The results can be readily extended to the three-dimensional situation.

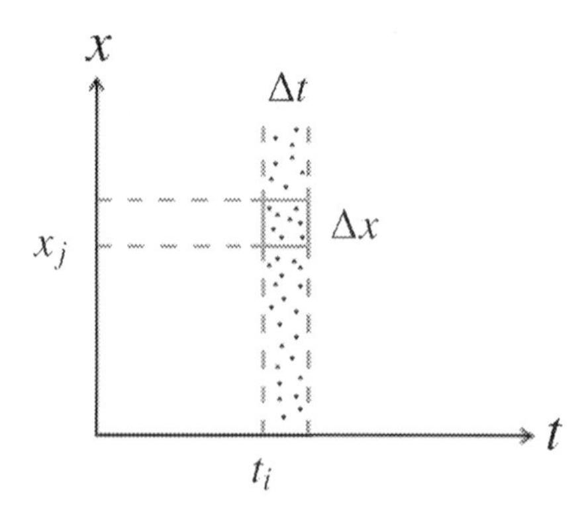

Fig. 3. Describing random discontinuous motion of a particle.

Consider the state of RDM of a particle in finite intervals Δt and Δx around a space-time point (t_i,x_j) as shown in Figure 3. The positions of the particle form a random, discontinuous trajectory in this square region.[k] We study the projection of this trajectory in the t-axis, which is a dense instant set in the time interval Δt. Let W be the discontinuous trajectory of the particle and Q be the square region $[x_j, x_j + \Delta x] \times [t_i, t_i + \Delta t]$. The dense instant set can be denoted by $\pi_t(W \cap Q) \in \Re$, where π_t is the projection on the t-axis. According to the measure theory, we can define the Lebesgue measure:

$$M_{\Delta x,\Delta t}(x_j, t_i) = \int_{\pi_t(W \cap Q) \in \Re} dt. \tag{1}$$

Since the sum of the measures of the dense instant sets in the time interval Δt for all x_j is equal to the length of the continuous time interval Δt, we have:

$$\sum_j M_{\Delta x,\Delta t}(x_j, t_i) = \Delta t. \tag{2}$$

Then we can define the measure density as follows:

$$\rho(x,t) = \lim_{\Delta x,\Delta t \to 0} M_{\Delta x,\Delta t}(x,t)/(\Delta x \cdot \Delta t). \tag{3}$$

We call $\rho(x,t)$ position measure density or position density in brief. This quantity provides a strict description of the position distribution of the particle in an infinitesimal space interval dx around position x during an infinitesimal interval dt around instant t, and it satisfies the normalization relation $\int_{-\infty}^{+\infty} \rho(x,t)dx = 1$ by (2). Note that the existence of the above limit relies on the precondition that the probability density that the particle appears in each position x at each instant t, which may be denoted by $\varrho(x,t)$, is differentiable with respect to both x and t. It can be seen that $\rho(x,t)$ is determined by $\varrho(x,t)$, and there exists the relation $\rho(x,t) = \varrho(x,t)$.

[k]Unlike deterministic continuous motion of particles, the discontinuous trajectory function, $x(t)$, no longer provides a useful description for RDM of particles. Recall that a trajectory function $x(t)$ is essentially discontinuous if it is not continuous at every instant t. A trajectory function $x(t)$ is continuous if and only if for every t and every real number $\varepsilon > 0$, there exists a real number $\delta > 0$ such that whenever a point t_0 has distance less than δ to t, the point $x(t_0)$ has distance less than ε to $x(t)$.

Since the position density $\rho(x,t)$ changes with time in general, we may further define the position flux density $j(x,t)$ through the relation $j(x,t) = \rho(x,t)v(x,t)$, where $v(x,t)$ is the velocity of the local position density. It describes the change rate of the position density. Due to the conservation of measure, $\rho(x,t)$ and $j(x,t)$ satisfy the continuity equation:

$$\frac{\partial \rho(x,t)}{\partial t} + \frac{\partial j(x,t)}{\partial x} = 0. \tag{4}$$

The position density $\rho(x,t)$ and position flux density $j(x,t)$ provide a complete description of the state of RDM of a particle.[1]

This description of the motion of a particle can be extended to the motion of many particles. At each instant a quantum system of N particles can be represented by a point in an $3N$-dimensional configuration space. During an arbitrarily short time interval or an infinitesimal time interval around each instant, these particles perform random discontinuous motion in three-dimensional space, and correspondingly, this point performs random discontinuous motion in the configuration space. Then, similar to the single particle case, the state of the system can be described by the position density $\rho(x_1, x_2, \ldots, x_N, t)$ and position flux density $j(x_1, x_2, \ldots, x_N, t)$ defined in the configuration space. There is also the relation $\rho(x_1, x_2, \ldots, x_N, t) = \varrho(x_1, x_2, \ldots, x_N, t)$, where $\varrho(x_1, x_2, \ldots, x_N, t)$ is the probability density that particle 1 appears in position x_1 and particle 2 appears in position x_2 ... and particle N appears in position x_N. When these N particles are independent with each other, the position density can be reduced to the direct product of the position density for each particle, namely $\rho(x_1, x_2, \ldots, x_N, t) = \prod_{i=1}^{N} \rho(x_i, t)$.

4.3. *Interpreting the wave function*

Although the motion of particles is essentially discontinuous and random, the discontinuity and randomness of motion are absorbed into the state of motion, which is defined during an infinitesimal time interval around a given instant and described by the position density and position flux density. Therefore, the evolution of the state of RDM of particles may obey

[1] It is also possible that the position density $\rho(x,t)$ alone provides a complete description of the state of RDM of a particle. Which possibility is the actual one depends on the laws of motion. As we will see later, quantum mechanics requires that a complete description of the state of RDM of particles includes both the position density and the position flux density.

a deterministic continuous equation. By assuming the nonrelativistic equation of RDM is the Schrödinger equation and considering the form of the resulting continuity equation, we can obtain the relationship between the position density $\rho(x,t)$, position flux density $j(x,t)$ and the wave function $\psi(x,t)$. $\rho(x,t)$ and $j(x,t)$ can be expressed by $\psi(x,t)$ as follows:[m]

$$\rho(x,t) = |\psi(x,t)|^2, \tag{5}$$

$$j(x,t) = \frac{\hbar}{2mi}[\psi^*(x,t)\frac{\partial\psi(x,t)}{\partial x} - \psi(x,t)\frac{\partial\psi^*(x,t)}{\partial x}]. \tag{6}$$

Correspondingly, the wave function $\psi(x,t)$ can be uniquely expressed by $\rho(x,t)$ and $j(x,t)$ or $v(x,t)$ (except for an overall phase factor):

$$\psi(x,t) = \sqrt{\rho(x,t)}e^{im\int_{-\infty}^{x} v(x',t)dx'/\hbar}. \tag{7}$$

In this way, the wave function $\psi(x,t)$ also provides a complete description of the state of RDM of a particle. A similar one-to-one relationship between the wave function and position density, position flux density also exists for RDM of many particles. For the motion of many particles, the position density and position flux density are defined in a $3N$-dimensional configuration space, and thus the many-particle wave function, which is composed of these two quantities, also lives on the $3N$-dimensional configuration space.

It is well known that there are several ways to understand objective probability, such as frequentist, propensity, and best-system intepretations (Hájek, 2012). In the case of RDM of particles, the propensity interpretation seems more appropriate. This means that the wave function in quantum mechanics should be regarded not simply as a description of the state of RDM of particles, but more suitably as a description of the instantaneous property of the particles that determines their random discontinuous motion at a deeper level. In particular, the modulus squared of the wave function represents the propensity property of the particles that determines the probability density that they appear in every possible group of positions

[m]Note that the relation between $j(x,t)$ and $\psi(x,t)$ depends on the concrete form of the external potential under which the studied system evolves, and the relation given below holds true for an external scalar potential. In contrast, the relation $\rho(x,t) = |\psi(x,t)|^2$ holds true universally, independently of the concrete evolution of the studied system.

in space.[n] In contrast, the position density and position flux density, which are defined during an infinitesimal time interval around a given instant, are only a description of the state of the resulting RDM of particles, and they are determined by the wave function. In this sense, we may say that the motion of particles is "guided" by their wave function in a probabilistic way.

4.4. *On momentum, energy and spin*

I have been analyzing RDM of particles in position space. Does the picture of random discontinuous motion exist for other observables such as momentum, energy and spin? Since there are also momentum wave functions etc in quantum mechanics, it seems tempting to assume that the above interpretation of the wave function in position space also applies to the wave functions in momentum space etc. This means that when a particle is in a superposition of the eigenstates of an observable, it also undergoes random discontinuous motion among the eigenvalues of this observable. For example, a particle in a superposition of momentum eigenstates also undergoes random discontinuous motion among all momentum eigenvalues. At each instant the momentum of the particle is definite, randomly assuming one of the momentum eigenvalues with probability density given by the modulus squared of the wave function at this momentum eigenvalue, and during an infinitesimal time interval around each instant the momentum of the particle spreads throughout all momentum eigenvalues.

Indeed, it can be seen that if an observable has a definite value at each instant, then it will also undergo random discontinuous change over time like position. The previous argument for RDM of particles, which is based on an analysis of the entangled states of a many-body system, applies to all observables of a quantum system which have a definite value at each instant. However, there is a well-known constraint on ascribing definite values to the observables of a quantum system. It is the Kochen-Specker theorem, which proves the impossibility of ascribing sharp values to all observables of a

[n] Note that the propensity here denotes single case propensity, as long run propensity theories fail to explain objective single-case probabilities. For a helpful analysis of the single-case propensity interpretation of probability in GRW theory see [15]. In addition, it is worth emphasizing that the propensities possessed by particles relate to their objective motion, not to the measurements on them. In contrast, according to the existing propensity interpretations of quantum mechanics, the propensities a quantum system has relate only to measurements; a quantum system possesses the propensity to exhibit a particular value of an observable if the observable is measured on the system [39, 40].

quantum system simultaneously, while preserving the functional relations between commuting observables [25].

There are two possible ways to deal with this constraint. The first way is to still ascribe sharp values to all observables of a quantum system and take all of them as physical properties of the system. This requires that the functional relations between commuting observables should not be always preserved for the values assigned to them.[o] There is an argument supporting this option. As Vink [42] has noticed, it is not necessary to require that the functional relations between commuting observables should hold for the values assigned to them for an arbitrary wave function; rather, the requirement must only hold for post-measurement wave functions (which are effectively eigenstates of the measured observable) and for observables that commute with the measured observable, which is enough to avoid conflicts with the predictions of quantum mechanics. In this way, the picture of RDM of particles may exist for all observables including momentum, energy and spin observables along all directions. But the functional relations between commuting observables will fail to hold for the values assigned to them in general.

The second way to deal with the constraint of the Kochen-Specker theorem is to ascribe sharp values to a finite number of observables of a quantum system and only take them as physical properties of the system. In this way, the functional relations between commuting observables can be preserved for the values assigned to them, while the picture of RDM exists only for a finite number of observables. The question is: Which observables?

First of all, since the proofs of the Kochen-Specker theorem do not prohibit ascribing sharp values to the position, momentum and energy (and their functions) of a quantum system simultaneously, the picture of RDM may exist also for momentum and energy. Next, it needs a more careful analysis whether the picture of RDM exists for spin. Consider a free quantum system with spin one. The Kochen-Specker theorem requires that only the system's squared spin components along a finite number of directions can be ascribed sharp values simultaneously (when the functional relations

[o]Note that incompatible observables can also have sharp values simultaneously and thus have a joint probability distribution when the functional relations between these observables are not preserved for the values assigned to them. The joint probability distribution of incompatible observables may be a product form, which fulfills the requirement that the marginal for every observable is the same given by quantum mechanics. The proofs of nonexistence of such a joint probability distribution require that the functional relations between these noncommuting observables should be preserved for the values assigned to them, which is a problematic assumption (see, e.g. [32, p. 95]).

between commuting observables are preserved). However, it can be argued that there is only one special direction for the spin of the system. It is the direction along which the spin of the system is definite. Thus it seems more reasonable to assume that only the spin observable along this direction can be ascribed a sharp value and taken as a physical property of the system.[p] Under this assumption, the picture of RDM does not exist for the spin of a free quantum system. But if the spin state of a quantum system is entangled with its spatial state due to interaction and the branches of the entangled state are well separated in space, the system in different branches will have different spin, and it will also undergo random discontinuous motion between these different spin states. This is the situation that usually happens during a spin measurement.

The key to decide which way is right is to determine whether the functional relations between commuting observables should be preserved for the values assigned to them for an arbitrary wave function. In my view, although this requirement is not necessary as Vink [42] has argued, there is no compelling reason to drop it either. Moreover, dropping this requirement and taking all observables including infinitely many related spin observables along various directions as physical properties of a quantum system seems to be superfluous in ontology and inconsistent with Occam's razor. Therefore, I think the second way is the right way to deal with the constraint of the Kochen-Specker theorem.

To sum up, I have argued that the picture of RDM also exists for some observables of a quantum system other than position, such as momentum and energy. But spin is a distinct property; the spin of a free quantum system is always definite along a certain direction, and it does not undergo random discontinuous motion.

---

[p]This analysis of spin also applies to projection operators on a system's Hilbert space (of dimension greater than or equal to 3). According to the Kochen-Specker theorem, only a finite number of projection operators can be ascribed sharp values simultaneously when the functional relations between commuting projection operators are preserved for the values assigned to them. Since there is no special direction for the set of orthogonal rays in the Hilbert space, all projection operators have the same status and no one is special. Therefore, it is arguable that the number of the projection operators which can be ascribed sharp values simultaneously should be either zero or infinity, depending on whether the functional relations between commuting observables are preserved for the values assigned to them. In particular, if this constraint for commuting observables is always satisfied, then projection operators on a system's Hilbert space do not correspond to properties of the system.

5. Particle ontology vs. field ontology

I have introduced a concrete ontological interpretation of the wave function in terms of RDM of particles. In some sense, the idea of RDM of particles can be regarded as a further development of Monton and Lewis's suggestion. The picture of RDM of particles clearly shows that interpreting the multi-dimensional wave function as representing the state of motion of particles in three-dimensional space *is* possible. In particular, the RDM of particles can explain quantum entanglement in a more vivid way, which thus reduces the force of the main motivation to adopt wave function realism. In any case, this new interpretion of the wave function in terms of RDM of particles provides an alternative to wave function realism.

Furthermore, the picture of RDM of particles has more explanatory power than wave function realism. It can more readily explain our three-dimensional impressions, which is still a difficult task for wave function realism. Moreover, it can also explain many fundamental features of the Schrödinger equation that governs the evolution of the wave function, which seem puzzling for wave function realism. For example, the existence of N particles in three-dimensional space for an N-body quantum system can readily explain why there are N mass parameters that are needed to describe the system, and why each mass parameter is *only* correlated with each group of three coordinates of the $3N$ coordinates on the configuration space of the system, and why each group of three coordinates of the $3N$ coordinates transforms under the Galilean transformation between two inertial coordinate systems in our three-dimensional space, etc.

Finally, it can be expected that the difference between particle ontology and field ontology may also result in different predictions that may be tested with experiments, and thus the two interpretations of the wave function can be distinguished in physics. The main difference between a particle and a field is that a particle exists only in one position in space at each instant, while a field exists throughout the whole space at each instant. It is a fundamental assumption in physics that a physical entity being at an instant has no interactions with itself being at another instant, while two physical entities may have interactions with each other. Therefore, a particle at an instant has no interactions with the particle at another instant, while any two parts of a field in space (as two local physical entities) may have interactions with each other. In particular, if a field is massive and charged, then any two parts of the field in space will have gravitational and electromagnetic interactions with each other.

Now consider a charged one-body quantum system such as an electron being in a superposition of two separated wavepackets. Since each wavepacket of the electron has gravitational and electromagnetic interactions with another electron, it is arguable that it is also massive and charged.[q] Besides this heuristic argument, protective measurements provide a more convincing argument for the existence of the mass and charge distributions of an electron in space [1, 2, 18]. As I have argued recently [19-21], when combined with a reasonable connection between the state of reality and results of measurements, protective measurements imply the reality of the wave function. For an electron whose wave function is $\psi(x)$ at a given instant, we can measure the density $|\psi(x)|^2$ in each position x in space by a protective measurement, and by the connection the density $|\psi(x)|^2$ is a physical property of the electron. Then, what density is the density $|\psi(x)|^2$? Since a measurement must always be realized by a certain physical interaction between the measured system and the measuring device, the density must be, in the first place, the density of a certain interacting charge. For instance, if the measurement is realized by an electrostatic interaction between the electron and the measuring device, then the density multiplied by the charge of the electron, namely $-|\psi(x)|^2 e$, will be the charge density of the electron in position x. This means that an electron has mass and charge distributions throughout space, and two separated wavepackets of an electron are massive and charged (see [19-21] for a more detailed analysis).

Let us now see how the difference between particle ontology and field ontology may result in different empirical predictions. According to the above analysis, if the wave function of an electron represents the state of RDM of a particle, then although two separated wavepackets of an electron are massive and charged, they will have no interactions with each other. This is consistent with the superposition principle of quantum mechanics and experimental observations. On the other hand, if the wave function of an electron represents a physical field, then since this field is massive

[q]One may object that if each wavepacket of the electron is massive and charged, then the two wavepackets will have gravitational and electromagnetic interactions with each other, but this is inconsistent with the superposition principle of quantum mechanics and experimental observations. But this objection is not valid. For if each wavepacket of the electron is not massive and charged, then how can it have gravitational and electromagnetic interactions with another electron? Rather, there should be a deeper reason why there are no interactions between the two wavepackets of the electron. As I will argue below, the reason is that what the wave function describes is not a field, but particles.

and charged, any two parts of the field will have gravitational and electromagnetic interactions with each other, which means that two separated wavepackets of an electron will have gravitational and electromagnetic interactions with each other. This is inconsistent with the superposition principle of quantum mechanics and experimental observations. Therefore, it is arguable that the interpretation of the wave function in terms of motion of particles, rather than wave function realism, is supported by quantum mechanics and experience.

6. Similar pictures of motion of particles

Although the picture of RDM of particles seems very strange, similar ideas have been proposed by other people for different purposes. In this section, I will briefly discuss these ideas.

6.1. *Epicurus's atomic swerve and Al-Nazzam's leap motion*

The ancient Greek philosopher Epicurus first considered the randomness of motion seriously. He presented the well-known idea of atomic "swerve" on the basis of Democritus's atomic theory [11]. Like Democritus, Epicurus also held that the elementary constituents of the world are atoms, which are indivisible microscopic bits of matter, moving in empty space. He, however, modified Democritus's strict determinism of elementary processes. Epicurus thought that occasionally the atoms swerve from their course at random times and places. Such swerves are uncaused motions. The main reason for introduing them is that they are needed to explain why there are atomic collisions. According to Democritus's atomic theory, the natural tendency of atoms is to fall straight downward at uniform velocity. If this were the only natural atomic motion, the atoms would never have collided with one another, forming macroscopic bodies. Therefore, Epicurus saw it necessary to introduce the random atomic swerves.

In order to solve Zeno's paradoxes, the 9th-century Islamic theologian Abu Ishaq Ibrahim Al-Nazzam proposed the idea of discontinuous motion, and called it the theory of "leap" [23, p. 258]. He argued that if a finite distance cannot be subdivided into a finite number of fractions but is subject to infinite divisibility, then an object in motion has to perform a leap (tafra) since not every imaginable point in space can be touched. In this way, "The mobile may occupy a certain place and then proceed to the third place without passing through the intermediate second place on the fashion of a leap." [23, p. 259]. Moreover, a leap from position A to position B consists of two interlocking sub-events; the original body in position A ceases to exist, and an "identical" body comes into being in position B.

It can be seen that neither Epicurus's atomic swerve nor Al-Nazzam's leap motion is exactly the RDM of particles; the atomic swerve is random but lack of explicit discontinuity, while the leap motion is discontinuous but lack of explicit randomness. In a sense, the picture of random discontinuous motion can be considered as an integration of Epicurus's random swerve and Al-Nazzam's discontinuous leap. However, the above reasons for introducing random and discontinuous motion are no longer valid as seen today.

6.2. *Bohr's discontinuous quantum jumps*

The picture of motion, which was proposed in modern times and more like random discontinuous motion, is Bohr's discontinuous quantum jumps.

In 1913, Bohr proposed what is now called the Bohr model of the atom [10]. He postulated that electrons as particles undergo two kinds of motion in atoms; they either move continuously around the nucleus in certain stationary orbits or discontinuously jump between these orbits. These discontinuous quantum jumps are supposed to be also truly random and uncaused. Indeed, Rutherford once raised the issue of causality. In a letter to Bohr dated 20 March 1913, he wrote:

> There appears to me one grave difficulty in your hypothesis, which I have no doubt you fully realize, namely, how does an electron decide what frequency it is going to vibrate at when it passes from one stationary state to the other? It seems to me that you would have to assume that the electron knows beforehand where it is going to stop. (Quoted in [36, p. 152-153])

Bohr admitted that "the dynamical equilibrium of the systems in the stationary states can be discussed by help of the ordinary mechanics, while the passing of the systems between different stationary states cannot be treated on that basis." [10, p. 7] However, he did not offer a further analysis of his discontinuous quantum jumps.

Although the electrons in Bohr's atomic model only undergo the random, discontinuous jumps occasionally (which is like Epicurus's atomic swerve), these jumps let us perceive the flavor of RDM of particles in the quantum world.

6.3. *Schrödinger's snapshot description*

In his 1927 Solvay congress report, Schrödinger gave a visualizable explanation of his interpretation of the wave function in terms of charge density.[r] He said,

> The classical system of material points does not really exist, instead there exists something that continuously fills the entire space and of which one would obtain a 'snapshot' if one dragged the classical system, with the camera shutter open, through *all* its configurations, the representative point in q-space spending in each volume element $d\tau$ a time that is proportional to the *instantaneous* value of $\psi\psi^*$. [6, p. 409]

If this description in terms of a snapshot is understood literally, then it seems to suggest that a quantum system is composed of point particles, and the continuous charge distribution of the system in space is formed by the ergodic motion of the system's point particles. Moreover, since the snapshot picture also requires that the spending time of the particles in each volume element is proportional to the instantaneous value of $\psi\psi^*$ there, it seems that the ergodic motion of the particles should be not continuous but discontinuous.

Indeed, what Schrödinger said later in his report seems to also support this analysis. He said,

> The pure field theory is not enough, it has to be supplemented by performing a kind of *individualisation* of the charge densities coming from the single point charges of the classical model, where however each single 'individual' may be spread over the whole of space, so that they overlap. [6, p. 415]

However, it is arguable that this picture of ergodic motion of particles was not the actual picture in Schrödinger's mind then, and he did not even mean by the above statement that the snapshot description is real. The reason is that Schrödinger later gave a clearer explanation of the above statement in the discussion of his report, which totally avoided the snapshot description. He said,

> It would seem that my description in terms of a snapshot was not very

[r]It is worth noting that Wallace and Timpson's [43] "spacetime state realism" can be regarded as a generation of Schrödinger's interpretation in some sense. Although this view may avoid the problems of wave function realism, it has the same problems as Schrödinger's interpretation. For a critical analysis of this view see [35].

> fortunate, since it has been misunderstood. Perhaps the following explanation is clearer. The interpretation of Born is well-known, who takes $\psi\psi^* d\tau$ to be the probability for the system being in the volume element $d\tau$ of the configuration space. Distribute a very large number N of systems in the configuration space, taking the above probability as 'frequency function'. Imagine these systems as superposed in real space, dividing however by N the charge of each point mass in each system. In this way, in the limiting case where $N = \infty$ one obtains the wave mechanical picture of the system. [6, p. 423]

Schrödinger never referred to his snapshot description, which he called a "somewhat naive but quite concrete idea", in his later publications.

It is interesting to note that like Schrödinger's snapshot description, Bell once gave a statement which may also suggest the picture of RDM of particles. In his well-known article "Against Measurement", after discussing Schrödinger's interpretation of the wave function in terms of charge density, Bell wrote:

> Then came the Born interpretation. The wavefunction gives not the density of stuff, but gives rather (on squaring its modulus) the density of probability. Probability of *what* exactly? Not of the electron *being* there, but of the electron being *found* there, if its position is 'measured'.
> Why this aversion to 'being' and insistence on 'finding'? The founding fathers were unable to form a clear picture of things on the remote atomic scale. [8]

It seems that by this statement Bell supported the following two claims: (1) the wave function should give (on squaring its modulus) the density of probability of the electron being in certain position in space, and (2) the wave function should also give the density of stuff. The key is to understand the meaning of probability in the first claim. If the probability means subjective probability, then the electron will be in a definite position which we don't know with certainty. But this result contradicts the second claim. The wave function of the electron giving the density of stuff means that the electron cannot be in only one position, but fill in the whole space in certain way with local density given by the modulus suqared of its wave function.

On the other hand, if the above probability means objective probability, then it seems that the two cliams can be reconciled and the resulting picture will be RDM of particles. The electron appears in every position in space with objective probability density given by the modulus squared of its wave function there, and during an arbitrarily short time interval such random discontinuous motion forms the density of stuff everywhere, which is also proportional to the modulus squared of the wave function of the electron.

6.4. *Bell's Everett (?) theory*

Although the above analysis may be an overinterpretation of Bell's statement, it can be argued that Bell's Everett (?) theory indeed suggests the picture of RDM of particles.

According to Bell [7], Everett's theory can be regarded as Bohm's theory without the continuous particle trajectories. Thus

> instantaneous classical configuration x are supposed to exist, and to be distributed in the comparison class of possible worlds with probability $|\psi|^2$. But no pairing of configuration at different times, as would be effected by the existence of trajectories, is supposed. [7]

Obviously, in Bell's Everett (?) theory, a quantum system is composed of particles, which have a definite position in space at each instant. Moreover, these particles jump among all possible configurations with probability $|\psi|^2$, and such jumps are random and discontinuous. This is clearly the picture of RDM of particles.

7. Conclusions

The physical meaning of the wave function is an important interpretative problem in the metaphysics of quantum mechanics. Notwithstanding more than eighty years' developments of the theory, it remains a hot topic of debate. In this chapter, I present an analysis of the meaning of the wave function, and the analysis leads to an ontological interpretation of the wave function in terms of random discontinuous motion of particles. According to this interpretation, quantum mechanics, like Newtonian mechanics, also deals with the motion of particles in space and time. Microscopic particles such as electrons are still particles with mass and charge, but they move in a discontinuous and random way. The wave function describes the state of the random discontinuous motion of particles, and in particular, the modulus squared of the wave function gives the probability density that the particles appear in every possible group of positions in space. At a deeper level, the wave function may represent the propensity property of the particles that determines their random discontinuous motion. Quantum mechanics, in this way, is essentially the laws of random discontinuous motion of particles. It is a further question whether the suggested particle ontology is complete in accounting for our definite experience and whether it needs to be revised in the relativistic domain.

Acknowledgements

This work is supported by the National Social Science Foundation of China (Grant No. 16BZX021).

References

1. Aharonov, Y., Anandan, J. and Vaidman, L. (1993). Meaning of the wave function. *Phys. Rev. A* 47, 4616.
2. Aharonov, Y. and Vaidman, L. (1993). Measurement of the Schrödinger wave of a single particle. *Phys. Lett. A* 178, 38.
3. Albert, D. Z. (1996), Elementary Quantum Metaphysics. in J. Cushing, A. Fine and S. Goldstein (eds.), Bohmian Mechanics and Quantum Theory: An Appraisal. Dordrecht: Kluwer, 277–284.
4. Albert, D. Z. (2013). Wave function realism. In Ney and Albert (2013), pp. 52–57.
5. Albert, D. Z. (2015). After Physics. Cambridge, MA: Harvard University Press.
6. Bacciagaluppi, G. and Valentini, A. (2009). Quantum Theory at the Crossroads: Reconsidering the 1927 Solvay Conference. Cambridge: Cambridge University Press.
7. Bell, J. S. (1981). Quantum mechanics for cosmologists. In C. J. Isham, R. Penrose, and D. W. Sciama (eds.), Quantum Gravity 2: A Second Oxford Symposium. Oxford: Oxford University Press, pp. 611–637.
8. Bell, J. S. (1990). Against 'measurement', in A. I. Miller (ed.), Sixty-Two Years of Uncertainty: Historical Philosophical and Physics Enquiries into the Foundations of Quantum Mechanics. Berlin: Springer, pp. 17–33.
9. Belot, G. (2012). Quantum states for primitive ontologists: A case study. *European Journal for Philosophy of Science* 2, 67–83.
10. Bohr, N. (1913). On the constitution of atoms and molecules. *Philosophical Magazine* 26, 1–25.
11. Englert, W. G. (1987). Epicurus on the Swerve and Voluntary Action. Atlanta: Scholars Press.
12. Forrest, P. (1988). Quantum Metaphysics. Oxford: Blackwell.
13. French, S. (2013). Whither wave function realism? In Ney and Albert (2013), pp. 76–90.
14. French, S. and Krause, D. (2006). Identity in Physics: A Historical, Philosophical, and Formal Analysis. Oxford: Oxford University Press.
15. Frigg, R and C. Hoefer (2007). Probability in GRW Theory. *Studies in the History and Philosophy of Modern Physics* 38, 371–389.
16. Gao, S. (2011). Meaning of the wave function. *International Journal of Quantum Chemistry* 111, 4124–4138.

17. Gao, S. (ed.) (2014a). Protective Measurement and Quantum Reality: Toward a New Understanding of Quantum Mechanics. Cambridge: Cambridge University Press.
18. Gao, S. (2014b). Reality and meaning of the wave function. In Gao (2014a), pp. 211–229.
19. Gao, S. (2015). An argument for ψ-ontology in terms of protective measurements. *Studies in History and Philosophy of Modern Physics* 52, 198–202.
20. Gao, S. (2017). The Meaning of the Wave Function: In Search of the Ontology of Quantum Mechanics. Cambridge: Cambridge University Press.
21. Gao, S. (2020). Protective Measurements and the Reality of the Wave Function. *The British Journal for the Philosophy of Science*, axaa004, https://doi.org/10.1093/bjps/axaa004 (2020).
22. Hájek, A. (2012). Interpretations of Probability, The Stanford Encyclopedia of Philosophy (Winter 2012 Edition), Edward N. Zalta (ed.), http://plato.stanford.edu/archives/win2012/entries/probability-interpret/.
23. Jammer, M. (1974). The Philosophy of Quantum Merchanics. New York: John Wiley and Sons.
24. Kline, M. (1990). Mathematical Thought from Ancient to Modern Times. Oxford: Oxford University Press.
25. Kochen, S. and Specker, E. (1967). The problem of hidden variables in quantum mechanics. *Journal of Mathematics and Mechanics* 17, 59–87.
26. Lewis, P. J. (2004). Life in configuration space. *British Journal for the Philosophy of Science* 55, 713–729.
27. Lewis, P. J. (2013). Dimension and illusion, in A. Ney, D. Albert (eds.), The Wave Function, Oxford: Oxford University Press, pp. 110–125.
28. Maudlin, T. (2013). The nature of the quantum state, in A. Ney, D. Albert (eds.), The Wave Function, Oxford: Oxford University Press, pp. 126–153.
29. Monton, B. (2002). Wave function ontology. *Synthese* 130, 265–277.
30. Monton, B. (2013). Against 3N-dimensional space. In A. Ney, D. Albert (eds.), The Wave Function, Oxford: Oxford University Press, pp. 154–167.
31. Myrvold, W. C. (2015). What is a wavefunction? *Synthese* 192, 3247–3274.
32. Nelson, E. (2001). Dynamical Theories of Brownian Motion. 2nd edition. Princeton: Princeton University Press. Available at https://web.math.princeton.edu/~nelson/books/bmotion.pdf.
33. Ney, A. (2012). The status of our ordinary three dimensions in a quantum universe. *Noûs* 46, 525–560.
34. Ney, A. and Albert, D. Z. (eds.) (2013). The Wave Function: Essays on the Metaphysics of Quantum Mechanics. Oxford: Oxford University Press.
35. Norsen, T. (2016). Quantum solipsism and non-locality. In M. Bell and S. Gao (eds.). Quantum Nonlocality and Reality: 50 Years of Bell's theorem. Cambridge: Cambridge University Press.
36. Pais, A. (1991). Niels Bohr's Times: In Physics, Philosophy, and Polity. Oxford: Oxford University Press.
37. Pusey, M., Barrett, J. and Rudolph, T. (2012). On the reality of the quantum state. *Nature Phys.* 8, 475–478.

38. Solé, A. (2013). Bohmian mechanics without wave function ontology. *Studies in History and Philosophy of Modern Physics* 44, 365–378.
39. Suárez, M. (2004). Quantum selections, propensities and the problem of measurement. *British Journal for the Philosophy of Science* 55(2), 219–255.
40. Suárez, M. (2007). Quantum Propensities. *Studies in the History and Philosophy of Modern Physics* 38, 418–438.
41. Teller, P. (1986). Relational Holism and Quantum Mechanics. *British Journal for the Philosophy of Science* 37, 71–81.
42. Vink, J. C. (1993). Quantum mechanics in terms of discrete beables. *Physical Review A* 48, 1808.
43. Wallace, D. and C. Timpson (2010). Quantum mechanics on spacetime I: spacetime state realism. *British Journal for the Philosophy of Science* 61, 697–727.

THE ELIMINATION OF METAPHYSICS THROUGH THE EPISTEMOLOGICAL ANALYSIS: LESSONS (UN)LEARNED FROM METAPHYSICAL UNDERDETERMINATION

RAONI W. ARROYO

Department of Philosophy, Communication and Performing Arts,
Roma Tre University,
Rome, Italy
Centre for Logic, Epistemology and the History of Science,
University of Campinas,
Campinas, São Paulo, Brazil
Fellow Researcher of the São Paulo Research Foundation (FAPESP)
Research Group in Logic and Foundations of Science (CNPq)
International Network on Foundations of Quantum Mechanics and Quantum Information
E-mail: rwarroyo@unicamp.br

JONAS R. BECKER ARENHART

Department of Philosophy,
Federal University of Santa Catarina,
Florianópolis, Santa Catarina, Brazil
Graduate Program in Philosophy,
Federal University of Maranhão,
São Luís, Maranhão, Brazil
Research Group in Logic and Foundations of Science (CNPq)
International Network on Foundations of Quantum Mechanics and Quantum Information
E-mail: jonas.becker2@gmail.com

DÉCIO KRAUSE

Graduate Program in Logic and Metaphysics,
Federal University of Rio de Janeiro,
Rio de Janeiro, Brazil
Research Group in Logic and Foundations of Science (CNPq)
International Network on Foundations of Quantum Mechanics and Quantum Information
E-mail: deciokrause@gmail.com

This chapter argues that the general philosophy of science should learn metaphilosophical lessons from the case of metaphysical underdetermination, as it occurs in non-relativistic quantum mechanics. Section 2 presents the traditional discussion of metaphysical underdetermination regarding the individuality and non-individuality of quantum particles. Section 3 discusses three reactions to it found in the literature: eliminativism about individuality; conservatism about individuality; eliminativism about objects. Section 4 wraps it all up with metametaphysical considerations regarding the epistemology of metaphysics of science.

Keywords: Individuality; metaphysical underdetermination; metaphysics of quantum mechanics; non-individuals; structuralism.

1. Introduction

The underdetermination of theory by data is a familiar topic in the general philosophy of science. It arises from the fact that there can be, in principle, more than one scientific theory that explains the same phenomena/data [73]. At first, this kind of underdetermination was thought to be hypothetical at best [72], but nowadays it is well-established that quantum mechanics exemplifies it well [18], viz. with the solutions to the measurement problem [60, 9] e.g. Bohmian mechanics, Everettian quantum mechanics, and collapse-based quantum theories [28]. Metaphysical underdetermination is a further problem. It appears when a scientific theory is compatible with more than one metaphysical profile, and it is widely known that quantum mechanics *also* exemplifies it well with the discussion concerning the fact that quantum objects can be metaphysically understood both as individuals *and* as non-individuals [30, 34].

The first kind of underdetermination clearly supports an anti-realist argument against scientific realism. After all, if scientific realism is roughly defined as the stance according to which scientific theories are (in some sense)[a] *true* descriptions of the world. Thus, in case of two (or more) competing theories to account for the same phenomena — often positing different entities in its ontological catalog and different dynamics with different state spaces and axioms [25, 69, 9] — it is not clear how to choose which one is true in a sense that is *of interest* for scientific realists. That is, the first kind of underdetermination prevents us from saying how the world looks like from the point of view of the ultimate furniture of the world.

[a]It is up for grabs whether 'truth' should be understood in a correspondentist sense [57] or in a partial/quasi-truth sense [26]. We shall not dwell on such matters here, however.

When it comes to the second kind of underdetermination, however, it is not obviously true that anti-realism may reap any benefit specifically from it. In this chapter, we hope to clarify the lessons that the philosophy of science may learn from such kind of underdetermination. Section 2 presents the traditional debate over quantum individuality and the birth of the term "metaphysical underdetermination". Section 3 discusses the reactions to metaphysical underdetermination: one that employs extra-empirical virtues to favor the non-individuality metaphysical profile [27, 51]; one that justifies the use of individuality [65]; finally, the one that considers metaphysical underdetermination to be an argument *for* scientific realism [36, 41], as long as the realist content concerns structures, not objects [58, 31, 32]. Finally, section 4 wraps it all up with epistemic considerations, viz. that once physics does not decide between metaphysical profiles, are we epistemically justified in adopting a philosophical attitude of (in)tolerance towards them [19]?

2. Whence metaphysical underdetermination?

Both classical and quantum particles are indistinguishable regarding state-independent properties (e.g., rest mass, electric charge). There is, however, a difference in how classical and quantum particles behave collectively. Such difference, captured by different statistical descriptions, is the source of the debate on quantum metaphysics that we address in this section. Since its earliest formulations, quantum-mechanical systems are known to obey the "indistinguishability postulate (IP)" defined as follows.

> Particles are indistinguishable if they satisfy the indistinguishability postulate (IP). This postulate states that all observables O must commute with all particle permutations $P : [O, P] = 0$.[b] Put informally, the IP is the requirement that no expectation value of any observable is affected by particle permutations. [47, p. 312].

Let us consider the case of a system composed of n indistinguishable particles, each particle $i \leq n$ having $\mathcal{H}_i$ as its state space. The state space of the whole system, described by $\mathcal{H}_n$, is written as:

$$\mathcal{H}_n \equiv \bigotimes_{i=1}^{n} \mathcal{H}_i \tag{1}$$

[b]N.B.: here there is a typical abuse of notation. Officially, $[O, P]$ is an operator, not a scalar.

Assume that $\mathcal{H}_n$ is closed under the permutation operator P_{ij} which exchanges the n factors of $\mathcal{H}_n$, i.e., ith and jth copies in $\mathcal{H}_n$. For $n = 2$:

$$P_{12}\big(|\varphi\rangle \otimes |\psi\rangle\big) = |\psi\rangle \otimes |\varphi\rangle \tag{2}$$

Moreover, the IP says that the tensor product is not commutative so that $\langle\psi|\hat{A}|\psi\rangle = \langle P\psi|\hat{A}|P\psi\rangle$. The theoretical consequences of the IP are usually presented through the counting process for particles by comparing the Maxwell–Boltzmann (classical) statistics with the Bose–Einstein and the Fermi–Dirac (quantum) statistics, which enables one to visualize what is at stake. Suppose one wishes to arrange two particles, A and B in two boxes, 1 and 2, as depicted in Table 1.

Table 1. Statistics for particles in boxes.

	Box 1	**Box 2**
1.	AB	
2.		AB
3.	A	B
4.	B	A

Within the Maxwell–Boltzmann statistics, there are four options to arrange the particles, to each of which we assign the equal probability weight of $\frac{1}{4}$. Regarding intuitiveness, so far so good. In the quantum case, due to IP, things are slightly different. Since quantum particles of the same kind are *indistinguishable*, one cannot assign any discerning role to the labels attributed to them — such as "A" and "B" — in any physically meaningful way. While the first and second options resemble the Maxwell–Boltzmann probability weight of $\frac{1}{4}$ for each, the third and fourth options do not. Due to IP, one can permute particles A and B without changing the system's state; in the case of fermions, the state changes signal, but its square, which stands for the probability, is the same. So, instead of attributing a probability weight of $\frac{1}{4}$, we write:

$$\frac{1}{\sqrt{2}}\Big(|A_1\rangle \otimes |B_2\rangle \pm |A_2\rangle \otimes |B_1\rangle\Big) \tag{3}$$

for cases 3 and 4. That's the theoretical difference between classical and quantum statistics in a nutshell — the addition being for bosons and the subtraction for fermions. To put it bluntly, what matters for our purposes is to state the fact that, from the experimental point of view, in the quantum case the permutation of particles does not give rise to a different physical situation. But in the classical case, it does: exchanging both particles A and B generates a different physical situation.

There are two standard ways of accommodating these matters in metaphysical terms regarding individuality. The first is to point out that in the classical realm, the difference in weighting probability in cases 3 and 4 implies that classical particles are individual objects in a metaphysical sense. Moreover, such individuality must be grounded in *something way beyond* the particles' state-independent properties, since they are indistinguishable with regard to those too. Even if they are absolutely indistinguishable, classical objects (e.g., A and B) are typically regarded as individuals; that is, there is some metaphysical feature that accounts for the fact that A is A, not B and vice versa. This would explain the statistical difference between 3. and 4. Due to Post [66], this 'something else' grounding individuality is known as "transcendental individuality" [33, 30], and it is typically framed in terms of an additional non-qualitative ingredient, like a substratum or an individual essence. Other options consist in attributing a unique space-time position to each particle and backing it with the Impenetrability Assumption, granting thus what is called 'space-time individuality' [30, chap. 2]. Quantum particles, by contrast, are objects that by virtue of assigning the same probability weight to cases 3 and 4 are said to have lost their individuality. To some founding fathers of quantum mechanics, such as Schrödinger [71, p. 197], this was enough to state that quantum particles are non-individual objects [75]. That is because quantum objects obey the IP; the metaphysical lesson to be learned is that the metaphysics of individuality should be revised (or abandoned) thoroughly. This is the standpoint of the "Received View" on quantum individuality [4], nowadays also called the "orthodoxy" [13]: quantum mechanics forces us to abandon the universal application of our notion of metaphysical individuality.

The second way to accommodate the odd statistics of quantum-mechanical systems is to resist the revisionary metaphysical maneuver. As argued by French [33, pp. 442–444] (see also [30]), quantum particles can *also* be seen as individuals, just like classical particles. In such a scenario, asymmetric states formed by permutations of particles' labels would not be counted not because they do not exist, but because they would not be available for the particles; such states are mathematically available, but not physically available, given that they violate the permutation symmetry required from quantum particles [30, p. 148]. Quantum statistics, in this view, would differ from the classical case, not because of objects lacking individuality principles, but in terms of accessibility of states, which is restricted to certain subspaces $\mathcal{H}$ — so the labels *can* reflect the individuality of quantum systems. In metaphysical terms, this individuality may be cashed out in terms of transcendental individuality, such as haecceity, primitive individuality, or by the substratum theory. As remarked by Caulton [21, pp. 581–582], there is a strict link *between haecceitism and anti-haecceitism* and *transcendental individuality (TI) and qualitative individuality (QI)*, both defined as follows in permuted states $\mathfrak{W}$ and the states' common domain D:

> (*Haecceitism*) Any two distinct states permutable in $\mathfrak{W}$ represent distinct possibilities.
> (*Anti-haecceitism*) Any two distinct states permutable in $\mathfrak{W}$ represent the same possibility.
> [...]
> (TI) Each label in D denotes some object of the target system, and that label denotes the same object in all states.
> (QI) The labels in D denote no object in particular of the target system. [21, pp. 581–582, original emphasis]

What is more relevant for us now is the fact that quantum theory, by itself, does not provide any kind of support for any of the options. Such a situation concerning the metaphysics of quantum objects enables French and Krause [30, p. xiii, see also chap. 4] to state that "[...] there exists a form of 'metaphysical underdetermination' between two conceptual packages which are both supported by the physics: particles-as-individuals and particles-as-non-individuals". What is important to stress, again, is that quantum mechanics is in no position to decide between them, i.e. the metaphysics of individuality is underdetermined by quantum mechanics. Furthermore, from now on we will use the taxonomy offered by Krause and Arenhart [50], according to which there are three key concepts in this debate, each related to a distinct philosophical dimension: identity (a logical relation), individuation (a matter of epistemology), individuality (a non-relation property, a concern for metaphysics). So one could say, for the sake of an example, that quantum objects can be dealt with by a logical system with *identity*, even though they cannot be *discerned* or *individuated* in certain physical circumstances (e.g. when in Bose–Einstein condensates), and still maintain on the top of that a metaphysical profile of individuality or non-individuality. This is what concerns us here. As the taxonomy employed makes it possible to discuss each of these components (which are often confused or, at least, used interchangeably in the literature), the subject of this chapter is exclusively metaphysical, viz. on the question of individuality. As discussed in Krause and Arenhart [50], individuality, the metaphysical dimension, may be related to identity and discernibility, so that we have the qualitative versions of individuality, or else it may not be so related, in case we have versions of transcendental individuality.

This characteristic of the metaphysics of quantum mechanics generated at least three types of philosophical reactions, which range from considering it to be a feature or a bug. Some anti-realists consider metaphysical underdetermination to be a bug, so we should say "good-bye to metaphysics" [74, p. 480]. Surely this is a stance one can voluntarily adopt [23]; one can even decide to be a realist about e.g. electrons but refuse to ask metaphysical questions about individuality [67]. This would put the scientific realist in the 'shallow' spectrum of scientific realism [59, 38, 2]. The problem with this attitude is that without metaphysics, it is said, one loses sight of what one is being a realist about [22], thus decreasing our overall understanding of the theory [36]. One alternative to it is to point out that one can *also* voluntarily bite the bullet and keep asking metaphysical questions concerning the individuality profile of quantum objects — a move which would put the scientific realist in the 'deep' part of the spectrum [59, 38, 2]. The particular debate on deep realism for (non-)individuality in the context of quantum mechanics is grounded in the notion put forth by Brading and Skiles [15] that objects must have their "individuality profile" spelled out, otherwise, we cannot make sense of objects whatsoever.

There are at least two realistic ways to proceed from there, which we'll cover in the next section.

3. Three reactions

The general message of metaphysical underdetermination for the philosophy of general science is this: as we are not able to empirically resolve the question of which metaphysical profile is the true one so that we can understand the nature of the entities with which scientific theories are committed, the question must appeal to non-empirical criteria.

Before doing that, however, one might be tempted to suggest leaving everything to the experimental part to decide. This is because the history of science has shown us that, through experiments, we can at least eliminate most of the possible metaphysics — a great example would be the local metaphysics associated with standard quantum mechanics after Bell's theorems and Aspect's experiments [10] —, so it is not vindictive to say that we should completely rule out empirical criteria for metaphysical theory decisions. The difficulty of this suggestion is twofold: on the one hand, the experience is compatible with the metaphysical theories discussed here, viz. from particles-as-individuals and particles-as-non-individuals; on the other hand, if we adopt only what is authorized by experience, we end up with no metaphysics at all, with the constructive empiricists. And as we recall, the whole point was based on providing a realistic perspective on the metaphysics of quantum mechanics. Indeed, perhaps future experiments will falsify some of these metaphysical alternatives. But the point is that this is not the case now: the metaphysics of individuals, non-individuals, and structures are on an experimental equal footing. So we have two alternatives: 1) adopt "quietism" [76] towards metaphysics until we have a final and fundamental physical theory [61]; 2) opt for non-empirical criteria to decide on metaphysical theories. As a working hypothesis, we will assume the second.

Generally, such criteria are pragmatic, aesthetic, and/or metaphysical, such as simplicity (pragmatic), indispensability (metaphysics), beauty (aesthetic), and compatibility with other scientific theories (epistemic). In this section, we'll analyze the following realist reactions to metaphysical determination: to consider it to be a bug, so we have to break the underdetermination in order to be a realist about them (about the particles-as-non-individuals horn or the particles-as-individuals horn); to consider it to be a feature, so we have to embrace it to move to another form of realism.

3.1. *Non-individuality horn*

Let us recall the claims made by the Received View: quantum mechanics forces us to revise our metaphysical notions. Instead of building a notion of super-individuals, we are suggested to abandon it for good. One may take this idea literally, and benefiting from the fact that identity and individuality are generally brought together in these contexts, suggest a revision of the logic of identity [4]. For instance, in classical logic, reflexivity of identity, viz. $x = x$ is trivially true, and if that is related to individuality, perhaps, failure of the latter requires failure of the former, no? But how can that be? As French and Krause stress:

> [...] the notion of non-individuality can be captured in the quantum context by formal systems in which self-identity is not always well-defined, so that the reflexive law of identity, namely, $\forall x(x = x)$, is not valid in general. [30, p. 13–14]

Logical systems in which the principle of reflexivity is relaxed are called "non-reflexive logics", which made, as French [39] put it, the notion of non-individuality "philosophically respectable"; even more: as French [36, p. 36] emphasizes, "[...] without them, this metaphysical position — of quantum particles as non-individuals — might not be treated as a viable 'horn' of an underdetermination argument at all".

Briefly, some self-avowed theoretical (albeit non-empirical) virtues of this formulation of the Received View (viz. particles-as-non-individuals) are:

Simplicity As it is well-known, the development of quasi-set theory [52, 53, 30] enabled one to treat indistinguishability and non-individuality *right from the start* in quantum mechanics. As an alternative to standard non-relativistic quantum mechanics, non-reflexive quantum mechanics [49] employs quasi-set theory as the theory's logical foundation. Regarding mathematical simplicity, we should favor the Received View.

Unity of science. Ionization is a chemical process by which an atom acquires a negative charge by gaining electrons and acquires a positive charge by losing electrons. As Krause [55, 54] emphasizes, chemistry wouldn't work properly with the assumption that the electron the atom loses is different from the electron the atom gains in the ionization process, viz. with the assumption that electrons are *individuals*. Regarding the unity of science, we should favor the Received View.

Intuitiveness. Due to the Kochen–Specker theorem, it is impossible to ascribe certain measurement/context-independent properties to quantum systems, which is a highly counter-intuitive result. As recently argued by de Barros, Holik, and Krause [27], one could attempt to argue that the Kochen–Specker theorem wouldn't hold if each measurement situation is considered to be indiscernible rather than identical, which is a feature that quasi-set theory and non-individuality can easily accommodate. So, regarding intuitiveness, we should favor the Received View.

3.1.1. *Logical underdetermination*

Regarding the abandonment of classical logic to cope with quantum non-individuality, Bigaj [13, p. 42, fn. 24] states that "[...] quantum mechanics does not force us to adopt such a radical view". To Bigaj,

> [...] quantum particles are definitely not individuals in the classical sense, [but] their non-individuality can hopefully be expressed without sacrificing classical logic that has served us well for two millennia. [13, p. 245]

The upshot is that it would be nice if we stick with classical logics, so we wouldn't need to change logical systems *because of* the impositions of quantum mechanics. However, as shown by Arenhart [6, p. 394], *classical* systems with "[...] first- and second-order languages may be provided with unintended interpretations of identity which may do the job of failing identity and play the role of an indiscernibility relation". So there's no *need* to change logics, strictly speaking, even if one defends the particles-as-non-individuals metaphysical package.

Bigaj's claim that quantum mechanics doesn't force us to adopt a non-classical logic can be considered on the following grounds: maybe he is right *if* the goal is just to get the probabilities of measurements, in a quite *instrumentalist* view in Bohr's [14] sense, that is, the use of classical logic and standard mathematics is compatible with the activities of the 'practical' physicist; any book of QM you can find in your library is based on standard mathematics, hence in classical logic. But the problem can be viewed from another point of view, that of *foundations*, mainly *philosophical foundations*. Here the goals are different, not that of the practitioner physicist, but of the philosopher interested in the meta-study of the theory itself, or better, of the cluster of theories we usually call "quantum mechanics". From this point of view, it is really fundamental to analyze the logical basis of the theory, and, as far as things go, apparently, the use of non-classical logics seems relevant and perhaps even necessary.

3.2. *Individuality horn*

Quantum objects can also be cashed out as individuals, in metaphysical terms. As we already mentioned, one can metaphysically dress their individuality profile as individuals either as primitives [62] or in haecceistic terms: a quantum object has the quintessential/transcendental feature of being itself an individual. In metaphysical debates, this is known as the "substratum theory" of individuality, which states that individuality is a feature that obtains for objects over and above their properties. The substratum theory is an example of the Transcendental Individuality account. However, another account of individuality is available in textbook metaphysics: the "bundle theory" of individuality. In it, an individual is characterized *solely* by its properties. So, if two objects share all their properties, they have to be one and the same. This, of course, assumes the validity of Leibniz's Principle of the Identity of Indiscernibles (PII). However, as we mentioned above, quantum objects are indiscernible with regard to their state-independent properties. In a very rough reading, one could argue that quantum objects *refutes* the PII, so bundle theories are not options to metaphysically interpret quantum mechanics [30]; alternatively, one could argue that, due to the PII, there is only one quantum object [29].

Somewhat recent developments in the foundations of quantum mechanics showed, however, that the PII can be tailored for quantum-mechanical purposes. Quantum objects are said to be "weakly discernible" [70, 64, 65, 48]; that means that quantum entities may be discerned because entering into irreflexive and symmetric relations (and that is what 'weakly discerning relations' mean). Given symmetry, a form of permutation is available: if x is R related to y, by symmetry, y is R related to x. Irreflexivity, on the other hand, accounts for the difference: if R is a weakly discerning relation and x is R related to y, then, due to irreflexivity, it is not the case that xRx and yRy, so that $x \neq y$. Such relations, then, are said to contribute to distinguishing two entities, saving a new version of the PII.

The traditional example concerns electrons' spin. Once electrons share the state-independent value of spin $\frac{1}{2}$, electrons cannot have an opposite spin value to themselves in a given axis. So, electrons A and B can be said to have opposite spin values from each other — which is a way to discern them; electron A has (say) opposite spin to electron B in such a manner that A and B don't share the same properties anymore. So, as the argument goes, one can apply the PII to them. The conclusion is then that quantum mechanics is compatible with the PII, so bundle theory is indeed a metaphysical option to the individuality metaphysical profile of quantum objects.

On that note, Caulton [20] and Bigaj [12] have recently argued that quantum particles are not indiscernible after all — they are indeed absolutely discernible. One might argue — in fact, it has been [13] — that this seems to be crucial for the metaphysical debate on quantum (non-) individuality, as it might be used to undermine the relevant form of underdetermination, viz. metaphysical underdetermination.

However, the *epistemic* notions of (in)discernibility do not need to be seen as determining the *metaphysical* issues concerning (non-)individuality. They can, but still, that connection itself is made in the discussion of the metaphysics; it does not come from quantum mechanics itself (see also [51]).

As we apply the terminology, nothing — except philosophical preferences — determines the metaphysical profile of individuality or non-individuality.

The issue is similar to the remarks already advanced in the cases of weak discernibility for quantum particles. While one may attempt to connect this kind of discernibility with some approach to individuality directly related to discernibility, one is not so obliged by quantum theory; that is a decision of metaphysical nature. As Arenhart [5, p. 110] puts it, "[...] there are many distinct and incompatible ways to put metaphysical flesh on the bare relational bones of weak discernibility". That is, even if one considers that the obtaining of weakly discerning relations may be derived from the quantum mechanical formalism, the metaphysical lesson about individuality does not result obvious. Let us mention how that would work. One could hold, for instance, that due to what 'individuality' actually means, weakly discerning relations are not enough to ground any kind of individuality worth its name. In a first approach, one could hold that individuality is conferred by some non-relational property. So, in the quantum case, although a weak version of PII could be seen as saved, individuality is still absent. For instance, due to the indistinguishability of quantum entities by their properties. In a second approach, one could still hold that individuality is granted by a transcendental principle of individuality, even though weak discernibility holds. In both cases, weakly discerning relations are not doing the expected metaphysical job. Also, in both cases, any metaphysical lesson drawn from weak discernibility must be added from outside of quantum mechanics.

In this way, the particles-as-individuals horn has its own metaphysical underdetermination: are quantum objects individuals-as-substrata or individuals-as-bundles? Hence, *even if* one assumes that there are undisputed cases of empirical criteria favoring the metaphysical package of particles-as-individuals, there is still not a clear picture in sight for us to understand what is the metaphysical profile for quantum objects' individuality.

3.3. *Structuralist horn*

Recall that the 'need' to ascribe a metaphysical profile concerning the (non-) individuality of quantum objects is to be *realist enough* about them. The claim, recall, is that without the metaphysical layer which floats free from what physics tells us about the nature of the entities with which it is ontologically committed to [7], one cannot be considered to be realist enough about them. As French [36, p. 50] argues, without the metaphysical import of individuality or non-individuality for objects, one cannot understand what objects *are*, hence one cannot adopt a legitimate realist stance towards it. The metaphysical profile of individuality or non-individuality is, then, indispensable for object-oriented scientific theories [15, 36]. It was van Fraassen [74, pp. 480–482] who first pointed out that the metaphysical underdetermination between individuality and non-individuality challenged the adoption of scientific realism; but as French [36, p. 37] argues that a particular form of scientific realism is challenged by metaphysical underdetermination: "object-oriented realism". What if one changes the ontological basis of scientific theories? Here's Ladyman [58]:

> It is an ersatz form of realism that recommends belief in the existence of entities that have such ambiguous metaphysical status. What is required is a shift to a different ontological basis altogether, one for which questions of individuality simply do not arise. [58, pp. 419–420]

In this passage, Ladyman [58] uses the terms 'ontology' and 'metaphysics' interchangeably, which is not advisable insofar as it is a source of several problems in the metaphysics of science [8]. However, the point is that we should take metaphysical underdetermination as a "motivational device" to a dash of realism based on structures rather than objects, viz. structural realism [35, 36, 41]. There are several forms of structural realism [56], our focus here being the ontological form of structural realism. But it wouldn't suffice to say that one believes in the structural components that remain through scientific changes; one must account for the metaphysical imports of what structures are. Otherwise, structuralists would remain in the shallow part of scientific realism, which is not advisable by their own standards of a genuine form of scientific realism [36]. One way to put it is to consider structures to be *fundamental* entities [43], to the point that objects are eliminable from an ontological point of view [40].

This, however, seems to be little informative about the metaphysical nature of structures, as Arenhart and Bueno [3] stressed. On this point, structuralists argue that the literature is not being fair to them by asking about the nature of structures. For instance, French writes that:

> [...] there's a certain asymmetry in the debate whereby structural realists are (constantly) asked 'what is structure?' but their non-structural friends and colleagues are almost never required to give an answer to the corresponding question 'what is an object?'. And of course it is not as if the answer to the latter is utterly straightforward. [42, pp. 4–5].

Let us recall, however, that the search for the individuality profile is the search for a metaphysical characterization of what objects are. Metaphysical underdetermination just happens to block such a way, but surely the individuality profile is not exhaustive concerning the metaphysical nature of objects. One can ask e.g. about *contrast* and/or *extension* [68]. When concerning metaphysics — or, as we are calling it here, the "metaphysical profile" —, Chakravartty [24, p. 12] already recalled that "[i]t is *always* possible to ask finer-grained questions [...]".

As Bianchi and Gianotti [11] emphasize, however, structuralists usually take structures to be the kind of entities metaphysically characterized extensionally: extrinsic properties, relations, symmetry groups, etc. To put it in another way, there is still metaphysical underdetermination concerning how should we understand 'structures' properly in metaphysical terms [42]. How's that different from object-oriented realism? It is unclear why we should favor structure-oriented realism over object-oriented realism *because of* metaphysical underdetermination, as both scientific-realist stances fall prey to metaphysical underdetermination.

4. Lessons (un)learned: The elimination of metaphysics

Given those discussions, the point is: as there are no empirical factors that can decide between these metaphysical views, all we have are pragmatic/aesthetic criteria for adopting one metaphysical profile or another. However, none of these solutions is final, as pragmatic values are not conducive to truth. Thus, what remains is to adopt an "irenic" attitude, à la Carnap, about such metaphysical proposals. As the famous "Principle of Tolerance" recommends:

> Let us grant to those who work in any special field of investigation the freedom to use any form of expression which seems useful to them; the work in the field will sooner or later lead to the elimination of those forms which have no useful function. [19, p. 40].

Thus, it seems that the maximum epistemic justification that one can have in the face of such metaphysical profiles is the acceptance of their empirical adequacy, and not the leading to the truth that such metaphysical profiles could supposedly have. To adopt such a stance, however, is something that constructive empiricists have always recommended.

But is that enough, given the purposes we started with? It seems that a stance accommodates the problem, but does not solve it. One way of perceiving the discomfort of such empiricist accommodation is through the notion of "understanding", as used by neo-Pyrrhonists such as Bueno:

> We obtain an understanding of the various possibilities that are available to make sense of the issues under consideration and the insights such possibilities offer even if neo-Pyrrhonists are unable in the end to decide which of them (if any) is ultimately correct. [17, p. 13].

It is difficult to see how the multiplicity of metaphysical options, however, gives us understanding rather than misunderstanding. After all, having several incompatible options for understanding what science tells us about the world leaves us in a situation of *confusion*, not enlightenment. In this way, perhaps, the biggest metametaphysical lesson that metaphysical underdetermination can bring us is the following: metaphysics, understood as an extra explanatory layer in relation to the scientific layer, must be avoided under the penalty of bringing more harm than gains.

Maybe van Fraassen [74] was right after all: metaphysical underdetermination means good-bye to metaphysics — or, at least, a farewell to its epistemic dignity for the time being. If metaphysics is too permissive from a methodological point of view, then it is easy to indicate the metaphysical profiles arbitrarily and to proliferate metaphysical profiles/options at any convenience. So we believe that "good-bye" is a farewell to the epistemic credentials of metaphysics — something that scientific metaphysicians/naturalists were keen to provide [57, 16, 46, 37]. That is, what is the epistemic value of a discipline in which (almost) anything goes? It seems to be very cheap. In any case, if the initial objective of justifying the introduction of a metaphysical layer of explanation was to increase our understanding of a certain domain of knowledge, well, it seems, the multiplicity of options ends up completely thwarting this ambition.

This, we think, is *the* most important metaphilosophical lesson unlearned from metaphysical underdetermination.

Acknowledgements

Order of authorship is alphabetical and does not represent any kind of priority; authors have contributed equally to this chapter. A previous version of this work was presented at the Hi-Phi International Conference, Lisbon, 2022. We would like to thank Miguel Ohnesorge and Paulo Castro, whose constructive suggestions contributed to the improvement of this text. We would also like to thank Christian de Ronde for inviting us to participate in this volume, and for the always thought-provoking conversations about the philosophy of physics and scientific realism. Raoni Wohnrath Arroyo acknowledges the support of grants #2021/11381-1 and #2022/15992-8, São Paulo Research Foundation (FAPESP); Jonas R. Becker Arenhart and Décio Krause acknowledge the partial support of the National Council for Scientific and Technological Development (CNPq).

References

1. Arenhart, J. R. B. & Arroyo, R. W. (2021). On physics, metaphysics, and metametaphysics. *Metaphilosophy* **52**, 175–199.
2. Arenhart, J. R. B. & Arroyo, R. W. (2021). The spectrum of metametaphysics: Mapping the state of art in scientific metaphysics. *Veritas* **66**, e41217.
3. Arenhart, J. R. B. & Bueno, O. (2015). Structural realism and the nature of structure. *European Journal for Philosophy of Science* **5**, 111–139.
4. Arenhart, J. R. B. (2017). The received view on quantum non-individuality: Formal and metaphysical analysis. *Synthese* **194**, 1323–1347.
5. Arenhart, J. R. B. (2017). Does weak discernibility determine metaphysics? *Theoria* **32**, 109–125.
6. Arenhart, J. R. B. (2018). New Logics for Quantum Non-individuals? *Logica Universalis* **12**, 375–395.
7. Arroyo, R. W. & Arenhart, J. R. B. (2020). *Floating free from Physics: The Metaphysics of Quantum Mechanics.* URL: http://philsci-archive.pitt.edu/18477
8. Arroyo, R. W. & Arenhart, J. R. B. (2021). Back to the question of ontology (and metaphysics). *Manuscrito* **44**, 1–51.
9. Arroyo, R. W. & da Silva, G. O. (2022). Against 'Interpretation': Quantum Mechanics Beyond Syntax and Semantics. *Axiomathes* **32**, 1243–1279.
10. Aspect, A. (2002). Bell's theorem: The naive view of an experimentalist. In: *Quantum (un)speakables: From Bell to quantum information.* Bertlmann, R. & Zeilinger, A. (eds.) Springer 119–153.
11. Bianchi, S. & Giannotti, J. (2021). Grounding ontic structuralism. *Synthese* **199**, 5205–5223.
12. Bigaj, T. (2021). On discernibility in symmetric languages: The case of quantum particles. *Synthese* **198**, 8485–8502.
13. Bigaj, T. (2022). *Identity and Indiscernibility in Quantum Mechanics.* Palgrave Macmillan.
14. Bohr, N. (1963). Essays 1958–1962 on Atomic Physics and Human Knowledge. John Wiley & Sons.
15. Brading, K. & Skiles, A. (2012). Underdetermination as a Path to Structural Realism. In: *Structural Realism: Structure, Object, and Causality.* Landy, E. M. & Rickles, D. P. (eds.) Springer, 99–116.
16. Bryant, A. (2020). Epistemic Infrastructure for a Scientific Metaphysics. *Grazer Philosophische Studien* **98**, 27–49.

17. Bueno, O. (2021). Neo-Pyrrhonism, Empiricism, and Scientific Activity. *Veritas* **66** e42184.
18. Callender, C. (2020). Can We Quarantine the Quantum Blight? In: *Scientific Realism and the Quantum.* French, S. & Saatsi, J. (eds.) Oxford University Press, 55–77.
19. Carnap, R. (1950). Empiricism, Semantics and Ontology. *Revue International de Philosophie* **4**.
20. Caulton, A. (2013). Discerning "Indistinguishable" Quantum Systems. *Philosophy of Science* **80**, 49–72.
21. Caulton, A. (2022). Permutations. In: *The Routledge Companion to Philosophy of Physics.* Knox, E. & Wilson, A. (eds.) Routledge, 578–594.
22. Chakravartty, A. (2007). *A metaphysics for scientific realism: Knowing the unobservable.* Cambridge University Press.
23. Chakravartty, A. (2017). *Scientific ontology: Integrating naturalized metaphysics and voluntarist epistemology.* Oxford University Press.
24. Chakravartty, A. (2019). Physics, metaphysics, dispositions, and symmetries – À la French. *Studies in Hisory and Philosophy of Science,* **74**, 10–15.
25. da Costa, N. C. A. & Bueno, O. (2011). Quasi-truth and Quantum Mechanics. In: *Brazilian Studies in Philosophy and History of Science.* Krause, D. & Videira, A. (eds.) Springer, 301–312.
26. da Costa, N. C. A. & French, S. (2003). *Science and partial truth: A unitary approach to models and scientific reasoning.* Oxford University Press.
27. de Barros, J. A., Holik, F. & Krause, D. (2017). Contextuality and Indistinguishability. *Entropy* **19**, 435–456
28. Dürr, D. & Lazarovici, D. (2020). *Understanding Quantum Mechanics: The World According to Modern Quantum Foundations.* Springer.
29. Feynman, R. P. (1965). The Development of the Space-Time View of Quantum Electrodynamics. *Nobel Lecture.* URL: www.nobelprize.org/prizes/physics/1965/feynman/lecture
30. French, S. & Krause, D. (2006). *Identity in physics: A historical, philosophical, and formal analysis.* Oxford University Press.
31. French, S. & Ladyman, J. (2003). Remodelling Structural Realism: Quantum Physics and the Metaphysics of Structure. *Synthese* **136**, 31–56.
32. French, S. & Ladyman, J. (2011). In Defence of Ontic Structural Realism. In: *Scientific Structuralism.* Bokulich, P. & Bokulich, A. (eds.) Springer, 25–42.
33. French, S. (1989). Identity and Individuality in Classical and Quantum Physics. *Australasian Journal of Philosophy* **67**, 432–446
34. French, S. (1998). On the Withering Away of Physical Objects. In: *Interpreting Bodies: Classical and Quantum Objects in Modern Physics.* Castellani, E. (ed.) Princeton University Press, 93–113.
35. French, S. (2011). Metaphysical underdetermination: Why worry *Synthese* **180**, 205–221.
36. French, S. (2014). *The structure of the world: Metaphysics and representation.* Oxford University Press.
37. French, S. (2018). Toying with the Toolbox: How Metaphysics Can Still Make a Contribution. *Journal for General Philosophy of Science* **49**, 211–230.

38. French, S. (2018). Realism and Metaphysics. In: *The Routledge Handbook of Scientific Realism.* Saatsi, J. (ed.) Routledge, 394–406.
39. French, S. (2019). Identity and Individuality in Quantum Theory. In: *The Stanford Encyclopedia of Philosophy.* Zalta, E. N. (ed.) Metaphysics Research Lab, Stanford University. URL: https://plato.stanford.edu/archives/win2019/entries/qt-idind
40. French, S. (2019). Defending eliminative structuralism and a whole lot more (or less). *Studies in History and Philosophy of Science* **74**, 22–29.
41. French, S. (2020). *Metaphysical Underdetermination as a Motivational Device.* URL: http://philsci-archive.pitt.edu/16922
42. French, S. (2020). *What is This Thing Called Structure? (Rummaging in the Toolbox of Metaphysics for an Answer).* URL: http://philsci-archive.pitt.edu/16921
43. French, S. (2022). Fundamentality. In: *The Routledge Companion to Philosophy of Physics.* Knox, E. & Wilson, A. (eds.) Routledge, 679–688.
44. French S. & McKenzie, K. (2012). Thinking outside the toolbox: Towards a more productive engagement between metaphysics and philosophy of physics. *European Journal of Analytic Philosophy* **8**, 42–59.
45. French S. & McKenzie, K. (2015). Rethinking Outside the Toolbox: Reflecting Again on the Relationship between Philosophy of Science and Metaphysics. In: *Metaphysics in Contemporary Physics.* Bigaj, T. & Wüthrich, C. (eds.) Brill/Rodopi, 25–54.
46. Hofweber, T. (2021). Is metaphysics special? In: *The Routledge Book of Metametaphysics.* Bliss, R. & Miller, J. T. M. (eds.) Routledge, 421–431.
47. Huggett, N. & Imbo, T. (2009). Indistinguishability. In: *Compendium of Quantum Physics: Concepts, Experiments, History and Philosophy.* Greenberger, D., Hentschel, K. & Weinert, F. (eds.) Springer, 311–317.
48. Huggett, N. & Norton, J. (2013). Weak Discernibility for Quanta, the Right Way. *The British Journal for the Philosophy of Science* **65**, 39–58.
49. Krause, D. & Arenhart, J. R. B. (2016). Presenting Nonreflexive Quantum Mechanics: Formalism and Metaphysics. *Cadernos de História e Filosofia da Ciência* **2**, 59–91.
50. Krause, D. & Arenhart, J. R. B. (2018). Quantum Non-individuality: Background Concepts and Possibilities. In: *The Map and the Territory: Exploring the Foundations of Science, Thought and Reality.* Wuppuluri, S. & Doria, F. A. (eds.) Springer, 281–305.
51. Krause, D. & Arenhart, J. R. B. (2020). *Identical particles in quantum mechanics: favouring the Received View.* URL: http://philsci-archive.pitt.edu/17116
52. Krause, D. (1991). Multisets, quasi-sets and Weyl's aggregates. *Journal of Non-Classical Logic* **8**, 9–39.
53. Krause, D. (1992). On a quasi-set theory. *Notre Dame Journal of Formal Logic* **33**, 402–411.
54. Krause, D. (2005). Structures and Structural Realism. *Logic Journal of the IGPL* **13**, 113–126.

55. Krause, D. (2019). Does Newtonian Space Provide Identity to Quantum Systems? *Foundations of Science* **24**, 197–215.
56. Ladyman, J. (2020). Structural Realism. In: *The Stanford Encyclopedia of Philosophy.* Zalta, E. N. (ed.) Metaphysics Research Lab, Stanford University. URL: https://plato.stanford.edu/archives/win2020/entries/structural-realism
57. Ladyman, J. & Ross, D. (2007). *Every Thing Must Go: Metaphysics Naturalized.* Oxford University Press.
58. Ladyman, J. (1998). What is structural realism? *Studies in History and Philosophy of Science* **29**, 409–424.
59. Magnus, P. D. (2012). *Scientific Enquiry and Natural Kinds: From Planets to Mallards.* Palgrave-Macmillan.
60. Maudlin, T. (1995). Three measurement problems. *Topoi* **14**, 7–15.
61. McKenzie, K. (2020). A Curse on Both Houses: Naturalistic Versus *A Priori* Metaphysics and the Problem of Progress. *Res Philosophica* **97**, 1–29.
62. Morganti, M. (2015). The metaphysics of individuality and the sciences. In: *Individuals Across the Sciences.* Guay, A. & Pradeu T. (eds.) Oxford University Press, 273–294.
63. Morganti, M. & Tahko, T. E. (2017). Moderately naturalistic metaphysics. *Synthese* **194**, 2557–2580.
64. Muller, F. A. & Saunders, S. (2008). Discerning Fermions. *The British Journal for the Philosophy of Science* **59**, 499–458.
65. Muller, F. A. (2011). Withering Away, Weakly. *Synthese* **180**, 223–233.
66. Post, H. (1963). Individuality and Physics. *The Listener* **70**, 534–537.
67. Psillos, S. (1999). *Scientific realism: How science tracks truth.* Routledge.
68. Rettler, B. & Bailey, A. M. (2017). Object. In: *The Stanford Encyclopedia of Philosophy.* Zalta, E. N. (ed.) Metaphysics Research Lab, Stanford University. URL: https://plato.stanford.edu/archives/win2017/entries/object
69. Ruetsche, L. (2018). Getting Real About Quantum Mechanics. *The Routledge Handbook of Scientific Realism.* Saatsi, J. (ed.) Routledge, 291–303.
70. Saunders, S. (2003). Physics and Leibniz's Principles. In: *Symmetries in Physics: Philosophical Reflections.* Brading, K. & Castellani, E. Cambridge University Press, 289–307.
71. Schrödinger, E. (1998). What is an Elementary Particle? *Interpreting Bodies: Classical and Quantum Objects in Modern Physics.* Castellani, E. (ed.) Princeton University Press, 197–210.
72. Stanford, K. (2021). Underdetermination of Scientific Theory In: *The Stanford Encyclopedia of Philosophy.* Zalta, E. N. (ed.) Metaphysics Research Lab, Stanford University. URL: https://plato.stanford.edu/archives/win2021/entries/scientific-underdetermination
73. van Fraassen, B. C. (1980). *The scientific image.* Oxford University Press.
74. van Fraassen, B. C. (1991). *Quantum mechanics: An Empiricist View.* Oxford University Press.
75. Weyl, H. (1931). *Gruppentheorie und Quantenmechanic.* Leipzig.
76. Wolff, J. (2019). Naturalistic quietism or scientific realism? *Synthese* **196**, 485–498.

THE (QUANTUM) MEASUREMENT PROBLEM IN CLASSICAL MECHANICS

CHRISTIAN DE RONDE
CONICET, Institute of Philosophy "Dr. A. Korn", Universidad de Buenos Aires
Universidad Nacional Arturo Jauretche, Argentina
Center Leo Apostel, Vrije Universiteit Brussel, Belgium
Universidade Federal de Santa Catarina, Brasil
E-mail: cderonde@vub.ac.be

In this work we analyze the deep link between the 20th Century positivist re-foundation of physics and the famous measurement problem of Quantum Mechanics (QM). We attempt to show why this is not an "obvious" nor "self evident" problem for the theory of quanta but a direct consequence of the empirical-positivist understanding of physical theories in terms of (binary) observations when applied to the orthodox linear formalism of QM. In contraposition, we discuss a representational realist account of both physical 'theories' and 'measurement' which goes back to the works of Einstein, Heisenberg and Pauli. After presenting a critical analysis of Bohr's definitions of 'measurement' we continue to discuss the way in which several contemporary approaches to QM — such as decoherence, modal interpretations and QBism — remain — still today — committed to Bohr's general methodology. Finally, in order to expose the inconsistencies present within the (empirical-positivist) presuppositions responsible for creating the quantum measurement problem we show how following these same set of presuppositions it is easy to derive a completely analogous paradox for the case of classical mechanics.

Keywords: Measurement problem; positivism; Bohr; quantum mechanics; classical mechanics.

1. The measurement problem in quantum mechanics

Still today, the measurement problem continues to be considered by many as *the* most important obstacle for a proper understanding of Quantum Mechanics (QM). The problem is strictly related to the the existence of superposed states within the quantum formalism and the seemingly contradicting fact that we never actually observe such superpositions in the lab. As argued by orthodoxy, instead of quantum superpositions we only observe single outcomes after a quantum measurement is performed. This apparent contradiction between the mathematical formalism — which in most cases provides an account of physical situations in terms of quantum superpositions — and observation gives rise to the famous measurement problem of QM, which can be stated in the following terms:

Quantum Measurement Problem (QMP): *Given a specific basis (or context), QM describes mathematically a quantum state in terms of a superposition of, in general, multiple states. Since the evolution described by QM allows us to predict that the quantum system will get entangled with the apparatus and thus its pointer positions will also become a superposition,*[a] *the question is why do we observe a single outcome instead of a superposition of them?*

An obvious way out of the QMP is to take an anti-realist viewpoint and simply dissolve it. By refusing to attach any representational reference to quantum superpositions anti-realists can easily escape the paradox right from the start. This path is clearly exemplified by Chris Fuchs and Asher Peres [46, p. 70] who argued that "quantum theory does not describe physical reality. What it does is provide an algorithm for computing probabilities for the macroscopic events ('detector clicks') that are the consequences of our experimental interventions." On the opposite realist corner, attempting to understand what is really going on beyond measurement outcomes, things look quite different. As recently pointed out by Matthias Egg [42]: "many metaphysicians (and some physicists) consider the abandonment of realism too high a price to pay and therefore insist that the measurement problem calls for a (realistic) solution rather than a dissolution along non-realist lines." Accordingly, Egg argues that there exist only two options: "either [...] to modify the physics (such as theories with additional variables or spontaneous collapses) or drastically inflate the empirically inaccessible content of reality (such as many-worlds interpretations)." As we shall discuss in detail, these — supposedly — realist approaches to QM assume uncritically the naive empiricist standpoint — which goes back to the positivist understanding of theories — according to which single observations can be considered as *givens* of experience, *prior* and independent of

[a]Given a quantum system represented by a superposition of more than one term, $\sum c_i|\alpha_i\rangle$, when in contact with an apparatus ready to measure, $|R_0\rangle$, QM predicts that system and apparatus will become "entangled" in such a way that the final 'system + apparatus' will be described by $\sum c_i|\alpha_i\rangle|R_i\rangle$. Thus, as a consequence of the quantum evolution, the pointers have also become — like the original quantum system — a superposition of pointers $\sum c_i|R_i\rangle$. This is why the measurement problem can be stated as a problem only in the case the original quantum state is described by a superposition of more than one term.

theoretical representation. Egg's map restricts the possibilities of analysis to implicit acceptance of the just mentioned naive empiricist presupposition. Taking distance from both anti-realism and empiricist versions of realism, we will argue that a truly realist account of QM necessarily implies a critical reconsideration of the role played by observation within physical theories. Against both naive realism and naive empiricism, observations in physics should not be regarded as "common sense" *givens* but, on the very contrary, as derived from theories themselves for as stressed repeatedly by both Einstein, Heisenberg and Pauli it is only the theory which decides what can be observed. And it is only through the creation and development of adequate physical concepts that we will be finally able to understand quantum phenomena.

The QMP takes for granted a series of empirical-positivist presuppositions which fail to address observations in a critical (realist) manner. In this work we attempt to expose its untenability showing that, taking as a standpoint the empirical-positivist presuppositions, one can also derive an analogous 'measurement problem' for classical mechanics. The article is organized as follows. In section 2 we consider the main scheme and cornerstones of the 20th positivist re-foundation of physics. In section 3 we revisit the deep relation between the QMP and the empirical-positivist understanding of physical theories. Section 4 restates in a contemporary manner the original Greek meaning of realism within physics as strictly founded on the possibility of a theoretical formal-conceptual invariant-objective representation of *physis*. In section 5, in order to expose the untenability of the QMP for realist perspectives, we show how an analogous problem can be also derived in the case of classical mechanics.

2. The 20th Century positivist re-foundation of physics

In the 17th Century metaphysics had escaped the constraints of experience and advanced wildly as a "mad dog" producing all sorts of amazing stories about existence and reality. There was no limit nor constrain for dogmatic metaphysics which debated not only about about the existence of God, but also about the properties of angels. In his fight against dogmatic metaphysics, the physicist and philosopher Immanuel Kant went back to the — apparently — more humble Sophistic viewpoint. On the one hand, following rationalists and empiricists he recognized that the standpoint of science needed to be human thought and experience; on the other, he accepted the finitude of man, and consequently, the impossibility to access the infinite reality-in-itself — i.e., objects as they are, independent of human perception. Kant reconfigured the foundation of knowledge itself grounding his new philosophy in a *transcendental subject* that captured through the table of (Aristotelian) *categories* and the *forms of intuition* (i.e., Newtonian space and time), the way in which we, humans, experience phenomena. From this standpoint, he was able to develop a new metaphysical architectonic in which the (transcendental) subject played the most fundamental role providing the conditions of possibility for objective experience itself; i.e., the experience about objects. In his *Critique of Pure Reason* he was then able to explain how Newtonian physics could be understood as providing *objective knowledge* about the phenomena observed by (empirical) subjects. By understanding the limits of human experience metaphysics would finally follow the secure path of science showing, at the same time, how (scientific) knowledge was possible. Unfortunately, very soon, the *a priori* forms of intuition (i.e., Newtonian absolute space and time) were taken as the fundamental unmovable cornerstones of all possible experience, restricting in this way all access to new phenomena and turning Kantian metaphysics itself into a new dogma. The categories and forms of intuition put forward by Kant had turned into the same he had striven to attack in the metaphysics of his time, *a priori* unquestionable elements of thought. As noted by van Fraassen [71, p. 2]: "Kant exposed the illusions of Reason, the way in which reason overreaches itself in traditional metaphysics, and the limits of what can be achieved within the limits of reason alone. But on one hand Kant's arguments were not faultless, and on the other there was a positive part to Kant's project that, in his successors, engaged a new metaphysics. About a century later the widespread rebellions against the Idealist tradition expressed the complaint that Reason had returned to its cherished Illusions, if perhaps in different ways." In the mid 19th Century, the cracks in the structure of the Kantian architectonic had begun to become visible through the work of Ernst Mach — another physicist and philosopher. Mach produced the most radical deconstruction of the main concepts of classical mechanics — which were also founding the Kantian scheme — by deriving a new positivist understanding of physics which led him to the conclusion that science is nothing but the systematic and synoptical recording of data of experience. In his famous book, *Analysis of Sensations*, he would write:

> Nature consists of the elements given by the senses. Primitive man first takes out of them certain complexes of these elements that present themselves with a certain stability and are most important to him. The first and oldest words are names for 'things'. [...] The sensations are no 'symbols of things'. On the contrary the 'thing' is a mental symbol for

> a sensation-complex of relative stability. Not the things, the bodies, but colors, sounds, pressures, times (what we usually call sensations) are the true elements of the world. [58]

Mach concluded that primary sensations constitute the ultimate building blocks of science, inferring at the same time that scientific concepts are only admissible if they can be defined in terms of sensations. In close analogy to Darwinistic ideas, Mach conceived the evolution of knowledge in physical theories as a process of "struggle for life" and "survival of the fittest". Although Mach had been himself a neo-Kantian, within his new positivist conception of science, he stated that we should reject every *a priori* element in the constitution of our knowledge about things. This move took experience and observation back to its naive pre-Kantian form. Scientific propositions should be empirically verifiable and science would be then nothing but a reflection of facts. Mach's pre-Kantian empiricist standpoint allowed him not only to criticize the physical concepts of classical mechanics, but also to produce a complete reformulation of the meaning and applicability of physical theories. This allowed the next generation of physicists — Einstein, Bohr, Heisenberg, Pauli and many others —, to use this new freedom in order to conceive not only a new experience — beyond classical notions — but also new mathematical formalisms. While Albert Einstein indicated on many occasions the importance of Mach's work for the creation of relativity theory, Heisenberg used the Machian principle according to which only observables should be considered within a theory in order to develop matrix mechanics — escaping right from the start the (classical) question regarding 'the trajectories of particles'.[b] At the beginning of the 20th Century, the Machian epistemological principle had finally broken the chains of classical Newtonian physics and Kant's metaphysics. A new quantum experience was disclosed, a new region of thought had been created.

Concomitant with the creation of relativity and QM, a new generation of positivists like Rudolph Carnap, Otto Neurath and Hans Hahn congregated in what they called 'the Vienna Circle'. In their famous and influential manifesto [16] they wrote the following: "In science there are no 'depths'; there is surface everywhere: all experience forms a complex network, which

[b]As noticed by Arthur Fine [19, p. 1195]: "Heisenberg's seminal paper of 1925 is prefaced by the following abstract, announcing, in effect, his philosophical stance: 'In this paper an attempt will be made to obtain bases for a quantum-theoretical mechanics based exclusively on relations between quantities observable in principle'."

cannot always be surveyed and, can often be grasped only in parts. Everything is accessible to man; and man is the measure of all things. Here is an affinity with the Sophists, not with the Platonists; with the Epicureans, not with the Pythagoreans; with all those who stand for earthly being and the here and now." Metaphysics was now understood as a dogmatic discourse about unobservable entities. According to them, science should focus in "statements as they are made by empirical science; their meaning can be determined by logical analysis or, more precisely, through reduction to the simplest statements about the empirically given." This new positivism sedimented Mach's pre-Kantian understanding of observation. In turn, this naive standpoint allowed them to draw a main distinction — that would guide the science of the new century — between "empirical terms", the empirically "given" in physical theories and "theoretical terms", their translation into "simple statements". An important consequence of this naive empiricist standpoint is that physical concepts become only supplementary elements, parts of a narrative that could be added, or not, to an already empirically adequate theory. When a physical phenomenon is understood as a self-evident *given* — independent of physical concepts and metaphysical presuppositions —, empirical terms configure an "objective" set of data which can be directly related to a formal scheme. Objective here does not relate to a conceptual *moment of unity* — like in the Kantian scheme — but instead, to something quite undefined like a "good" or "honest" observation made by a subject.

Empirical Observable Data ———— Theoretical Terms

(Addition of a Supplementary 'Interpretation' or 'Fictional Narrative')

Since an empirically adequate theory is already able to predict the observations it talks about, there seems to be no essential need to add a "story" or "interpretation" about the external reality lying beyond the observable realm. The role of concepts becomes then completely accessory to those needing to believe in narratives about how the world could be according to the theory. But as remarked by van Fraassen [70, p. 242]: "However we may answer these questions, believing in the theory being true or false is something of a different level." According to the (naive) empiricist viewpoint, the world is unproblematically described in terms of our "common sense" access to experience and an adequate empirical theory can perfectly account for experiments without the need of any interpretation.

The project of articulating the empirical-formal relation through the distinction between *theoretical terms* and *observational terms* never accomplished the promise of justifying their independence — specially with respect to the categorical or metaphysical definition of concepts. The fundamental reason had already been discussed in Kant's *Critique of Pure Reason*: the observation of a phenomenon cannot be considered without previously taking into account a categorical-metaphysical scheme. The description of phenomena always necessarily presupposes — implicitly or explicitly — certain metaphysical principles. 'Identity' or 'non-contradiction' are not 'things' that we see walking in the street, but rather the very *conditions of possibility* of classical experience itself. As David Hume would remark with respect to causality we never find these principles in experience; rather, we presuppose them in order to make sense of phenomena. This categorical systematization, allowing for a theoretical-conceptual representation, is in itself metaphysical. As the philosopher from Königsberg would have said, it is the representational framework of the transcendental subject, articulating categories and forms of intuition, that which allows for an objective experience of the empirical subject.[c] In fact, this was according to Einstein the really significant philosophical achievement of Kant:

> From Hume Kant had learned that there are concepts (as, for example, that of causal connection), which play a dominating role in our thinking, and which, nevertheless, can not be deduced by means of a logical process from the empirically given (a fact which several empiricists recognize, it is true, but seem always again to forget). What justifies the use of such concepts? Suppose he had replied in this sense: Thinking is necessary in order to understand the empirically given, *and concepts and 'categories' are necessary as indispensable elements of thinking.* [43, p. 678] (emphasis in the original)

Maybe due to the general failure of the program [53], it is difficult to find today anyone willing to call herself a positivist. However, regardless of its failure, the basic ideas present in early positivism have reached our days and — more importantly — continue to configure the problems we discuss in the present. The limits of positivism itself are of course difficult to encapsulate, but there are four main cornerstones which might be regarded to conform the main positivist program. These pillars have played an essential role in the deep 20th Century postmodern re-foundation of physics.

[c]In the context of analytic philosophy, this metaphysical aspect of observation was rediscovered by Norwood Rusell Hanson in the mid 20th Century [49]. However, the main positivist scheme of thought remained exactly the same as before, taking for granted a naive empiricist standpoint.

I. Naive Empiricism: observation is a self evident *given* of "common sense" experience.

II. Physics as an Economy of Experience: physical theories are mathematical machineries (algorithmic formalisms) which produce predictions about observable measurement outcomes.

III. Anti-Metaphysics: metaphysics, understood as an interpretation or narrative about *unobservable entities*, is not essentially required within empirically adequate theories.

IV. Intersubjective Justification: the *intersubjective* communication of theoretically predicted observable data between the members of a scientific community allows to regard a theory as "objective".

All 'isms' in science imply a specific *praxis* defined by the field of its problems. Scientific problems are not independent of the understanding of what the scientific enterprise is really about. In this respect, the influence of the just mentioned pillars — regardless of the fact positivism has become a quite unpopular term — continue to play — still today — the most fundamental role within the discussions and debates that conform the specialized physical and philosophical literature about QM. It is important to realize that in order to be part of a program, it doesn't matter if you truly believe or not in it; what really matters is what you actually do. If you follow a program, then — quite regardless of your personal beliefs — you will engage with its problems and discuss what is important for the program. The actuality of the positivist program within QM is exposed by the centrality of the infamous measurement problem within today's philosophical and foundational debates.

3. The positivist origin of the quantum measurement problem

As we discussed above, positivist ideas were deeply influential for the creation and development of QM. In 1925, Werner Heisenberg was finally able to find the key to develop the closed mathematical formalism of matrix mechanics through the Machian positivist observability principle. Positivism allowed him to abandon the focus in finding the classical trajectories of particles and restate the problem from a completely different angle, in formal-operational terms. One year later, following the work by Louis de Broglie on matter waves, Erwin Schrödinger would propose a differential "wave-looking" equation which introduced the famous quantum wave function, Ψ. Regardless of the initial expectations to restore the classical space-time representation, it was soon realized that the equation was written in *configuration space*, precluding its classical understanding in terms of a space-time 3-dimensional wave. That same year Max Born presented his probabilistic interpretation of Ψ, which only made reference to the prediction of 'clicks' in detectors consequence of elementary particles. This interpretation — which Einstein strongly rejected —, also made implicit use of the positivist *Zeitgeist* according to which theories, instead of describing a real states of affairs, had to be understood as "tools" for predicting observations. As remarked by Osnaghi et al. in [62]: "During the 1920s and 1930s, the ideas which were to be identified with the 'orthodox view' of quantum mechanics became quite popular. The positivist flavor of the approach developed by Heisenberg, Jordan, Born and Pauli was not only in tune with the cultural climate of continental Europe between the two wars, but was also well suited to cope with the change of paradigm that atomic phenomena seemed to demand." After the theoretical developments of Heisenberg and Schrödinger, Paul Dirac, a young English engineer and mathematician, attempted to provide a sound axiomatic mathematical presentation of the new theory. In 1930 he presented his book, *The Principles of Quantum Mechanics*, where from an explicit positivist perspective he stressed that it is "important to remember that science is concerned only with observable things and that we can observe an object only by letting it interact with some outside influence. An act of observation is thus necessarily accompanied by some disturbance of the object observed." Following Bohr's teachings, he also remarked that [41, pp. 3-4]: "we have to assume that *there is a limit to the finiteness of our powers of observation and the smallness of the accompanying disturbance — a limit which is inherent in the nature of things and can never be surpassed by improved technique or increased skill on the part of the observer.*" Dirac was maybe the first to realize the importance of the superposition principle in the quantum formalism which implied the existence of strange quantum superpositions. Dirac realized that this introduced a serious obstacle for his own positivist reading in terms of the certain prediction of single (binary) outcomes. In order to bridge this gap, making use of an example of polarized photons, Dirac introduced for the first time the now famous "collapse" of the quantum wave function:

> When we make the photon meet a tourmaline crystal, we are subjecting it to an observation. We are observing wither it is polarized parallel or perpendicular to the optic axis. The effect of making this observation is to force the photon entirely into the state of parallel or entirely into the state of perpendicular polarization. It has to make a sudden jump from

> being partly in each of these two states to being entirely in one or the other of them. Which of the two states it will jump cannot be predicted, but is governed only by probability laws. [41, p. 7]

Even though the explanation did not succeed in providing a consistent 'picture' of what was going on, Dirac recalled his readers that "the main object of physical science is not the provision of pictures, but the formulation of laws governing phenomena and the application of these laws to the discovery of phenomena. If a picture exists, so much the better; but whether a picture exists of not is a matter of only secondary importance." An empirically adequate theory might possess an 'interpretation', but this is not essential, the scheme only "becomes a precise physical theory when all the axioms and rules of manipulation governing the mathematical quantities are specified and when in addition certain laws are laid down connecting physical facts with the mathematical formalism." Two years later, in 1932, the famous Hungarian mathematician John von Neumann published his famous *Mathematische Grundlagen der Quantenmechanik* where he attempted to provide an even more rigorous mathematical foundation for QM.[d] In this case, the need of an axiomatization of the theory might be easily linked to von Neumann's empiricist understanding of the method of the physical sciences (see also [15]).

> To begin, we must emphasize a statement which I am sure you have heard before, but which must be repeated again and again. It is that the sciences do not try to explain, they hardly ever try to interpret, they mainly make models. By a model is meant a mathematical construct which, with the addition of some verbal interpretations describes observed phenomena. The justification of such a mathematical construct

[d]Von Neumann was strongly influenced by Hilbert regarding the axiomatization program. In a joint paper by Hilbert, Nordheim and von Neumann from 1926 (translated by Redei and Stöltzner) they claim: "In physics the axiomatic procedure alluded to above is not followed closely, however; here and as a rule the way to set up a new theory is the following. One typically conjectures the analytic machinery before one has set up a complete system of axioms, and then one gets to setting up the basic physical relations only through the interpretation of the formalism. It is difficult to understand such a theory if these two things, the formalism and its physical interpretation, are not kept sharply apart. This separation should be performed here as clearly as possible although, corresponding to the current status of the theory, we do not want yet to establish a complete axiomatics. What however is uniquely determined, is the analytic machinery which — as a purely mathematical entity — cannot be altered. What can be modified — and is likely to be modified in the future — is the physical interpretation, which contains a certain freedom and arbitrariness."

> is solely and precisely that it is expected to work — that is correctly to describe phenomena from a reasonably wide area. [65, p. 9]

This empiricist standpoint assumed by both Dirac and von Neumann seemed to confront the existence of superpositions in QM which did not seem to describe what was being actually observed. But after Dirac had introduced the "collapse", von Neumann [68, p. 214] was ready to transform it into a *postulate* of the theory itself: "Therefore, if the system is initially found in a state in which the values of $\mathcal{R}$ cannot be predicted with certainty, then this state is transformed by a measurement M of $\mathcal{R}$ into another state: namely, into one in which the value of $\mathcal{R}$ is uniquely determined. Moreover, the new state, in which M places the system, depends not only on the arrangement of M, but also on the result of M (which could not be predicted causally in the original state) — because the value of $\mathcal{R}$ in the new state must actually be equal to this M-result." The *projection postulate* secured the observation of 'clicks' in detectors and 'spots' in photographic plates, but this addition came with a great cost. Like a Troyan horse, the postulate hosted an unwanted guest. Apart from the unitary deterministic evolution guided by Schrödinger's equation of motion, applicable when we do not observe what is going on, there was now a new non-unitary indeterministic evolution that took place every time we measured (or observed) a quantum state. Measurement in QM seemed to be responsible in producing a real physical process which was not represented by the theory itself. This new "quantum jump", different to the one made popular by Bohr,[e] took place between the mathematical formulation and *hic et nunc* observations made by subjects. The need of introducing an agent capable of observing, implied the breakdown of an objective (subject-independent) theoretical representation. Without any success in the physics community, Einstein and Schrödinger would raise their voices against this unsatisfactory evolution of the theory commanded by what could be seen as a silent alliance between Bohr and positivists. As Einstein [62], would mention to Everett some years later, he "could not believe that a mouse could bring about drastic changes in the universe simply by looking at it." Also Schrödinger criticized explicitly, not only the collapse, but also the explicit subjectivity

[e]It is important to remark however that the jumps discussed by Bohr took place within the atomic representation. Electrons jumped from one orbit to the next without any clear explanation. So while Bohr's jumps made reference to a limit of the representation within the theory, Dirac's jumps made reference to the shift from the mathematical representation of superpositions to the empirical realm of observation.

involved within the process:

> But jokes apart, I shall not waste the time by tritely ridiculing the attitude that the state-vector (or wave function) undergoes an abrupt change, when 'I' choose to inspect a registering tape. (Another person does not inspect it, hence for him no change occurs.) The orthodox school wards off such insulting smiles by calling us to order: would we at last take notice of the fact that according to them the wave function does not indicate the state of the physical object but its relation to the subject; this relation depends on the knowledge the subject has acquired, which may differ for different subjects, and so must the wave function. [62, p. 9]

The triumph of the Bohrian-positivist alliance silenced the critical remarks made by Einstein and Schrödinger and the orthodox textbook account of QM would become the following: quantum states behave in a *deterministic* manner when there is no measurement, but there is also an *indeterministic* evolution which takes place each time we attempt to observe what is really going on. Of course, as noticed by Dennis Dieks [38, p. 120]: "Collapses constitute a process of evolution that conflicts with the evolution governed by the Schrödinger equation. And this raises the question of exactly when during the measurement process such a collapse could take place or, in other words, of when the Schrödinger equation is suspended. This question has become very urgent in the last couple of decades, during which sophisticated experiments have clearly demonstrated that in interaction processes on the sub-microscopic, microscopic and mesoscopic scales collapses are never encountered." In the last decades, the experimental research seems to confirm there is nothing like a "real collapse" taking place when measurements are performed. In a more recent paper, Dieks [40] acknowledges the fact that: "The evidence against collapses has not yet affected the textbook tradition, which has not questioned the status of collapses as a mechanism of evolution alongside unitary Schrödinger dynamics." It is clear there is no experimental nor formal justification of the addition of the collapse. And yet it continues to play the most essential role in both physical and philosophical debates about QM.

We believe that in order to disentangle the mess created by the application of the positivist pillars to QM, It is essential not only to understand the anti-realist root of the measurement problem, it might be also wise to go back to the representational realist understanding of physics through the writings of Einstein, Heisenberg and Pauli.

4. Realism in physics: Theoretical representation and experience

The origin of physics is intrinsically related to the ancient Greek attempt to study and understand, in theoretical terms, what physicists and philosophers called *physis* — later on translated as reality or nature. Physics begins by presupposing the existence of *physis* as the fundament of all existence and the possibility of knowing it through the development of *theories*. These are the fundamental pillar of physical (or realistic) thought.

Physical (Formal-Conceptual) Theory —— *Physis/Reality*

Physics and realism go hand in hand, simply because physicists are mainly interested in reality, which is just a translation of *physis* [17]. One of the main presuppositions of physical realism is that *physis* is not chaotic, it has an internal order, a *logos* and that theories are able — in some way — to express the *logos* of *physis*, making in this way scientific knowledge possible. Physics and realism take as a basic standpoint that there exists a *relation* between theory and reality. However, this does not imply the general widespread naive (pre-Kantian) claim according to which the relation between theory and reality should be regarded as a one-to-one *correspondence relation*, the idea that theories "mirror" reality.[f] Of course, one can think of many different more interesting *relations* between theory and *physis* [22,23]. But the characterization of realism does not end here, it continues with a particular account of the relation between theoretical representation and experience. For realists — contrary to anti-realists — theoretical representation comes always first, experience is necessarily second. Observation is derived from theoretical representation itself which is conformed by an interrelated formal-conceptual framework. As Einstein would tell to a very young Heisenberg: "It is only the theory which decides what can be observed." Theoretical representation is first, observation and perception are necessarily second. This marks the kernel point of disagreement between empiricism — from which positivism, instrumentalism and even scientific realism were developed — and realism. According to the latter, willingly or not, we physicists, are always producing our praxis *within* a specific representation. And in this respect, there is no such thing as 'internal' and

[f]This idea, according to which the theory describes *reality as it is* can be only sustained by extremely naive forms of realism such as scientific realism; a formulation created by empiricists much closer to the positivists pillars than to the *praxis* of either science or realism.

'external' realities. Physical analysis takes always as a standpoint the theoretical representation of a state of affairs. And thus, for physics there can exist no discourse "outside" the gates of theoretical representation itself.

Realism: The working presupposition that there exists a *relation* between *theory* and *physis* and that it is possible to develop *theoretical (formal-conceptual) representations* which are independent of reference frames and the actions of empirical subjects.

Representation is not only mathematical (or formal), it is also conceptual (or metaphysical). And this marks the second point of distinction between realism and anti-realism, namely, the meaning and role played by metaphysics with respect to observation and experience — which is not that presented by positivists.[g] While for anti-realists metaphysics is a mere discourse or story about the un-observable realm, a narrative that might be added to an empirically adequate theory capable to predict unproblematic observable *givens*; for realists, metaphysics is the systematic creation of a relational net of concepts which constitute the condition of possibility for thought and experience themselves. As remarked by Einstein:

> I dislike the basic positivistic attitude, which from my point of view is untenable, and which seems to me to come to the same thing as Berkeley's principle, *esse est percipi*. 'Being' is always something which is mentally constructed by us, that is, something which we freely posit (in the logical sense). The justification of such constructs does not lie in their derivation from what is given by the senses. Such a type of derivation (in the sense of logical deducibility) is nowhere to be had, not even in the domain of pre-scientific thinking. The justification of the constructs, which represent 'reality' for us, lies alone in their quality of making intelligible what is sensorily given. [43, p. 669]

It is only through such generalizations that thought becomes possible, that we can even think and imagine an experience which was never actually observed. As Heisenberg [52, p. 264] made the point: "The history of physics is not only a sequence of experimental discoveries and observations, followed by their mathematical description; it is also a history of concepts. For an understanding of the phenomena the first condition is the introduction of adequate concepts. Only with the help of correct concepts can we really

[g]For a detailed discussion of the orthodox role played by metaphysics in philosophy of physics we refer to [3].

know what has been observed." But this distinction regarding the meaning of metaphysics is also a distinction regarding the meaning and role played by concepts. While for anti-realists concepts *refer* to 'things' in the world, in the case of realism concepts are always related to other concepts, they are intrinsically relational. A concept in not something that exists alone. A concept is always part of a conceptual net, a whole in which each element is supported by its neighbor — and viceversa. As remarked by Heisenberg [50, p. 94]: "the connection between the different concepts in the system is so close that one could generally not change any one of the concepts without destroying the whole system." The very notion of 'object' (e.g., a 'table') is a good example of the way in which logical and metaphysical principles, namely, the principles of existence, non-contradiction and identity allow to create a conceptual *moment of unity* capable to account for experience. An object looses its meaning when it is disconnected from its principles (e.g., identity). Conceptual nets allow us to think — in a specific manner — about physical reality. This is the reason why the discovery of a new field of experience implies always the creation of new conceptual frameworks and mathematical formalisms. As Heisenberg [11] explains: "the transition in science from previously investigated fields of experience to new ones will never consist simply of the application of already known laws to these new fields. On the contrary, a really new field of experience will always lead to the crystallization of a new system of scientific concepts and laws". In physics, theoretical representations are not only conceptual, they are also mathematical. Only together, a theory is capable of producing a qualitative and quantitative representative understanding of experience. While the conceptual part provides a *qualitative* type of understanding, the formal part provides a *quantitative* understanding. In order to conform a whole, metaphysical concepts must be consistently and coherently related to mathematical notions — and viceversa. It is this relation which provides the missing link between mathematics and metaphysics. Once again, the notion of 'object' provides a very good example of such intrinsic relation since the principles of existence, non-contradiction and identity play not only a metaphysical role, but also a logical one. In fact, these principles determine classical (Aristotelian) logic itself, which in turn, is also related to a specific mathematical formalism which is consistent with these principles (for a detailed discussion see [27,28]).

Another essential point of the realist understanding of physical (theoretical) representations is that the construction of a scheme requires a complete independence with respect to particular viewpoints considered in terms of reference frames or (empirical) subjects. Agents and their particular observations described from particular reference frames can play no essential role within the physical representation of a state of affairs. As Einstein [34, p. 175] made the point: "[...] it is the purpose of theoretical physics to achieve understanding of physical reality which exists independently of the observer, and for which the distinction between 'direct observable' and 'not directly observable' has no ontological significance." While the formal aspect allowing to detach theoretical representation from (empirical) subjects is provided via the *invariance* of the mathematical formalism, the conceptual component is articulated by the Kantian notion of *objectivity.* The notions of invariance and objectivity are intrinsically related; one being the counterpart of the other (see for a detailed analysis [29]). These two definitions, combined, allow physics to represent things as a *moment of unity* independently of subjects and their particular observations.[h] Physics does not deny the existence of (empirical) subjects, it simply considers them *within* Nature. The scientific enterprise considers subjects, not more nor less but *as* important as any other existent. This marks a third point of disagreement between realists and anti-realists. While for realists scientific analysis begins always from a theoretical standpoint, from the *physis*-perspective; for anti-realists, assuming the *relative* perspective of an individual subject, the analysis begins from "common sense" observations, perceptions or measurements. While the latter argues that the observations made by subjects (or agents) provide the ground from which theories are built, realists claim that the systematic understanding of experience can be only produced through the creation of a general theoretical (mathematical and metaphysical) framework. As Wolfgang Pauli would make clear to a young Heisenberg:

> 'Understanding' probably means nothing more than having whatever ideas and concepts are needed to recognize that a great many different phenomena are part of coherent whole. Our mind becomes less puzzled once we have recognized that a special, apparently confused situation is merely a special case of something wider, that as a result it can be formulated much more simply. The reduction of a colorful variety of phenomena to a general and simple principle, or, as the Greeks would

[h]Let us remark that the notion of *object* defined in categorical terms is one of such possible conceptual *moments of unity.* However, objectivity can be regarded as a more general idea, one which is not necessarily restricted to the reference to classical objects. As we have argued, there can also exist objectivity without objects; i.e. objective schemes with conceptual moments of unity which are non-classical [23,25].

> have put it, the reduction of the many to the one, is precisely what we mean by 'understanding'. The ability to predict is often the consequence of understanding, of having the right concepts, but is not identical with 'understanding'. [51, p. 63]

However, the most important point of disagreement between realists and anti-realists is marked by the different type of problems — grounded on essentially different pillars — they both address in each case. What is important for the realist is not important for the anti-realist, and viceversa. In the context of QM, as we have discussed in [23,25] these problems are simply orthogonal. On the one hand, the anti-realist, assuming that observations in the lab are "unproblematic" still has to confront the fact that representation and interpretation seem to be playing an essential role in order to connect empirical observations with the mathematical formalism. On the other hand, the realist, having a mathematical formalism that is capable of operational predictions must still produce a conceptual (subject-independent) representation which is structurally connected to the formalism and is also able to provide an *anschaulich* (intuitive) content to the theory. This realist problem is extremely difficult to tackle in QM due to the extreme departure which the theory of quanta seems to imply with respect to our (classical) metaphysical representation of (classical) reality in terms of objects moving in space-time. As Heisenberg [50, p. 3], made the point: "the change in the concept of reality manifesting itself in quantum theory is not simply a continuation of the past; it seems to be a real break in the structure of modern science". Thus, while the anti-realist attempts to build a bridge between the quantum formalism and our "common sense" classical or manifest image of the world, the realist is confronted with the need to create a consistent theoretical scheme which provides a formal-conceptual representation in invariant-objective terms. This implies in the case of QM, the development of a new (non-classical) conceptual scheme which can be structurally linked to the mathematical formalism of the theory allowing the development of a consistent representation of a state of affairs. These different problems are intrinsically related to two very different understandings of 'physics', either as providing theoretical (formal-conceptual) representations of an objective-invariant reality, or, as a mathematical algorithm capable of predicting the observations made by agents.

Tim Maudlin [59, p. xii] has recently argued that "physical theories are neither realist or anti-realist [...] It is a person's attitude toward a physical theory that is either realist or antirealist." On the contrary, we have argued that the realist and anti-realist schemes of understanding of theories are essentially different. While an anti-realist theory is a mathematical formalism grounded on observations, a realist theory is constituted by the entrenched relation between an objective conceptual framework an invariant mathematical formalism and a field of thought-experience. Unlike for the anti-realist, an operational algorithm capable of predictions — such as QM — with no consistent conceptual framework cannot be — yet — regarded as a (closed) theory. It still requires an objective conceptual representation which provides an intuitive (*anschaulicht*) access to the experience it talks about. According to Maudlin: "The scientific realist maintains that in at least some cases, we have good evidential reasons to accept theories or theoretical claims as true, or approximately true, or on-the-road-to-truth. The scientific antirealist denies this. These attitudes come in degrees: You can be a mild, medium, or strong scientific realist and similarly a mild, medium, or strong scientific antirealist." Like in many contemporary postmodern characterizations of realism, Maudlin ends up equating realism with the naive unjustified belief made by an agent. Against this definition, realism should be understood as a working presupposition in which the production of objective-invariant (subject-independent) representations is a kernel goal to achieve for the development of theories. Being a realist is not a matter of faith, it is rather a question which regards the choice of the problems someone chooses to confront. In this respect, being a realist is, so to say, a matter of *praxis*.

Realist: Those who assume a *praxis* according to which there exists a *relation* between *theory* and *physis*. Those who engage and work on realist problems which confront the production of objective-invariant (subject-independent) theoretical representations.

To sum up, while realism implies a perspective from *physis* itself, the anti-realist always assumes a perspective relative either to an individual *subject* or a community of subjects. Realism and anti-realism also imply different pillars from which two completely different fields of problems arise in each case. This distinction implies a different *praxis*. To know if you are a realist, you simply have to find out — regardless of your personal beliefs — which are the pillars that support the problems you are working on. Wolfgang Pauli, argued in this respect that the main problem that realists confront might imply a major revolution in our way of thinking:

> When the layman says 'reality' he usually thinks that he is speaking about something which is self-evidently known; while to me it appears

> to be specifically the most important and extremely difficult task of our time to work on the elaboration of a new idea of reality. [57, p. 193]

5. Measurement: From theoretical experience to actual observation

Exposing the fact that theories do not make exclusive reference to observations, the realist analysis escapes the empiricist standpoint right from the start. A very good example of this is provided by the Kochen-Specker theorem [56] which states — taking as a standpoint the orthodox mathematical formalism of QM — that projection operators cannot be related to a *Global Binary Valuation* [30]. This formal result has a deep consequence at the conceptual level, namely, it presents a limit to the possibility of interpreting projection operators in terms of definite binary-valued properties. It is interesting to notice that the theorem goes explicitly beyond the consideration of measurements and observations, in fact it discusses a situation which by definition cannot be subject of measurement (see for a detailed analysis [26]). This of course does not mean that for physicists (or realists) measurements are unimportant. On the contrary, measurements are an essential way to check out if a theory is capable of *expressing* an aspect of reality (or not). At some point, the thought-experience created by a theory in terms of a net of concepts and a mathematical formalism needs to be tested through measurements in the lab. As stressed by Einstein [34, p. 175], even though observability does not play a role within the theory itself, "the only decisive factor for the question whether or not to accept a particular physical theory is its empirical success." Measurement is a way to connect theoretical experience with *hic et nunc* observation, it provides the link between an objective theoretical representation and subjective empirical observability. Given a theoretical representation of a state of affairs we can imagine the possible experiences contained within the theory. And this thinking is not restricted by our technical or instrumental capabilities nor by our previous observations, it is only restricted by our theoretical (mathematical and conceptual) schemes of thought. This is the true power of physics. Not only the possibility to escape the here and now creating a thought-experience without the need to actually produce it in the lab, but to escape the technical limitations of our time and advance through thought and representation in our understanding of un-observed theoretical experience. It is *Gedankenexperiments* which show explicitly that, in physics, theoretical experience comes always before actual measurement and observability. In fact, this is exactly what was done in 1935 by both

Einstein and Schödinger in their famous EPR and 'cat' *Gedankenexperiments* where, going beyond the technical restrictions of their epoch, they discussed the experimental and representational consequences of the theory of quanta. It took half a century — partly due to the unwillingness of positivists, Bohrians and later on instrumentalists — to actually investigate the existence of quantum superpositions and entanglement. Two kernel notions which, in turn, made possible the ongoing technological revolution of quantum information processing.[i]

Theoretical experience needs to make contact with *hic et nunc* subjective observation. While the field of thought-experience is strictly limited by the theory itself, observation is a purely subjective conscious action which cannot be theoretically represented. It is 'measurement' which stands just in the middle between the theoretical representation of physical experience and empirical observation. Measurements, at least for the realist, are conscious actions performed by human subjects which are able to select, reproduce and understand a specific type of phenomenon. This is of course a very complicated process created by humans which interrelates practical, technical and theoretical knowledge. Anyone attempting to perform a measurement must be able to think about a specific problem, she must be also able to construct a measurement arrangement, she must be capable to analyze what might be going on within the process, and finally, she must be qualified to observe, interpret and understand the phenomenon taking place when the measurement is actually performed. All these requirements imply human capacities and, in particular, consciousness. The table supporting the measurement set up does not understand what complicated process is taking place above itself. Tables and chairs cannot construct a measuring set-up. The chair that stands just beside the table cannot observe a measurement result, and the light entering the lab through the window cannot interpret what is going on. It is only a conscious (empirical) subject (or agent) who is capable of performing a measurement. And

[i]As remarked by Jeffrey Bub [14], "[...] it was not until the 1980s that physicists, computer scientists, and cryptographers began to regard the non-local correlations of entangled quantum states as a new kind of non-classical resource that could be exploited, rather than an embarrassment to be explained away." The reason behind this shift in attitude towards *entanglement* is an interesting one. As Bub continues to explain: "Most physicists attributed the puzzling features of entangled quantum states to Einstein's inappropriate 'detached observer' view of physical theory, and regarded Bohr's reply to the EPR argument (Bohr, 1935) as vindicating the Copenhagen interpretation. This was unfortunate, because the study of entanglement was ignored for thirty years until John Bell's reconsideration of the EPR argument (Bell, 1964).".

these actions have nothing to do with "common sense". Measurement is a controlled technical-theoretical activity performed by subjects.

At this point it becomes of outmost importance to clearly distinguish between the theoretical representation of the interactions of 'systems' which are already part of a conceptual picture and the empirical observation which requires a conscious agent capable of understanding the theory and interpreting her observation as a measurement.[j]

THEORETICAL EXPERIENCE: *The process of interaction and evolution of a state of affairs represented in theoretical terms via the consistent conjunction of both conceptual and mathematical frameworks.*

EMPIRICAL OBSERVATION BY SUBJECTS (OR AGENTS): *The here and now conscious act of observation of a subject (or agent).*

The theory of electromagnetism comprised through Maxwell's equations and the concepts of charge, field, electricity, magnetism, etc., can be only learned in a classroom. As a student in physics one does not go around looking for fields or electromagnetic waves without understanding beforehand what the notion of 'field' or 'wave' really means; how they relate to other concepts of the theory or how to mathematically compute the evolution of a specific state of affairs. The understanding of physical notions is always relational, in the sense that a notion always refers to other other notions, like a net in which each node is supported by its neighbor. Only once you grasp the theory of electromagnetism as a whole, it makes sense to go into the lab and try to measure an electromagnetic field. The observation of a field can be only discussed as a complex interplay between theoretical knowledge — constituted by mathematical equations and concepts — and what actually happens *hic et nunc*. Theories do not come with a user's manual which explains how to measure the 'things' the theory talks about. Physics simply does not work like that. Theories do not begin with observations, they end with observations as corroborations of a theoretical representation. It is only once you have a theory that you can understand what is observed. Theoretical measurements can be only performed by subjects (or agents) who understand the theory and the technical devices used in the lab in order to reproduce *hic et nunc* an already represented theoretical experience. A physical measurement implies thus the most difficult

[j]It is important not to confuse a 'conscious subject' with a 'system' described by a theory.

balance between theory and observation. Theoretical experience requires conceptual and mathematical frameworks, but theories are nothing without an adequate contact to empirical observation. This is what physical (or theoretical) measurement is all about.

PHYSICAL (OR THEORETICAL) MEASUREMENT: *The point of contact between the objective theoretical representation of a specific situation and the hic et nunc conscious act of observation performed by a subject (or agent) who knows the theory and interprets both the experiment and the phenomena accordingly.*

At this point it is also important to clearly distinguish between a theoretical measurement and an operational procedure which can also end up in the prediction of an observation. Like theories, some mathematical models are also capable of predicting specific observations. However, in the case of models there is in general no conceptual-formal unity, and consequently, no consistency achieved by the representation. A very good example of this is Bohr's quantum model of the atom which is a set of "magical" rules which allow to compute the spectral lines of the Helium atom. Bohr himself accepted in many discussions with Heisenberg, Pauli and even Schrödinger that his model was not a theory; it was not only inconsistent, it simply did not provide a tenable representation of how to think about the atom. It was all these recognized problems that led the efforts of the time to develop a theory with a closed mathematical formalism that would allow to explain and understand the observed phenomena. An operational observation does not necessarily imply a formal conceptual representation. Unlike physical theories which describe 'states of affairs' through formal and conceptual moments of unity provided in invariant and objective terms, algorithmic models are only capable of making reference to measurement outcomes, like 'clicks' in detectors and 'spots' in photographic plates which refer to nothing beyond themselves.

OPERATIONAL OBSERVATION: *The observation of a measurement result, like a 'click' in a detector or a 'spot' in a photographic plate, which is predicted in operational terms through a an algorithmic model with no formal nor conceptual unity.*

An operational observation does not require the reference to a physical concept that captures, as a moment of unity, the phenomena in question. In fact, the reference to 'clicks' in detectors in the context of QM, exposes the complete lack of conceptual representation.[k] This is the reason why QM must be still regarded as a proto-theory which, even though possesses a closed mathematical formalism capable of operational quantitative predictions, is still in need of a conceptual representation capable to explain qualitatively the theoretical experience it talks about.[l] This should be regarded as one of the main realist (or physical) problems within QM; i.e., the development of a conceptual representation that provides not only a conceptual *moment of unity* to the mathematical formalism, but also an *anschaulich* (intuitive) content to the theory.

6. On what 'measurement' is not: From Bohr to Neo-Bohrians

During the re-foundation of physics in the 20th Century, one of the most influential alterations was produced by Niels Bohr's redefinition of the notion of 'measurement' in the context of QM. This new characterization of the meaning of measurement goes hand in hand with what could be called 'Bohr's pendular scheme of argumentation' which consisted on a balanced oscillation between an operational (or instrumentalist) reference to 'clicks' in detectors and an atomist (metaphysical) narrative grounded on fictional notions such as quantum particles and jumps, both of them connected via the *ad hoc* introduction of what he called the *correspondence principle*. This allowed Bohr to create a circular justification that remained constantly in motion without ever reaching any reference nor understanding. After many small battles, it was in 1935, that Bohr was able to popularize his pendular scheme within science when most physicists accepted uncritically his reply

[k]It is not true that 'clicks' predicted by the quantum formalism can be regarded — as Bohr and positivists did — as being "classical". It is Boole-Bell inequalities which have proven that such 'clicks' cannot be considered within any classical theory which represents reality as an actual state of affairs (for a detailed discussion see [30]). The predictions implied by the theory of quanta simply cannot be regarded as arising from the classical presuppositions implied by the theories of Newton and Maxwell. In short, Boole-Bell inequalities must be understood as providing the very conditions of possible classical experience [63]. Aspect' operational measurements have proven that the 'clicks' arising in an EPR type experiment cannot be regarded as 'classical clicks' (see [2]). It is interesting to notice that the Bohrian reference to 'clicks' in quantum experiments as being "classical" is the kernel point that demonstrates that — after all — Bohr was much closer to empiricism than to Kantian philosophy.

[l]This is what Tim Maudlin has characterized as a "recipe" which in the context of QM allows to make predictions from the mathematical formalism. "What is presented in the average physics textbook, what students learn and researchers use, turns out not to be a precise physical theory at all. It is rather a very effective and accurate recipe for making certain sorts of predictions. What physics students learn is how to use the recipe."

to the famous EPR paper [8] as the final triumph of quantum theory over Einstein's critical remarks. Without making any fuzz of the many lacunas, ambiguities and even inconsistencies within the paper, Bohr was congratulated and applauded by the physicist community as the new champion. In his reply, the Danish physicist had begun by immediately shifting the focus of analysis from EPR's theoretical definition of physical reality to the discussion about the applicability of classical measurement apparatuses. Once the classical representation was introduced within the discussion with extreme detail, he created a story — with no relation to the mathematical formalism nor to any operational test — through the introduction of irrepresentable *fictional notions* such as 'quantum particles', 'quantum jumps', 'quantum individuality', etc. In this way, according to Bohr's narrative, the measurement in QM was caused by the interaction of elementary quantum particles which affected through quantum jumps the classical measuring apparatuses in a manner that, due to the quantum of action, could not be described by the theory. As he argued [8, p. 701]: "The impossibility of a closer analysis of the reactions between the particle and the measuring instrument is indeed no peculiarity of the experimental procedure described, but is rather an essential property of any arrangement suited to the study of the phenomena of the type concerned, where we have to do with a feature of [quantum] individuality completely foreign to classical physics." Bohr had already applied this "solution" in a discussion with Schrödinger about the existence of "quantum jumps" within the atom. During a meeting in Copenhagen in 1926 under the attentive gaze of Heisenberg, Schrödinger presented several arguments exposing not only the lack of explanation but also the contradictions reached when introducing these a-causal jumps. The lack of theoretical support allowed Schrödinger [51, p. 73] to conclude that "the whole idea of quantum jumps is sheer fantasy." But while the Austrian physicist was not willing to accept the complete lack of theoretical representation within the atom — a critical position shared only by Einstein —, Bohr was ready to blame the lack of a consistent representation to the theory of quanta itself. As he explained to Schrödinger:

> What you say is absolutely correct. But it does not prove that there are no quantum jumps. It only proves that we cannot imagine them, that the representational concepts with which we describe events in daily life and experiments in classical physics are inadequate when it comes to describing quantum jumps. Nor should we be surprised to find it so, seeing that the processes involved are not the objects of direct experience. [51, p. 74]

According to Bohr, QM went beyond our classical image of the world, and thus — he argued — it was no surprise that our (classical) concepts were incapable of explaining what was really going on in the quantum domain. QM referred to a microscopic realm that simply could not be represented. There was nothing to be done but accept the physical and technical limitations we had finally reached within science. According to Bohr, these epistemological restrictions were ontologically imposed by Nature herself. In this way, Bohr turned his own incapacity to develop a consistent representation of QM into a proof of the theory's own difficulties and limits.[m] As a leader of the community, Bohr [72, p. 7] forbid physicists to go beyond the limits imposed by classical concepts. No one was allowed to look for a different solution. As he warned everyone: "it would be a misconception to believe that the difficulties of the atomic theory may be evaded by eventually replacing the concepts of classical physics by new conceptual forms." Of course, as we have just seen, Bohr did not respect his own dictum and repeatedly introduced many non-classical concepts (such as quantum particles, quantum jumps, quantum waves, etc.) which allowed him to create illusions which, in turn, also allowed him to avoid explanations. These newly introduced notions didn't really mean anything, they were not related to the mathematical formalism nor could be operationally tested. However, they did play an essential role within Bohr's scheme, namely, to stop difficult questions. In this respect, David Deutsch, one of the few contemporary physicists who has criticized in depth Bohr's scheme of thought, has remarked the following:

> Let me define 'bad philosophy' as philosophy that is not merely false, but actively prevents the growth of other knowledge. In this case, instru-

[m]Recognizing the danger of Bohr's scheme Karl Popper [64] strongly criticized Bohr's complementarity solution. Popper hoped that physicists would recognize the untenability of Bohr's proposal: "I trust that physicists will soon come to realize that the principle of complementarity is *ad hoc*, and (what is more important) that its only function is to avoid criticism and to prevent the discussion of physical interpretations; though criticism and discussion are urgently needed for reforming any theory. They will then no longer believe that instrumentalism is forced upon them by the structure of contemporary physical theory." Unfortunately, Popper's expectations were not fulfilled. Some decades after Deustch [33, pp. 309-310] characterized the effect of Bohr's scheme: "For decades, various versions of all that were taught as fact — vagueness, anthropocentrism, instrumentalism and all — in university physics courses. Few physicists claimed to understand it. None did, and so students' questions were met with such nonsense as 'If you think you've understood quantum mechanics then you don't.' Inconsistency was defended as 'complementarity' or 'duality'; parochialism was hailed as philosophical sophistication. Thus the theory claimed to stand outside the jurisdiction of normal (i.e. all) modes of criticism — a hallmark of bad philosophy."

> mentalism was acting to prevent the explanations in Schrödinger's and Heisenberg's theories from being improved or elaborated or unified. The physicist Niels Bohr (another of the pioneers of quantum theory) then developed an 'interpretation' of the theory which later became known as the 'Copenhagen interpretation'. It said that quantum theory, including the rule of thumb, was a complete description of reality. Bohr excused the various contradictions and gaps by using a combination of instrumentalism and studied ambiguity. He denied the 'possibility of speaking of phenomena as existing objectively' — but said that only the outcomes of observations should count as phenomena. He also said that, although observation has no access to 'the real essence of phenomena', it does reveal relationships between them, and that, in addition, quantum theory blurs the distinction between observer and observed. As for what would happen if one observer performed a quantum-level observation on another, he avoided the issue [...] [33, p. 308]

The Bohrian technique of argumentation consisted in discussing on two disconnected parallel levels jumping back and forth from one to the other, avoiding in this way questions that could not be answered. The methodology is quite simple, if a question cannot be answered in one level, simply shift to the other. Moving from our "common sense" manifest image of the world to fictitious stories about a quantum realm that could not be represented, Bohr was able to create the illusion of understanding. His circular analysis begun with classically described experimental arrangements and ended up in the prediction of 'clicks' in detectors consequence of unseen and irrepresentable quantum particles. As remarked by Deutsch, Bohr's scheme might reminds us of conjuring tricks rather than scientific explanation.

> Some people may enjoy conjuring tricks without ever wanting to know how they work. Similarly, during the twentieth century, most philosophers, and many scientists, took the view that science is incapable of discovering anything about reality. Starting from empiricism, they drew the inevitable conclusion (which would nevertheless have horrified the early empiricists) that science cannot validly do more than predict the outcomes of observations, and that it should never purport to describe the reality that brings those outcomes about. This is known as instrumentalism. It denies that what I have been calling 'explanation' can exist at all. It is still very influential. In some fields (such as statistical analysis) the very word 'explanation' has come to mean prediction, so that a

> mathematical formula is said to 'explain' a set of experimental data. By 'reality' is meant merely the observed data that the formula is supposed to approximate. That leaves no term for assertions about reality itself, except perhaps 'useful fiction'. [33, p. 15]

In this way, theoretical explanation became confused with the *ad hoc* postulation of rules and inconsistent interpretations became confused with theoretical explanation. Bohr created a fictional quantum narrative which allowed physicists to believe in something they did not understand. The power of Bohr's pendular scheme was that it allowed a dual discourse. While the realist believer could claim that quantum particles truly existed and were responsible for the macroscopic world we observe, the anti-realist was allowed to maintain their sceptic position according to which QM only made reference to the prediction of 'clicks' in detectors or 'spots' in photographic plates. Both assertions co-existed in Bohr's inconsistent rhetorics. The replacement of theoretical (formal-conceptual) representation by *ad hoc* rules and fictional stories meant also the abandonment of the methodology and goals implied by physical research (section 4). Bohr's program, in line with the positivist *Zeitgeist* of the 20th Century, was then ready to take a step further and redefine the meaning of 'physics' itself:

> Physics is to be regarded not so much as the study of something a priori given, but rather as *the development of methods of ordering and surveying human experience.* In this respect our task must be to account for such experience in a manner independent of individual subjective judgement and therefor *objective*[n] *in the sense that it can be unambiguously communicated in ordinary human language.*'[10] (emphasis added)

This shift — which was already part of the positivist scheme of thought — turned physics away from *physis* and closer to the (empirical) subject's "common sense" observations and measurements. Since the concept of measurement had always made reference to physical reality, the redefinition of a new 'physics' detached from *physis* implied also the need to redefine the notion of 'measurement' accordingly. Making use of his pendular rhetorics, Bohr provided two confronting definitions of 'measurement'. On the one hand, he [9, p. 209] related measurement to the intersubjective communication of observations by (empirical) subjects arguing that "by the word 'experiment' we refer to a situation where we can tell others what we have

[n]'Objective' for Bohr means in fact 'intersubjective'. See section 6.3.

done and what we have learned and that, therefore, the account of the experimental arrangement and of the results of the observations must be expressed in unambiguous language with suitable application of the terminology of classical physics."

MEASUREMENT COMMUNICATION: *The here and now conscious act of observation of a subject (or agent) who is capable to communicate what she observed to other subjects (or agents).*

On the other hand, Bohr also made reference to an understanding of 'measurement' in terms of 'interacting systems'. Shifting to a metaphysical level of description, it was now assumed that a measurement was essentially a process in which a 'click' in a detector or the 'imprint' in a photographic plate was produced due to the interaction between a quantum particle and a classical apparatus: "In fact to measure the position of one of the particles can mean nothing else than to establish a correlation between its behavior and some instrument rigidly fixed to the support which defines the space frame of reference." Once again, the existence of this process was constructed in a fictional manner without providing any consistent link to the mathematical formalism of the theory.

MEASUREMENT INTERACTION: *The interaction between a system and an apparatus produced by an incontrollable interaction (due to the quantum of action) which allows a correlation between them. This process, independent of subjects, ends up in the imprint of a 'spot' in a photographic plate or the sound of a 'click' in a detector.*

Adding to the confusion, in his pendular fashion, Bohr argued:

> As we have seen, any observation necessitates an interference with the course of the phenomena, which is of such a nature that it deprives us of the foundation underlying the causal mode of description. The limit, which nature herself has thus imposed upon us, of the possibility of speaking about phenomena as existing objectively finds its expression, as far as we can judge, just in the formulation of quantum mechanics. The discovery of the quantum of action shows us, in fact, not only the natural limitation of classical physics, but, by throwing a new light upon the old philosophical problem of the objective existence of phenomena independently of our observations, confronts us with a situation hitherto unknown in natural science. [7, p. 115]

Quite irrespectively of the internal contradictions, inconsistencies and ambiguities, Bohr's pendular scheme has continued to play an essential role within the contemporary debates about QM. Today, there are many interpretations which expose the impact of Bohr's ideas, arguments and forms of reasoning within the specialized contemporary foundational and philosophical literature. In the following, we are interested in discussing three different approaches which develop the understanding of measurement in QM in a neo-Bohrian fashion: Zurek's model of decoherence, modal interpretations and QBism.

6.1. *A Neo-Bohrian "technical solution": Decoherence*

From a positivist viewpoint which tends to focus in the analysis of formal (mathematical and logical) models which are able to account for empirical observations, one of the main problems in the writings of Bohr is the complete lack of any reference to the orthodox mathematical formalism of QM. Bohr systematically evaded such reference and shifted his attention — and that of the community — to the analysis of classical measurement situations. He justified himself by explaining that after all [72, p. 7]: "[...] the unambiguous interpretation of any measurement must be essentially framed in terms of classical physical theories, and we may say that in this sense the language of Newton and Maxwell will remain the language of physicists for all time." At the beginning of the 1970s, attempting to fill this formal void, Dieter Zhe [73] discussed the meaning of measurement in QM in a paper that would pioneer what would be later on known as decoherence. Decoherence was conceived by Zeh as a model that would finally bridge the gap between QM and our classical macroscopic world; explaining how weird microscopic quantum superpositions could end up transforming themselves into tables and chairs. During the 1980s, Wojciech Zurek popularized these ideas relating them more explicitly to Bohr's understanding(s) of measurement and the quantum to classical limit [74,75]. There is an essential circularity in the proposal since the application of particle metaphysics to the quantum formalism is what needs to be explained, instead of presupposed. One can easily understand the way in which "small particles" are able to constitute "big objects", what is difficult is to explain is the way in which quantum superpositions can be represent "small particles" or how entangled states can become a table.

Using an analogy with classical thermodynamics with no obvious link to the mathematical formalism of QM a new fictional notion was added to the scheme of decoherence, namely, that of *quantum harmonic oscillator.*

By considering an infinite sum of such "harmonic oscillators" the "coherence" of quantum particles was expected to decrease. Adding to the confusion, Everett's relative state formulation — closely connected to Bohr's ideas on contextuality — was another essential condiment of the original scheme. However, regardless of the efforts to create a narrative, according to the formalism decoherence never actually took place. It was found that a sum of quantum states retain their entanglement independently of how many of them are considered. Even an infinite sum of quantum states do not de-cohere. At this kernel point, in a complete Bohrian fashion, it was argued that what was in fact needed was the addition of a (classical) "continuum bath" also called "environment". After all — it was argued —, what surrounds the quantum microscopic realm is our classical macroscopic one. This shift from a (quantum) *discrete* representation — consequence of Planck's quantum postulate — to a (classical) *continuous* representation might seem to the attentive reader like a desperate attempt to impose a solution — instead of finding it. Indeed, an infinite numerable sum of Hamiltonians of elementary harmonic oscillators with natural frequencies[o] is not the same as an integral of the Hamiltonians of a continuum of oscillators with real frequencies. It is this mathematical addition of the continuous which hides the *ad hoc* imposition a classical representation. Thus, that which needed to be physically explained in conceptual and formal terms (i.e., the appearance of classical systems from the quantum formalism) was simply presupposed. The introduction of the "continuum bath" (of harmonic oscillators) was justified through the "commonsensical idea" that an "open system" (i.e., a great number of interacting systems) is "more real", or "less idealized", than a "closed system" (i.e., a completely isolated system). The argument points to the seemingly obvious fact that what actually happens in our world is that systems are always in interaction with other systems. Due to technical reasons it might be impossible to "close a system" completely and that is the reason why quantum states become classical. After going beyond representation and treating systems as 'things' in the real world, the argument also makes use of the Bohrian-positivist presupposition according to which an 'object', in order to bear existence, must be observed by someone or something. As Bohr stressed repeatedly, a quantum system cannot be described independently of its context of existence, it must be always related to a classical apparatus or — in the case of decoherence —

[o]It should be remarked that the notion of harmonic oscillator has a clear meaning within classical physics; however, its extension to quantum mechanics is far from evident. If QM does not describe 'particles' nor 'waves', what is then oscillating?

to the environment. However, as we have argued above, a theoretical representation characterizes the existents it talks about and the interactions between them are just part of the representation; a representation cannot be considered as "open" or "closed". The representation of a "single particle", is as abstract as that of "many particles". In fact, one cannot think of "many particles" without presupposing the representation of a "single particle". Thermodynamics is a generalization of classical mechanics, but since it makes only sense by presupposing it, it would be ridiculous to claim that thermodynamics is "more real" than classical mechanics. It is simply ridiculous to claim that if we consider many particles the representation becomes "more real" than if we consider only one particle — since the first presupposes the latter. The problem is there is in QM an intrinsic discreteness due to the quantum postulate inherent within the mathematical formalism which makes the description of a single particle untenable. In fact, Max Planck created QM by replacing an integer of *continuous* energy by a sum of quantum packages of *discrete* energies. That is the whole point about the *quantum*, namely, that it is *discrete*. Making use of the "common sense" atomist idea that it is impossible to "close a system" completely, the discrete sum of superpositions is blamed for being part of an "inaccurate" representation of reality. As if the idea of "open system" — which is nothing but many interacting (closed) systems — was not part of the same representation, the addition of the continuum becomes then a necessary condition for a more accurate representation of things *as they really are.* Decoherence not only shows the complete lack of comprehension regarding the scope of theoretical representation and measurement, it also exposes the inconsistency of the program itself for if quantum superpositions are "less real" than the environment — which is "closer to reality" — why should we even bother in finding a limit?

In addition to these unjustifiable formal jumps and *ad hoc* maneuvers in the conceptual and formal descriptions, there are many other technical aspects which also expose the failure of the program. The fact that the diagonalization is not complete, since "very small" is obviously not "equal to zero",[p] the fact that the diagonalization can recompose itself into un-diagonalized mixtures if enough time is considered [1,18] and the fact that the principle turns (non-diagonal) improper mixtures into ("approxi-

[p]Notice that within an epistemological account, "very small" might be considered as superfluous when compared to "very big"; however, this is clearly not the case from an ontological account. From an ontological perspective there is no essential difference between "very big" and "very small", they both have exactly the same importance.

mately" diagonal) improper mixtures which still cannot be interpreted in terms of ignorance are just a few of the many failures of the decoherence program.[q] After many criticisms within the specialized literature, and just to make things even less clear, Zurek [76, p. 22] — in a truly Bohrian spirit — decided to introduce also a subjective account of measurement by arguing that: "Quantum state vectors can be real, but only when the superposition principle — a cornerstone of quantum behavior — is 'turned off' by einselection. Yet einselection is caused by the transfer of information about selected observables. Hence, the ontological features of the state vectors — objective existence of the einselected states — is acquired through the epistemological 'information transfer'."

It was due to the insistent critical analysis coming mainly from philosophers of QM, that the failure of the decoherence program to explain the quantum to classical limit and the measurement problem had to be recognized by its supporters.[r] But in an amazing rhetorical move — which reminds us Bohr's method —, making use of the fact that some operational results and models had been already constructed, decoherent theorists argued that even though decoherence did not provide a theoretical explanation of the quantum to classical limit nor of the measurement problem, it did provide a solution "For All Practical Purposes" (a "FAPP solution"). By creating a new type of "solution", the process of decoherence justified its own existence and became part — together with "quantum jumps" and "quantum particles" — of the contemporary postmodern narrative of quantum physics. Today, the widespread acceptance of the decoherence shows the influence of Bohr's fictional-instrumentalist legacy. The creation of a new science which does not need to provide a consistent theoretical account of what it talks about and allows to justify itself through the application of operational models supported by a fragmented inconsistent fictions and illusions.

[q]The late recognition of this fact by Zurek has lead him to venture into many worlds interpretation in which case there are also serious inconsistencies threatening the project [20]. Ruth Kastner has even pointed out quite clearly why — even if these many points would be left aside — the main reasoning of the decoherence program is circular [54].

[r]Regardless of this, as remarked by Guido Bacciagaluppi [5]: "[some physicists and philosophers] still believe decoherence would provide a solution to the measurement problem of quantum mechanics. As pointed out by many authors, however (e.g. Adler 2003; Zeh 1995, pp. 14-15), this claim is not tenable. [...] Unfortunately, naive claims of the kind that decoherence gives a complete answer to the measurement problem are still somewhat part of the 'folklore' of decoherence, and deservedly attract the wrath of physicists (e.g. Pearle 1997) and philosophers (e.g. Bub 1997, Chap. 8) alike."

6.2. *A Neo-Bohrian "formal solution": Modal interpretations*

Almost concomitant with the creation of decoherence, at the beginning of the 1980s Bas van Fraassen, one of today's most influential contemporary empiricists, proposed another Neo-Bohrian interpretation of QM. Van Fraassen's main idea was to introduce modal logics in order to provide a consistent account of the theory in empiricist terms [69, pp. 202-203] which meant for him: "to withhold belief in anything that goes beyond the actual, observable phenomena, and to recognize no objective modality in nature. To develop an empiricist account of science is to depict it as involving a search for truth only about the empirical world, about what is actual and observable." Following Bohr's ideas regarding the purely algorithmic understanding of the quantum wave function, van Fraassen [70, p. 288] argued that: "[the emergence of a result is] *as if the Projection Postulate were correct.* For at the end of a measurement of **A** on system X, it is indeed true that **A** has the actual value which is the measurement outcome. But, of course, the Projection Postulate is not really correct: there has been a transition from possible to actual value, so what it entailed about values of observables is correct, but that is all. There has been no acausal state transition." Since the *possible* was addressed in the terms of modal logics, van Fraassen's interpretation became known as the "modal interpretation". Following the same line of reasoning, Dennis Dieks has recently tried to explain why the quantum measurement problem was never explicitly considered by the Danish physicist:[s]

> [...] measuring devices, like all macroscopic objects around us, can and must be described classically. It is an immediate consequence of this that measurements necessarily have only one single outcome. Pointers can only have one position at a time, a light flashes or does not flash, and so on — this is all inherent in the uniqueness of the classical description. Because of this, Bohr's interpretation does not face the "measurement problem" in the form in which it is often posed in the foundational literature, namely as the problem of how to explain — in the face of the

[s]Regarding the measurement problem, Petersen [62, p. 249] in his letter of 1957 to Hugh Everett argued in the same line as Dieks: "There can on [Bohr's] view be no special observational problem in quantum mechanics in accordance with the fact that the very idea of observation belongs to the frame of classical concepts. The aim of [Bohr's] analysis is only to make explicit what the formalism implies about the application of the elementary physical concepts. The requirement that these concepts are indispensable for an unambiguous account of the observations is met without further assumptions [...]."

> presence of superpositions in the mathematical formalism — that there is only one outcome realized each time we run an experiment. For Bohr this is not something to be explained, but rather something that is given and has to be assumed to start with. It is a primitive datum, in the same sense that the applicability of classical language to our everyday world is a brute fact to which the interpretation of quantum mechanics necessarily has to conform. An interpretation that would predict that pointers can have more than one position, that a cat can be both dead and alive, etc., would be a non-starter from Bohr's point of view. So the measurement problem in its usual form does not exist; it is dissolved. [39, p. 24]

This dissolution of the measurement problem simply re-states the naive empiricist standpoint present in the empirical-positivist and Bohrian understanding of theories in general, and of QM in particular. That's fair. But going beyond the empiricist reference to actualities, Dieks himself advanced in the late 1980s a "realist" version of the modal interpretation in which quantum particles were explicitly addressed. Rather than making reference to measurement outcomes, QM should be better understood "in terms of properties possessed by physical systems, independently of consciousness and measurements (in the sense of human interventions)" [37]. Following Bohr's ideas on contextuality and Simon Kochen's proposal [55], Dieks took the Schmidt (bi-orthogonal) decomposition and preferred basis as a necessary standpoint for interpreting QM.[t]

Theorem 6.1. *Given a state* $|\Psi_{\alpha\beta}\rangle$ *in* $\mathcal{H} = \mathcal{H}_\alpha \otimes \mathcal{H}_\beta$. *The Schmidt theorem assures there always exist orthonormal bases for* $\mathcal{H}_\alpha$ *and* $\mathcal{H}_\beta$, $\{|a_i\rangle\}$ *and* $\{|b_j\rangle\}$ *such that* $|\Psi_{\alpha\beta}\rangle$ *can be written as*

$$|\Psi_{\alpha\beta}\rangle = \sum c_j |a_j\rangle \otimes |b_j\rangle.$$

The different values in $\{|c_j|^2\}$ *represent the spectrum of the state. Every* λ_j *represents a projection in* $\mathcal{H}_\alpha$ *and a projection in* $\mathcal{H}_\beta$ *defined as* $P_\alpha(\lambda_j) = \sum |a_j\rangle\langle a_j|$ *and* $P_\beta(\lambda_j) = \sum |b_j\rangle\langle b_j|$, *respectively. Furthermore, if the* $\{|c_j|^2\}$

[t]It is interesting to notice that Carl Friedrich von Weizsäcker and Theodor Görnitz [48, p. 357] referred specifically to Kochen's proposal in a paper entitled "Remarks on S. Kochen's Interpretation of Quantum Mechanics". In this paper they state: "We consider it is an illuminating clarification of the mathematical structure of the theory, especially apt to describe the measuring process. We would, however feel that it means not an alternative but a continuation to the Copenhagen interpretation (Bohr and, to some extent, Heisenberg)."

are non degenerate, there is a one-to-one correlation between the projections $P_\alpha = \sum |a_j\rangle\langle a_j|$ *and* $P_\beta = \sum |b_j\rangle\langle b_j|$ *pertaining to subsystems* $\mathcal{H}_\alpha$ *and* $\mathcal{H}_\beta$ *given by each value of the spectrum.* □

If we assume non-degeneracy the modal interpretation based on the Schmidt decomposition establishes a one-to-one correlation between the reduced states of 'system' and 'apparatus'. As noted by Kochen [55, p. 152]: "Every interaction gives rise to a unique correlation between certain canonically defined properties of the two interacting systems. These properties form a Boolean algebra and so obey the laws of classical logic." The bi-orthogonal decomposition provides in this way a one-to-one correlation between apparatus and (quantum) system according to the following interpretation: *The system* α *possibly possesses one of the properties* $\{|a_j\rangle\langle a_j|\}$*, and the actual possessed property* $|a_k\rangle\langle a_k|$ *is determined by the device possessing the reading* $|b_k\rangle\langle b_k|$.[u] However, there is an essential drawback. Tracing over the degrees of freedom of the system, one obtains an *improper mixture* which cannot be interpreted in terms of ignorance [31]. Consequently, the path from the possible to the actual cannot be regarded as making reference to an underlying preexistent system with definite valued properties. At this crucial point, in a truly Bohrian spirit, Dieks returned to the safety of an instrumentalist account of the theory which — leaving 'quantum systems' behind — goes back to the possibility of predicting measurement outcomes.[v] The path from the mathematical formalism to the actual observation was then postulated by Dieks in terms of an *ad hoc* algorithmic rule which is completely equivalent to the projection or measurement postulate:

> I now propose the following interpretational rule: as soon as there is a unique decomposition of the form $[|\Psi_{\alpha\beta}\rangle = \sum c_j|\phi_j\rangle \otimes |R_j\rangle]$, the partial system represented by the $|\phi_k\rangle$, taken by itself, can be described as possessing one of the values of the physical quantity corresponding to the set $|\phi_k\rangle$, with probability $|c_k|^2$.

---

[u] As Bacciagaluppi [4] points out with respect to Kochen's interpretation: "It seems that he conceives [the states in the Schmidt decomposition] rather as states that are relative to each other. It seems that he espouses an Everettian view that systems have states only relative to each other but that he considers the ascription of relative states (in Everett sense) only in the symmetrical situation in which not only the $|a_j\rangle$ are relative to the $|a_i\rangle$, but at the same time the $|a_j\rangle$ are relative to the $|a_i\rangle$."

[v] In order to do so, while van Fraassen's distinguishes between *dynamical states* and *value states* [70], Dieks and Vermaas do exactly the same by distinguishing between *mathematical states* and *physical states* [67].

> This rule is intended to have the following important consequence. Experimental data that pertain only to the object system, and that say it possesses the property associated with, e.g., $|\phi_1\rangle$, not only count as support for the theoretical description $|\phi_1\rangle|R_1\rangle$ but also as empirical support for the theoretical description (2). [35, p. 39]

Instead of using the mathematical formalism in order to describe systems with definite properties and the way they end up producing 'clicks', the formalism was suddenly applied in an instrumentalist fashion in order to predict the actual observation of measurement outcomes. Dieks [34, p. 182] argued that "[...] there is no need for the projection postulate. On the theoretical level the full superposition of states is always maintained, and the time evolution is unitary. One could say that the 'projection' has been shifted from the level of the theoretical formalism to the semantics: it is only the empirical interpretation of the superposition that the component terms sometimes, and to some extent, receive an independent status." The *projection postulate* was renamed as an "interpretational rule" which was accepted at the level of operational prediction but rejected at the level of the realist interpretation. Suddenly, the reference to 'systems' and 'properties' disappeared and Dieks [34, p. 177] went back to an analysis about measurement outcomes: "[...] an irreducible statistical theory only speaks about possible outcomes, not about the actual one; this predicts only probability distributions of all outcomes, and says nothing about the result which really will be realized in a single case. In brief, such a theory is not about what is real and actual but only about what could be the case." As it becomes clear, this instrumentalist interpretation of quantum modalities makes no contact with an underlying realist interpretation about 'systems'. Dieks — following the Bohrian scheme — goes back and forth between a realist interpretation of the mathematical formalism which supposedly describes quantum systems (independently of measurement outcomes) and an empiricist pragmatic interpretation of the same formalism used as a black box in order to predict actual measurement outcomes (see [38]). The failure of the project extends itself to the *preferred* Schmidt basis which becomes only essential in creating a new formal narrative about quantum systems that interact with classical apparatuses.

> [...] there should be a one-to-one correlation between the definite properties of a system and the definite properties of its environment [...] can be seen as a way to generalize (and make rigorous) a significant part of Bohr's interpretation of quantum mechanics. According to Bohr the

> applicability of concepts depends on the type of macroscopic measuring device that is present; given a 'phenomenon' there is a one- to-one correspondence between the properties of the measuring device and those of the object system. Our second requirement implements this idea also in situations in which there is no macroscopic measuring device, but only a correlation with the (possibly microscopic) environment.
> The idea that there is a correspondence between properties of a system and those of its environment is also physically motivated by other approaches to the interpretation of quantum mechanics, especially the decoherence approach (the 'monitoring' of a system by its environment, see Ref. [5] and references therein). [36, p. 368]

Independently of the initial lacunas and ambiguities, during the 1990s Dieks' modal realistic reference to systems was confronted to the mathematical formalism of the theory itself. As a deeply provocative result, many no-go theorems specifically designed for modal interpretations begun to expose the already known difficulties to link the orthodox mathematical formalism with the notion of 'system' — i.e., a physical entity composed of definite valued properties.[w] Like in the case of Bohr, Dieks was then forced to assume an even more extreme form of relativism — already implicit in the self imposed restriction of modal interpretations to the Schmidt preferred basis. Dieks perspectivalist proposal co-authored with Gyula Bene, added a further non-invariant reference through the choice of a new factorization of the total Hilbert space [6]. As in the case of decoherence, the essential difficulty of the modal interpretation is that even though there might seem to exist an account of correlated systems, there is no independent representation of the constituents. So even though the 'interpretation' attempts to evade an account of 'single systems' — through the reference to the bi-orthogonal decomposition and factorizations — it nevertheless talks about composite ones creating the illusion of a reference which simply does not exist. The realist version of the modal interpretation fails to explain the basics. What is a 'quantum system'? What is a 'classical apparatus'? How does a 'quantum system' interact with a 'classical system' or how is a single 'click' generated by a particle? A realist interpretation should be able to answer these questions instead of providing an instrumentalist reply which makes only reference to the prediction of outcomes.

[w] We refer the interested reader to the detailed analysis presented in [21]. For an analysis of Kochen-Specker contextuality and its implications for the interpretation of projection operators see [26].

6.3. *A Neo-Bohrian "subjectivist solution": QBism*

Bohr's understanding of the mathematical formalism of QM as making exclusive reference to measurement outcomes had as a consequence a radical shift from the *objective* character of theoretical representation — in terms of conceptual moments of unity — to the *intersubjective* communication of individual observations between empirical subjects. In this way, the particular observations made by different agents were detached from a common *objective* reference and representation.[x] This also implied a silent replacement of the notion of *object* — categorically constituted through the general principles of existence, non-contradiction and identity — by that of *event* (e.g., 'clicks' and 'spots') — which has no categorical nor conceptual constitution.[y] Bohr shifted from objectivity to intersubjectivity, but he was not willing to abandon the term 'objective'. Instead, he simply renamed 'intersubjective statements' as 'objective statements'. Stressing the claim that his account of QM was as objective as classical physics he [32, p. 98] argued that: "The description of atomic phenomena has [...] a perfectly objective character, in the sense that no explicit reference is made to any individual observer and that therefore... no ambiguity is involved in the communication of observation." Bernard D'Espagnat explains this quote in the following manner: "That Bohr identified objectivity with intersubjectivity is a fact that the quote above makes crystal clear. In view of this, one cannot fail to be surprised by the large number of his commentators, including competent ones, who merely half-agree on this, and only with ambiguous words. It seem they could not resign themselves to the ominous fact that Bohr was not a realist."

With the arrival of the new millennium and in tune with the anti-realist postmodern *Zeitgeist* of the 20th Century, Quantum Bayesianism (QBism for short) was developed by a group of researchers as a new neo-Bohrian approach to QM. Following Bohr, QBism [46, p. 70] took as a standpoint the unspoken divorce between quantum theory and the representation of reality: "[...] quantum theory does not describe physical reality. What it does is provide an algorithm for computing probabilities for the macroscopic events ('detector clicks') that are the consequences of experimental

[x]In his book, [32], D'Espagnat clearly distinguishes between *objective statements* and Bohr's *intersubjective statements*, which he calls: *weakly objective statements*.
[y]While the object acts as a moment of unity that is able to account for the multiplicity of experience, the 'click' makes only reference to a fragmented experience with no internal unity.

interventions." As made explicit by Chris Fuchs [47]: "QBism agrees with Bohr that the primitive concept of experience is fundamental to an understanding of science."[z] It is in this context that the Bayesian subjectivist interpretation of probability is introduced.

> QBism explicitly takes the 'subjective' or 'judgmental' or 'personalist' view of probability, which, though common among contemporary statisticians and economists, is still rare among physicists: probabilities are assigned to an event by an agent and are particular to that agent. The agent's probability assignments express her own personal degrees of belief about the event. The personal character of probability includes cases in which the agent is certain about the event: even probabilities 0 and 1 are measures of an agent's (very strongly held) belief. [47, p. 750]

At least in a first stage, QBism allows to make explicitly clear the subjectivist standpoint implied by anti-realist approaches and avoids the pendular reference imposed by Bohr's inconsistent rhetorics.[aa] The QBist understanding of measurement is then exclusively related to a purely conscious act of observation [*Op. cit.*, p. 750]: "A measurement in QBism is more than a procedure in a laboratory. It is any action an agent takes to elicit a set of possible experiences. The measurement outcome is the particular experience of that agent elicited in this way. Given a measurement outcome, the quantum formalism guides the agent in updating her probabilities for subsequent measurements." Indeed, as QBist make explicitly clear: "A measurement does not, as the term unfortunately suggests, reveal a pre-existing state of affairs." Measurements are personal, individual and QM is a "tool" for the "user" — as Mermin prefers to call the "agent" [61]. Just like a mobile phone or a laptop, QM is a tool that we subjects use in order to organize our experience.

> QBist takes quantum mechanics to be a personal mode of thought — a very powerful tool that any agent can use to organize her own experience. That each of us can use such a tool to organize our own experience with spectacular success is an extremely important objective fact about the world we live in. But quantum mechanics itself does not deal directly

[z]In recent years David Mermin has become also part of the QBist team, publishing several papers which not only support, but also make clear the connection of QBism to the Bohrian interpretation of QM. See: [60,61].

[aa]Recently, Chris Fuchs has changed his anti-realist claims and presented something called "participatory realism".

> with the objective world; it deals with the experiences of that objective world that belong to whatever particular agent is making use of the quantum theory. [*Op. cit.*, p. 751]

According to Fuchs, measurement is an *irrepresentable* "act of creation":

> The research program of Quantum Bayesianism (or QBism) is an approach to quantum theory that hopes to show with mathematical precision that its greatest lesson is the world's plasticity. With every quantum measurement set by an experimenter's free will, the world is shaped just a little as it takes part in a moment of creation. So too it is with every action of every agent everywhere, not just experimentalists in laboratories. Quantum measurement represents those moments of creation that are sought out or noticed. [45, p. 2114]

QBism — following Bohr — accepts the failure of science to provide a representation of the world we live in.

> [...] a measurement apparatus must be understood as an extension of the agent herself, not something foreign and separate. A quantum measurement device is like a prosthetic hand, and the outcome of a measurement is an unpredictable, undetermined 'experience' shared between the agent and external system. Quantum theory, thus, is no mirror image of what the world is, but rather a 'user's manual' that any agent can adopt for better navigation in a world suffused with creation: The agent uses it for her little part and participation in this creation. [45, p. 2042]

Once representation is erased, the problem is also erased [66, p. 147]: "The measurement problem — from our view — is a problem fueled by the fear of thinking that quantum theory might be just the kind of user's manual theory for individual agents (contemplating the consequences of their individual interactions with quantum systems) that we have described in the previous answers. Take the source of the paradox away, we say, and the paradox itself will go away." Of course, the question that rises is the following: if a physical theory like QM does not provide understanding of the world and Nature, what discipline is supposed to do this? And if the stories told by science are not making reference to reality, to particles, fields or other type of entities, should we simply accept that all we observe as subjects is part of an illusion? A fictitious story we created ourselves?

7. The (quantum) measurement problem in classical mechanics

As we discussed above, the QMP has been created by forcing within the orthodox quantum formalism the idea that a theory must account for what is observed in the lab. This naive empiricist presupposition according to which there can be a "direct access" to the world that surround us by "simply observing what is going on" was fantastically addressed — and ironically criticized — by the Argentine writer Jorge Luis Borges in a beautiful short story called *Funes the memorious* [12].[bb] Borges recalls his encounter with Ireneo Funes, a young man from Fray Bentos who after having an accident become paralyzed. Since then, Funes' perception and memory became infallible. According to Borges, the least important of his recollections was more minutely precise and more lively than our perception of a physical pleasure or a physical torment. However, as Borges also remarked: "He was, let us not forget, almost incapable of general, platonic ideas. It was not only difficult for him to understand that the generic term dog embraced so many unlike specimens of differing sizes and different forms; he was disturbed by the fact that a dog at three-fourteen (seen in profile) should have the same name as the dog at three fifteen (seen from the front). [...] Without effort, he had learned English, French, Portuguese, Latin. I suspect, however, that he was not very capable of thought. To think is to forget differences, generalize, make abstractions. In the teeming world of Funes there were only details, almost immediate in their presence." Using the story as a *Gedankenexperiment* Borges shows why, for a radical empiricist as Funes, there is no reason to believe in the (metaphysical) identity of 'the dog at three-fourteen (seen in profile)' and 'the dog at three fifteen (seen from the front)'. For Funes, being capable of apprehending experience beyond conceptual presuppositions, there is no meaning to the word 'dog'. The experience of Funes is one that stands beyond language, beyond the (metaphysical) moments of unity required to make reference to *the same* through time. As Borges explains: "Locke, in the seventeenth century, postulated (and rejected) an impossible language in which each individual thing, each

[bb]The first cornerstone of positivism which was also partially accepted by Bohr's claim that experience is always "classical". Even though Bohr's neo-Kantian approach recognized the fact that classical experience was not a *given* but rather a complex constitution provided by the subjects transcendental categories and forms of intuition, the prohibition to consider any other experience placed observations in the same place as positivists: as the fundament of all possible scientific knowledge. This coincidence is what allowed to seal the silent alliance between positivists and neo-Kantians.

stone, each bird and each branch, would have its own name; Funes once projected an analogous language, but discarded it because it seemed too general to him, too ambiguous. In fact, Funes remembered not only every leaf of every tree of every wood, but also every one of the times he had perceived or imagined it." It is through Aristotle's application of the ontological and logical principles of existence, identity and non-contradiction, that science was able to provide a conceptual (metaphysical) solution to the problem of movement. This is the architectonic which allows to connect the 'the dog at three-fourteen (seen in profile)' and 'the dog at three fifteen (seen from the front)' in terms of a *sameness*, the same concept. Only by presupposing principles can we think in terms of 'individual systems' — such as, for example, a 'dog'. Borges shows why these principles are not self-evident *givens* of experience, and neither is a 'dog'. And this is the reason why Borges also suspected that Funes "was not very capable of thought." Funes understood that every observation is just a fragmented conglomerate of sensations with no causal relation to a subsequent one. Now, returning to our discussion, what is important to understand is that the QMP does not make reference to theoretical measurements, it makes reference to operational observations of 'clicks' in detectors and 'spots' in photographic plates — which is not what QM talks about. Such 'clicks' and 'spots' play exactly the same role as Funes' pure fragmented observations. Like Funes' static observations of pure sensation a 'click' has no reference beyond itself. A 'click' is not a conceptual *moment of unity*, it has no metaphysical constitution, and consequently it escapes theoretical representation right from the start. No wonder, it becomes then impossible to relate it to a physical concept or to the mathematical formalism.

We are now able to expose the untenability of the so called measurement problem which, in fact, should be called "the observability problem". If we assume that observations (e.g., a 'click' in a detector) have no need of a conceptual reference (i.e., a moment of unity), then an analogous measurement problem can be derived also in the case of classical mechanics. If, following the empirical-positivist understanding, a theory must describe observations, then it is easy to show that classical mechanics fails to do so exactly in the same way as QM. Let us explain this. Classical mechanics describes theoretically a 'table' in terms of a *rigid body*; i.e., a 3-dimensional object in which a number of atoms constitutes the table as an individual unity. Now, if we enter a lab in which there is a table, classical mechanics simply fails to tell us which profile of the table will be observed by us in each instant of time. Classical mechanics describes the table and its evolution

— in case there are forces acting on it —, but remains silent regarding the particular empirical observations made by agents. Classical mechanics simply does not talk about observations or measurements. Instead, it provides a (subject-independent) representation in terms of objects in space-time. If Funes would be taught the positivist understanding of theories and then go into the lab in order to observe a 'table', he would certainly not relate the table seen from the front at 3.15 with the table seen from above at 3.20. Observing the 'table' from different perspectives Funes would be right to conclude that classical mechanics simply fails to describe what he observed in each instant of time, namely, the profiles of the table — and not the whole table. He would be also completely justified in deriving the following measurement problem for classical mechanics.

Classical Measurement Problem (CMP): *Given a specific system of reference, classical mechanics describes mathematically a rigid body in terms of a system constituted in $\mathcal{R}^3$ by a set of different properties. Since classical mechanics predicts the existence of all properties and profiles simultaneously, the question is why do we observe a single profile instead of all the profiles at the same time?*

Classical mechanics simply doesn't tell us which profile we will actually observe, it only describes the object as a whole — independently of us observing it. So what is the path from the theoretical description provided by classical mechanics to the profile we observe? From an empirical-positivist perspective there must be also something wrong with Newtonian mechanics since it is not able to account for what we actually observe in each instant of time, namely, the particular profile of the table.

Acknowledgements

I want to thank Juan Vila for comments and discussions related to an earlier version of this paper. This work was partially supported by the following grants: Project PIO-CONICET-UNAJ (15520150100008CO) "Quantum Superpositions in Quantum Information Processing".

References

1. Anglin, J.R., Laflamme, R. & Zurek, W.H., 1995, "Decoherence, re-coherence, and the black hole information paradox", *Physical Review D*, **52**, 2221.
2. Aspect A., Grangier, P. & Roger, G., 1981, "Experimental Tests on Realistic Local Theories via Bell's Theorem", *Physical Review Letters*, **47**, 725–729.
3. Arenhart, J., 2019, "Bridging the Gap Between Science and Metaphysics, with a Little Help from Quantum Mechanics", in *Proceedings of the 3rd*

Filomena Workshop, p. 9–33, J.D. Dantas, E. Erickson and S. Molick (Eds.), PPGFIL, Natal.
4. Bacciagaluppi, G., 1996, *Topics in the Modal Interpretation of Quantum Mechanics*, Doctoral dissertation, University of Cambridge, Cambridge.
5. Bacciagaluppi, G., 2012, "The Role of Decoherence in Quantum Mechanics", *The Stanford Encyclopedia of Philosophy (Winter 2012 Edition)*, Edward N. Zalta (ed.), http://plato.stanford.edu/archives/win2012/entries/qm-decoherence/.
6. Bene, G. & Dieks, D., 2002, "A Perspectival Version of the Modal Interpretation of Quantum Mechanics and the Origin of Macroscopic Behavior", *Foundations of Physics*, **32**, 645–671.
7. Bohr, N., 1934, *Atomic theory and the description of nature*, Cambridge University Press, Cambridge.
8. Bohr, N., 1935, "Can Quantum Mechanical Description of Physical Reality be Considered Complete?", *Physical Review*, **48**, 696–702.
9. Bohr, N. 1949. "Discussions with Einstein on Epistemological Problems in Atomic Physics", in *Albert Einstein: Philosopher-Scientist*, edited by P. A. Schilpp, 201–41. La Salle: Open Court. Reprinted in (Bohr 1985, 9–49).
10. Bohr, N., 1960, *The Unity of Human Knowledge*, In *Philosophical writings of Neils Bohr*, vol. 3., Ox Bow Press, Woodbridge.
11. Bokulich, A., 2004, "Open or Closed? Dirac, Heisenberg, and the Relation between Classical and Quantum Mechanics", *Studies in History and Philosophy of Modern Physics*, **35**, 377–396.
12. Borges, J.L., 1989, *Obras completas: Tomo I*, María Kodama y Emecé (Eds.), Barcelona. Translated by James Irby from *Labyrinths*, 1962.
13. Bub, J., 1997, *Interpreting the Quantum World*, Cambridge University Press, Cambridge.
14. Bub, J., 2017, "Quantum Entanglement and Information", *The Stanford Encyclopedia of Philosophy (Spring 2017 Edition)*, Edward N. Zalta (ed.), URL = https://plato.stanford.edu/archives/spr2017/entries/qt-entangle/.
15. Bueno, O., 2016, "Von Neumann, Empiricism and the Foundations of Quantum Mechanics" in *Probing the Meaning of Quantum Mechanics: Superpositions, Semantics, Dynamics and Identity*, pp. 192–230, D. Aerts, C. de Ronde, H. Freytes and R. Giuntini (Eds.), World Scientific, Singapore.
16. Carnap, H., Hahn, H. & Neurath, O., 1929, "The Scientific Conception of the World: The Vienna Circle", *Wissendchaftliche Weltausffassung.*
17. Cordero, N.L., 2014, *Cuando la realidad palpitaba*, Buenos Aires, Biblos.
18. Cormick, C. & Paz J.P., "Decoherence of Bell states by local interactions with a dynamic spin environment", *Physical Review A*, **78**, 012357.
19. Curd, M. & Cover, J.A., 1998, *Philosophy of Science. The Central Issues*, Norton and Company (Eds.), Cambridge University Press, Cambridge.
20. Dawin, R. & Thébault, K., 2015, "Many worlds: Incoherent or decoherent?", *Synthese*, **192**, 1559–1580.
21. de Ronde, C., 2011, *The contextual and modal character of quantum mechanics: A formal and philosophical analysis in the foundations of physics*, Doctoral dissertation, Utrecht University, Utrecht.

22. de Ronde, C., 2014, "The Problem of Representation and Experience in Quantum Mechanics", in *Probing the Meaning of Quantum Mechanics: Physical, Philosophical and Logical Perspectives*, pp. 91–111, D. Aerts, S. Aerts and C. de Ronde (Eds.), World Scientific, Singapore.
23. de Ronde, C., 2016, "Representational Realism, Closed Theories and the Quantum to Classical Limit", in *Quantum Structural Studies*, pp. 105–135, R.E. Kastner, J. Jeknic-Dugic and G. Jaroszkiewicz (Eds.), World Scientific, Singapore.
24. de Ronde, C., 2016, "Probabilistic Knowledge as Objective Knowledge in Quantum Mechanics: Potential Immanent Powers instead of Actual Properties.", in D. Aerts, C. de Ronde, H. Freytes and R. Giuntini (Eds.), pp. 141–178, *Probing the Meaning of Quantum Mechanics: Superpositions, Semantics, Dynamics and Identity*, World Scientific, Singapore.
25. de Ronde, C., 2018, "Quantum Superpositions and the Representation of Physical Reality Beyond Measurement Outcomes and Mathematical Structures", *Foundations of Science*, **23**, 621–648.
26. de Ronde, C., 2020, "Unscrambling the Omelette of Quantum Contextuality (Part I): Preexistent Properties or Measurement Outcomes?", *Foundations of Science*, **25**, 55–76.
27. de Ronde, C. & Bontems, V., 2011, "La notion d'entité en tant qu'obstacle épistémologique: Bachelard, la mécanique quantique et la logique", *Bulletin des Amis de Gaston Bachelard*, **13**, 12–38.
28. de Ronde, C., Freytes, H. & Domenech, G., Quantum Logic in Historical and Philosophical Perspective, *The Internet Encyclopedia of Philosophy (IEP)*, URL: https://www.iep.utm.edu/qu-logic/
29. de Ronde, C. & Massri, C., 2017, "Kochen-Specker Theorem, Physical Invariance and Quantum Individuality", *Cadernos da Filosofia da Ciencia*, **2**, 107–130.
30. de Ronde, C. & Massri, C., 2018, "The Logos Categorical Approach to Quantum Mechanics: I. Kochen-Specker Contextuality and Global Intensive Valuations.", *International Journal of Theoretical Physics*, DOI: 10.1007/s10773-018-3914-0.
31. D'Espagnat, B., 1976, *Conceptual Foundations of Quantum Mechanics*, Benjamin, Reading MA.
32. D'Espagnat, B., 2006, *On Physics and Philosophy*, Princeton University Press, Princeton.
33. Deutsch, D., 2004, *The Beginning of Infinity. Explanations that Transform the World*, Viking, Ontario.
34. Dieks, D., 1988, "The Formalism of Quantum Theory: An Objective description of reality", *Annalen der Physik*, **7**, 174–190.
35. Dieks, D., 1988, "Quantum Mechanics and Realism", *Conceptus XXII*, **57**, 31–47.
36. Dieks, D., 1995, "Physical motivation of the modal interpretation of quantum mechanics", *Physics Letters A*, **197**, 367–371.
37. Dieks, D., 2007, "Probability in the modal interpretation of quantum mechanics", *Studies in History and Philosophy of Modern Physics*, **38**, 292–310.

38. Dieks, D., 2010, "Quantum Mechanics, Chance and Modality", *Philosophica*, **83**, 117–137.
39. Dieks, D., 2016, "Niels Bohr and the Formalism of Quantum Mechanics", in *Niels Bohr in the 21st Century*, J. Faye and H. Folse (Eds.), forthcoming.
40. Dieks, D., 2018, "Quantum Mechanics and Perspectivalism", preprint. (quant-ph:1801.09307)
41. Dirac, P.A.M., 1974, *The Principles of Quantum Mechanics*, 4th Edition, Oxford University Press, London.
42. Egg, M., 2018, "Dissolving the Measurement Problem Is Not an Option for the Realist", *Studies in History and Philosophy of Modern Physics*, forthcoming.
43. Einstein, A., 1949, "Remarks concerning the essays brought together in this co-operative volume", in *Albert Einstein. Philosopher-Scientist*, P.A. Schlipp (Eds.), pp. 665–689, MJF Books, New York.
44. Einstein, A., Podolsky, B. & Rosen, N., 1935, "Can Quantum-Mechanical Description be Considered Complete?", *Physical Review*, **47**, 777–780.
45. Fuchs, C.A., 2014, *My Struggles with the Block Universe*, preprint. (quant-ph: 1405.2390)
46. Fuchs, C.A. & Peres A., 2000, "Quantum theory needs no 'interpretation"', *Physics Today* **53**, 70–71.
47. Fuchs, C.A., Mermin, N.D. & Schack, R., 2014, "An introduction to QBism with an application to the locality of quantum mechanics", *American Journal of Physics*, **82**, 749. (quant-ph:1311.5253)
48. Gornitz, Th. & Von Weiszäcker, C.F., 1987, "Remarks on S. Kochen's Interpretation of Quantum Mechanics", in *Symposium on the Foundations of Modern Physics 1987*, pp. 365–367, P. Lathi and P. Mittelslaedt (Eds.), World Scientific, Singapore.
49. Hanson, R., 1958, *Patterns of Discovery: An Inquiry into the Conceptual Foundations of Science*, Cambridge University Press, Cambridge.
50. Heisenberg, W., 1958, *Physics and Philosophy*, World perspectives, George Allen and Unwin Ltd., London.
51. Heisenberg, W., 1971, *Physics and Beyond*, Harper & Row, New York.
52. Heisenberg, W., 1973, "Development of Concepts in the History of Quantum Theory", in *The Physicist's Conception of Nature*, pp. 264–275, J. Mehra (Ed.), Reidel, Dordrecht.
53. Hempel, C.G., 1958, "The theoretician's dilemma: A study in the logic of theory", *Minnesota Studies in the Philosophy of Science*, **2**, 173–226.
54. Kastner, R., 2014, "'Einselection' of pointer observables: The new H-theorem?", *Studies in History and Philosophy of Modern Physics*, **48**, 56–58.
55. Kochen, S., 1985, "A New Interpretation of Quantum Mechanics", in *Symposium on the Foundations of Modern Physics 1985*, 151–169, P. Lathi and P. Mittelslaedt (Eds.), World Scientific, Joensuu.
56. Kochen, S. & Specker, E., 1967, "On the problem of Hidden Variables in Quantum Mechanics", *Journal of Mathematics and Mechanics*, **17**, 59–87. Reprinted in Hooker, 1975, 293–328.

57. Laurikainen, K.V., 1998, *The Message of the Atoms, Essays on Wolfgang Pauli and the Unspeakable*, Springer Verlag, Berlin.
58. Mach, E., 1959, *The Analysis of Sensations*, Dover Edition, New York.
59. Maudlin, T., 2019, *Philosophy of Physics. Quantum Theory*, Princeton University Press, Princeton.
60. Mermin, D., 2004, "Copenhagen Computation: How I Learned to Stop Worrying and Love Bohr", *IBM Journal of Research and Development*, 48–53. (quant-ph:0305088)
61. Mermin, D., 2014, "Why QBism is not the Copenhagen interpretation and what John Bell might have thought of it", Preprint. (quant-ph:1409.2454)
62. Osnaghi, S., Freitas, F. & Freire, O., 2009, "The origin of the Everettian heresy" *Studies in History and Philosophy of Modern Physics*, **40**, 97–123.
63. Pitowsky, I., 1994, "George Boole's 'Conditions of Possible Experience' and the Quantum Puzzle", *The British Journal for the Philosophy of Science*, **45**, 95–125.
64. Popper, K.R., 1963, *Conjectures and Refutations: The Growth of Scientific Knowledge*, Routledge Classics, London.
65. Redei, M. & Stöltzner, M., "Soft Axiomatisation: John von Neumann on Method and von Neumann's Method in the Physical Sciences", Kluwer Academic Publishers, Dordrecht.
66. Schlosshauer, M. (Ed.), 2011, *Elegance and Enigma. The Quantum Interviews*, Springer-Verlag, Berlin.
67. Vermaas, P.E. & Dieks, D., 1995, "The Modal Interpretation of Quantum Mechanics and Its Generalization to Density Operators", *Foundations of Physics*, **25**, 145–158.
68. Von Neumann, J., 1955, *Mathematical Foundations of Quantum Mechanics.* Trans. Robert T. Geyer. Princeton, Princeton University Press, Princeton.
69. Van Fraassen, B.C., 1980, *The Scientific Image*, Clarendon, Oxford.
70. Van Fraassen, B.C., 1991, *Quantum Mechanics: An Empiricist View*, Oxford: Clarendon.
71. Van Fraassen, B.C., 2002, *The Empirical Stance*, Yale University Press, New Haven.
72. Wheeler, J.A. & Zurek, W.H. (Eds.) 1983, *Theory and Measurement*, Princeton University Press, Princeton.
73. Zhe, H.D., 1970, "On the Interpretation of Measurement in Quantum Theory", *Foundations of Physics*, **1**, 69–76.
74. Zurek, W., 1981, "Pointer Basis of Quantum Apparatus: Into What Mixture Does the Wave Packet Collapse?", *Physical Review D*, **24**, 1516.
75. Zurek, W., 1982, "Environment-Induced Superselection Rules", *Physical Review D*, **26**, 1862.
76. Zurek, W., 2002, "Decoherence and the Transition from Quantum to Classical—Revisited", *Los Alamos Science*, **27**, 2–25.

JOHN BELL AND THE GREAT ENTERPRISE*

ANTHONY SUDBERY

Department of Mathematics, University of York,
Heslington, York, England YO10 5DD
E-mail: tony.sudbery@york.ac.uk

I outline Bell's vision of the "great enterprise" of science, and his view that conventional teachings about quantum mechanics constituted a betrayal of this enterprise. I describe a proposal of his to put the theory on a more satisfactory footing, and review the subsequent uses that have been made of one element of this proposal, namely Bell's transition probabilities regarded as fundamental physical processes.

Keywords: Bell; probability; beables; histories.

1. Introduction: The great enterprise

John Bell was a scientist. That was a vocation that he followed with great respect, devotion and sense of responsibility. For him, to be a scientist was to participate in the "great enterprise" [1] of understanding the world we live in; in particular, to be a physicist was to pursue the grand vision of describing the physical world in terms of its ultimate constituents and delineating how those constituents behave. The great enterprise is undertaken according to the scientific method: first, carefully observe and experiment to see how what happens in the world; second, imaginatively construct theories to explain these observations; third, rigorously test these theories by calculating what they predict for the results of further experiments. If these predictions are successful, we can feel, diffidently and tentatively, that we have made progress towards our original goal of truly describing and understanding the world.

When Bell embarked on his career as a physicist, the furthest advances towards the goal of physics were represented by quantum mechanics, as

*This article is reprinted from *Quanta* **7** (2018), 68–73.
http://dx.doi.org/10.12743/quanta.v7i1.79

developed and expounded by Bohr, Born and Heisenberg. All the professional training he received followed the teaching of these great men: not only their discoveries, but also their pronouncements on how these discoveries should be regarded, and how future physics should be conducted. Bell was puzzled and dismayed. He felt that everything he was taught constituted a betrayal of the great enterprise: a surrender to the difficulties of the pursuit, and an insistence that there was no alternative but to join the leaders of the field in retreat.

The doctrine which he found all physicists were expected to accept seemed to him to be a distortion of the scientific method. It dismissed, or forgot, the purpose of the method, and held up the method itself as if the very essence of science was contained in the third of these steps: the purpose of physics is to predict the results of experiments. It was a central feature of quantum mechanics, according to the founding fathers, that it could *only* describe the results of experiments. The aim of describing the world apart from experiments was totally and explicitly abandoned. This doctrine was criticised, in a text-book influenced by Bell, as follows:

> It cannot be true that the sole purpose of a scientific theory is to predict the results of experiments. Why on earth would anyone want to predict the results of experiments? Most of them have no practical use; and even if they had, practical usefulness has nothing to do with scientific enquiry. Predicting the results of experiments is not the *purpose* of a theory, it is a *test* to see if the theory is true. The purpose of a theory is to understand the physical world. ([2], p. 214)

In Bell's own words,

> To restrict quantum mechanics to be exclusively about piddling laboratory operations is to betray the great enterprise.

Bell's unhappiness with this situation led him to examine the possibility of explaining the results of quantum mechanics in terms of "hidden variables" — some way of describing the actual disposition of the material world, regardless of whether any experiments were being done. In the early 1960s it was known that this could be done, following David Bohm's revival in 1952 of a model proposed by Louis de Broglie in 1927. However, although this was known, it was not widely known; indeed, it was generally thought to be impossible because of Pauli's early opposition, to which de Broglie himself surrendered, and a supposed proof by the respected mathematician John von Neumann. But as Bell wrote [3], "In 1952 I saw the impossible done". In [3] he analysed the reasons why it continued to be the accepted opinion that hidden variables were impossible, and acknowledged that there were good reasons not to like the de Broglie/Bohm model; Einstein, for example, whom Bell followed in his dissatisfaction with quantum mechanics, found this solution "cheap". It made sense, therefore, for Bell to look at the full range of possible hidden-variable models; and in doing so he discovered that one particular reason for disliking the de Broglie/Bohm model was unavoidable: any such model would necessarily exhibit *nonlocality*, the feature that distant parts of the model would affect each other instantaneously. This discovery is what Bell is famous for. In this paper, however, I want to focus on one of his later contributions to the project of rescuing the great enterprise. But I should emphasise that there is no substitute for reading Bell's contributions in his own wonderfully elegant and entertaining sentences [4].

2. Beables

For Bell, the feature of nonlocality, or action at a distance, was no reason to reject a theory. It might be surprising, it might be difficult to reconcile with special relativity, and it might, as it did for Einstein, defy one's presupposition of what a scientific description of the world should look like; but this is outweighed by the virtue of actually giving a description of the world, independent of human beings and "piddling laboratory operations" [1]. In conversation, Bell would emphasise that he would encourage anyone working on a theory with this overriding virtue. He put aside his own opinions as to whether the work was likely to be successful; the important thing was to get physicists thinking in a healthy, "professional" [5] way. And he was enormously helpful and supportive: I remember, at a conference in 1987, diffidently giving him the manuscript of a paper at the end of one afternoon. Despite attending an alcoholic reception that evening, he sought me out the next morning to give me detailed comments.

Bell's hostility to the official version of quantum mechanics, as preached in nearly all university physics courses, is emblazoned in the two words of the title of his paper *Against "measurement"* [1]. Another key word to which he took exception is "observable". This is the only word available in the official theory to refer to properties of physical objects; it insists that the only quantities that can have any place in a physical theory are those which can be measured, or *observed*. In line with his conviction that a theory should describe physical objects themselves, regardless of their relation to human observers, Bell proposed [6] to replace the word "observable" by a new coinage of "beable" to emphasise the autonomous existence of the quantities in question.

In the Bohm/de Broglie theory of non-relativistic many-particle quantum mechanics, the beables are the positions of the particles. This might be one of the reasons why the theory is not universally liked. A fundamental feature of the conventional theory is that all the quantities of classical mechanics, i.e. functions on phase space, have quantum counterparts which enter the theory on an equal footing, as they do in Hamiltonian mechanics. This symmetry under canonical transformations, becoming unitary symmetry in quantum mechanics, is widely regarded as very attractive. It is explicitly broken in de Broglie/Bohm theory. Bell defended this, arguing that all actual observations in experimental physics come down to measurements of position (of pointers in instruments, for example). This strikes me as dubious, and anyway it represents a reliance on experimental considerations which is curiously inappropriate in the author of *Against "measurement"*. However, in the paper on which I now want to focus [5], Bell addressed a more serious problem: de Broglie/Bohm theory is non-relativistic, and is a theory of a finite number of particles. On the other hand, the fundamental theories which Bell was seeking as an elementary particle physicist would have to be relativistic, and they would have to be theories of fields, not particles. Hence his title: *Beables for quantum field theory.*

In this paper Bell tackled only one of these desiderata. He proposed a quantum theory of fields with a clearly defined set of beables, quantities with privileged ontological status, and with dynamics which reproduced the predictions of orthodox quantum theory, just as the Bohm/de Broglie equation of motion reproduces the predictions of the Schrödinger equation. But his theory is not relativistic. He assumes an absolute distinction between space and time, in which time is continuous but space is taken to be a discrete three-dimensional lattice $\mathcal{L}$. The beables of the theory are the total numbers of fermions at the points of the lattice. These fermion numbers are the eigenvalues of the field operators

$$B(x) = \sum_i \psi_i(x)^\dagger \psi_i(x), \qquad x \in \mathcal{L}$$

where ψ_i is a particular variety of Dirac field, labelled by i, and the sum is over all such varieties. Thus the actual situation of the world, is given by the set of integers $\{F(x) : x \in \mathcal{L}\}$ representing numbers of fermions at all points of the lattice. Bell does not consider it necessary or even possible for the numbers of different types of fermion to be beables, since interacting fields will not all commute. He regards microscopic details such as these as "entirely redundant. What is essential is to be able to define the positions of things", by which he seems to mean things that we would recognise in our macroscopic world. His only elucidation of these "things" is that they should include "positions of instrument pointers or (the modern equivalent) of ink on computer output". This is not intended to be exhaustive: it is not that these are the only beables but that the beables must *at least* include the positions of instrument pointers.

3. Transition probabilities

In Bell's model the state of the world at each time t is given by an element $\mathbf{n}(t)$ of the discrete set of functions $\mathbf{n} : \mathcal{L} \to \mathbb{N}$ ($\mathbb{N}$ being the set of non-negative integers), so each $\mathbf{n}$ is a set of non-negative integers, one for each lattice point. The change of this state with time is governed, as in de Broglie/Bohm theory, by a time-dependent element $|\Psi(t)\rangle$ of a Hilbert space spanned by eigenvectors of the field operators. We are used to calling $|\Psi(t)\rangle$ a "state vector", but in this theory that terminology is misleading: $|\Psi(t)\rangle$ does not describe a state of the world, but something that governs change in the state of the world. Let us call it the "pilot vector", in memory of the pilot wave of de Broglie/Bohm theory. However, the value of $|\Psi(t)\rangle$ is a fact about the world, and Bell therefore considered $|\Psi(t)\rangle$, as well as $\mathbf{n}(t)$, to be a beable. The complete specification of the world at time t is then given by the pair $(\mathbf{n}(t), |\Psi(t)\rangle)$.

The way that the world changes in time is an adaptation of the evolution equations of de Broglie/Bohm theory. As in that theory, the pilot vector $|\Psi(t)\rangle$ evolves according to the Schrödinger equation with a Hamiltonian H:

$$i\hbar\frac{\mathrm{d}}{\mathrm{d}t}|\Psi(t)\rangle = H|\Psi(t)\rangle.$$

Because the possible values of $\mathbf{n}(t)$ are a discrete set, however, the change from one value to another is stochastic: Bell postulates that if, at time t, the value of the total fermion number distribution is $\mathbf{m}$, then the probability that at time $t + \delta t$ its value has changed to $\mathbf{n}$ is $w_{\mathbf{mn}}\delta t$ where the transition probability $w_{\mathbf{mn}}$ is given by

$$w_{\mathbf{mn}} = \begin{cases} \dfrac{2\mathrm{Re}[(i\hbar)^{-1}\langle\mathbf{n}|H|\mathbf{m}\rangle\overline{c_{\mathbf{n}}}c_{\mathbf{m}}]}{\langle\Psi(t)|P_{\mathbf{m}}|\Psi(t)\rangle} & \text{if this is } \geq 0 \\ 0 & \text{if it is negative} \end{cases} \tag{1}$$

where $P_{\mathbf{m}}$ is the projection onto the subspace of simultaneous eigenstates of the local occupation number operators $B(x)$ in which the eigenvalue of $B(x)$ is $\mathbf{m}(x)$. Bell then shows that this joint time development is consistent with the probabilities given by the Born rule in the same sense as in de Broglie/Bohm theory: if the Born rule holds at some initial time, i.e. the value of the total fermion number distribution at that time is given probabilistically so that the probability of the distribution $\mathbf{m}$ is $\langle\Psi(0)|P_{\mathbf{m}}|\Psi(0)\rangle$, then this remains true at all subsequent times.

Bell found the stochastic nature of this time development "unwelcome"; he suspected that it was purely a consequence of his artificial assumption of a discrete lattice of points of space, and that it would "go away in some sense in the continuum limit". Indeed, it was shown by Vink [7] and myself [8], working independently, that a stochastic model of a particle on a one-dimensional lattice, modelled on this theory of Bell's, did become the deterministic de Broglie/Bohm theory in the continuum limit. Bell's unease arose from his respect for the time-reversal invariance of both quantum and classical mechanics, in the forms of Schrödinger's equation and Newton's equations of motion. Others, however, have welcomed both stochastic elements in fundamental theory, as reflecting our actual experience of quantum phenomena, and non-invariance under time reversal, as reflecting the true nature of time. ("Others" here possibly means just myself [9].)

The transition probabilities introduced by Bell can be used in a wide variety of theories, not restricted to those which postulate a special class of quantities which are "beable". In general, consider a theory which supposes that there is a true description of the world by means of a vector $|\Psi(t)\rangle$ which evolves according to a Schrödinger equation with Hamiltonian H, and that there is some reason to give special consideration to one of the components of this vector in a decomposition given by special subspaces $\mathcal{S}_n$ of the Hilbert space $\mathcal{H}$, known as *viable* subspaces. Thus $\mathcal{H}$ is the orthogonal direct sum of the subspaces $\mathcal{S}_n$, and any $|\Psi\rangle \in \mathcal{H}$ can be written

$$|\Psi(t)\rangle = \sum_n |\psi_n(t)\rangle \qquad \text{with} \qquad |\psi_n(t)\rangle \in \mathcal{S}_n.$$

Then $|\psi_n(t)\rangle = P_n|\Psi(t)\rangle$ where P_n is the orthogonal projection onto the subspace $\mathcal{S}_n$. As $|\Psi(t)\rangle$ changes in accordance with the Schrödinger equation, the components $|\psi_n(t)\rangle = P_n|\Psi(t)\rangle$ will also change inside their respective viable subspaces; but in addition to this, the spotlight shining on the component with special status will also move stochastically from one subspace to another. This stochastic change is given by Bell's transition probabilities: if the special component is in subspace $\mathcal{S}_m$ at time t, then the probability that it has moved to subspace $\mathcal{S}_n$ by time $t + \delta t$ is $w_{mn}\delta t$ where w_{mn} is given by (1). If the viable subspaces $\mathcal{S}_n$ are themselves changing with time, then there is an extra term in this equation [9,10].

I will refer to theories with this structure as "generalised Bell-type theories". All such theories share the property proved by Bell for his version of quantum field theory: the transition probabilities (1) guarantee the continuing validity of the Born rule if it is valid initially. They are not uniquely determined by this requirement: there is a range of possible transition probabilities with the same property [10]. However, Bell's formula is uniquely natural in applications to decay [9] and measurement processes [11,12]: it ensures that the underlying direction of change in such processes is always forwards, without intermittent reversals (decay products, for example, recombining to reconstitute the unstable decaying state).

In many such theories the special status of the highlighted component $|\psi_n(t)\rangle$ is ontological; only this component describes the actual state of the world, and the function of the overall vector $|\Psi(t)\rangle$ is to act as a pilot, guiding the discontinuous quantum transitions of the world. Such theories are liable to face a *preferred basis* problem: what defines the viable subspaces $\mathcal{S}_n$? Bell formulated the concept of *beables* precisely to give an answer to this problem: the viable subspaces are the eigenspaces of beables. In the original de Broglie/Bohm theory, the beables are the particle positions. We have already noted on the one hand the unease this arouses because of its violation of symmetry under canonical transformations, and on the other hand Bell's defence of it on the grounds that ultimately all observations are of position.

In Bell's theory in [5], the beables are the total fermion numbers at each point in the lattice of space. At first sight it seems reasonable that these should have fundamental status, but this is thrown into doubt by the Unruh effect, according to which the number of particles present in a region of space depends on a frame of reference as soon as one moves away from frames in constant relative motion.

This preferred-basis problem is also often thought to arise in the "many-worlds" interpretation of quantum mechanics. That theory has only the universal state vector $|\Psi(t)\rangle$ and does not single out a component of that vector as describing the actual world. Nevertheless, if the vector $|\Psi(t)\rangle$ is regarded as describing many worlds, all of which are real, then some commentators, including Bell [13], demand that there should be a specification of which vectors or subspaces can describe "worlds". However, this is not a problem in Everett's original version [14] of this interpretation but only arises when too much weight is placed on the expository terminology of "worlds" [2, p. 221]. Even with this terminology, the components of the state vector $|\Psi(t)\rangle$ which describe worlds can be determined by the structure of $|\Psi(t)\rangle$ itself [15] and need no independent definition.

Other generalised Bell-type theories have no preferred-basis problem, defining the viable subspaces purely in terms of the pilot vector $|\Psi(t)\rangle$. In the (now largely discarded) modal interpretation, in which the universe is divided into two systems so that the Hilbert space $\mathcal{H}$ is a tensor product of two factors, the viable components are defined by the Schmidt decomposition of $|\Psi(t)\rangle$ with respect to this structure.

The format of a generalised Bell-type theory is appropriate to describe the changing experience of a sentient subsystem of the physical universe [16]. This can be done in the context of Everettian theory, in which the universe is described by a single time-dependent state vector $|\Psi(t)\rangle$, and nothing else. We know that the universe has sentient subsystems, each of which is capable of experiences relating to the rest of the universe. I am myself such a subsystem. I have various possible experiences, for each of which there is a set of physical states of my body in which I have the experience. Since I can distinguish between the experiences (if I couldn't they wouldn't be different experiences), it seems to be in keeping with quantum mechanics that the corresponding states form a set of orthogonal subspaces $\mathcal{S}_n^{\mathrm{me}}$ of my Hilbert space $\mathcal{H}_{\mathrm{me}}$. These subspaces then define a set of subspaces $\mathcal{S}_n = \mathcal{S}_n^{\mathrm{me}} \otimes \mathcal{H}_{\mathrm{rest}}$ of the universal Hilbert space $\mathcal{H} = \mathcal{H}_{\mathrm{me}} \otimes \mathcal{H}_{\mathrm{rest}}$. The changes in my experience then constitute transitions between these subspaces. Bell's formula (1) gives the probabilities of these transitions, subject to a universal state vector and a universal Hamiltonian.

This formalism can also be used [11,12] to model the progress of a quantum measurement.

4. Histories

Once a significant set of preferred subspaces has been identified for each time t, a generalised Bell-type theory makes it possible to calculate probabilities for *histories* of the system. It is usual, and convenient, to consider only a discrete set of times $t_1, \ldots, t_f$; then a *history* of the system is a sequence $(\mathcal{S}_1, \ldots, \mathcal{S}_f)$ where each $\mathcal{S}_i$ is a closed subspace of the Hilbert space of the system, or equivalently a sequence of projection operators $h = (\Pi_1, \ldots, \Pi_f)$ where Π_i is the projection onto $\mathcal{S}_i$. Such histories are the fundamental concepts in the *consistent histories* interpretation of quantum mechanics [17], in which the probability of the history h is taken to be

$$P(h) = \mathrm{tr}[\widetilde{\Pi}_1 \dots \widetilde{\Pi}_{f-1} \widetilde{\Pi}_f \widetilde{\Pi}_{f-1} \dots \widetilde{\Pi}_1] \tag{2}$$

where

$$\widetilde{\Pi}_i = \mathrm{e}^{iHt/\hbar} \Pi_i \mathrm{e}^{-iHt/\hbar}.$$

This probability can be obtained from the Copenhagen interpretation of quantum mechanics by assuming that at each time t_i there is a measurement of an observable whose eigenspaces include $\mathcal{S}_i$, and applying the projection postulate after each measurement. Then $P(h)$ is the probability that this sequence of measurements has the results corresponding to the subspaces $\mathcal{S}_1, \dots \mathcal{S}_n$. It can be written

$$p(h) = \mathrm{tr}[C_h C_h^\dagger]$$

where C_h is the *history operator*

$$C_h = \widetilde{\Pi}_1 \cdots \widetilde{\Pi}_f. \tag{3}$$

In general, these probabilities will not be consistent with the following natural requirement. Suppose two histories h_1 and h_2 differ only in the subspaces $\mathcal{S}_i^{(1)}$ and $\mathcal{S}_i^{(2)}$ which they assign to time t_i, and that these subspaces are orthogonal. We can consider a third history $h_1 \vee h_2$ in which the subspace at time t_i is the direct sum $\mathcal{S}_1 \oplus \mathcal{S}_2$. In terms of measurement, this describes a result of the measurement at time t_i which was either the result corresponding to $\mathcal{S}_1$ or that corresponding to $\mathcal{S}_2$; so $h_1 \vee h_2$ relates that the history of the system was either h_1 or h_2. We expect that the corresponding probabilities should satisfy

$$P(h_1 \vee h_2) = P(h_1) + P(h_2).$$

In particular, if $\mathcal{S}_1 \oplus \mathcal{S}_2 = \mathcal{H}$, we expect that $\mathcal{S}_1$ and $\mathcal{S}_2$ are an exhaustive set of possibilities at time t_i and so

$$P(h_1) + P(h_2) = P(h')$$

where h' is the history which coincides with h_1 and h_2 at all times except t_i, but does not say anything about time t_i. However, these equations will not in general be true. A condition which guarantees them is

$$\mathrm{tr}[C_{h_1} C_{h_2}^\dagger] = 0. \tag{4}$$

A set of histories is said to be *consistent* (or *decoherent*) if this condition holds true for every pair of different histories in the set.

This is not an issue in generalised Bell-type theories. It would be an issue if the transition probabilities (1) were supposed to apply for transitions to any subspace in the Boolean algebra generated by the subspaces $\mathcal{S}_i$, but that would not be in accord with the basic presuppositions of such a theory. To take a linear sum of the preferred subspaces as having the same status as those subspaces would be to assume that the system could exist in a superposition of states from the preferred subspaces, whereas the philosophy of these theories is that such superpositions are not actual states. In the theory of sentient experience, for example, a sum of experience states describing different experiences is not an experience state (a sum of eigenvectors with different eigenvalues is not an eigenvector). Thus the state of the system can be in a subspace $\mathcal{S}_1 \oplus \mathcal{S}_2$ only if it is in the subset $\mathcal{S}_1 \cup \mathcal{S}_2$, and the appropriate probability is

$$P(\mathcal{S}_1 \oplus \mathcal{S}_2) = P(\mathcal{S}_1) + P(\mathcal{S}_2).$$

In [16], where probabilities are identified with truth values, and a history is formed by logical operations from single-time propositions, the probability of a history was taken to be given by the usual formula (2). For the development of a satisfactory logic, it was then found to be necessary to make the consistent-histories assumption (4) (in a somewhat weaker version). I now think that this was a mistake. If Bell's transition probabilities had been used to define the truth value (= probability) of a history rather than (2), there would have been no need for a subsidiary assumption, and the logic could have been developed in much greater generality.

5. Conclusion

John Bell never lost sight of the great enterprise of science. He rejected the narrow scepticism and pessimism in the reaction of the founding fathers of quantum mechanics to the difficulties which they encountered, and the instrumentalist view of physics which became the dogma in which all physics students were indoctrinated. His own most famous and influential work only served to emphasise the difficulties in the way of understanding quantum mechanics as he thought physical theories should be understood. Nevertheless, he persevered in the search for such an understanding. His concept of "beables" has become a standard tool for those seeking to understand and develop quantum mechanics, and deserves deeper philosophical analysis. The transition probabilities that he formulated as a component of theories of such beables are a lasting legacy of his search, and have proved valuable even to those who do not share his vision of what a satisfactory physical theory should be like.

References

1. J. S. Bell, *Physics World* **3**, 33 (August 1990).
2. A. Sudbery, *Quantum Mechanics and the Particles of Nature* (Cambridge University Press, 1986).
3. J. S. Bell, *Found. Phys.* **12**, p. 989 (1982). Reprinted in [4].
4. J. S. Bell, *Speakable and unspeakable in quantum mechanics* (Cambridge University Press, 1987).
5. J. S. Bell, Beables for quantum field theory, in *Quantum Implications*, eds. B. J. Hiley and F. D. Peat (Routledge and Kegan Paul, London, 1984) pp. 227–234. Reprinted in [4].
6. J. S. Bell, Subject and object, in *The Physicist's Conception of Nature*, ed. J. Mehra (Reidel, Dordrecht, 1973) pp. 687–690. Reprinted in [4].
7. J. C. Vink, *Phys. Rev. A* **48**, p. 1808 (1993).
8. A. Sudbery, *J. Phys. A* **20**, p. 1743 (1987).
9. A. Sudbery, *Stud. Hist. Phil. Mod. Phys.* **33B**, 387 (2002), `arXiv:quant-ph/0011082`.
10. G. Bacciagaluppi and M. Dickson, *Found. Phys.* **29**, p. 1165 (1999), `arXiv:quant-ph/9711048`.
11. A. Sudbery, *Found. Phys.* **47**, 658 (2017), arXiv:1608.05873.
12. T. J. Hollowood (2018), arXiv:1803.04700.
13. J. S. Bell, Six possible worlds of quantum mechanics, in *Possible Worlds in Arts and Sciences*, ed. S. Allén, Proceedings of the Nobel Symposium, Vol. 65 (Nobel Foundation, Stockholm, 1986). Reprinted in [4].
14. H. Everett III, *Rev. Mod. Phys.* **29**, 141 (1957).
15. D. Wallace, *The Emergent Multiverse* (Oxford University Press, 2012).
16. A. Sudbery, *Synthese* **194**, 4429 (2017), arXiv:1409.0755.
17. R. B. Griffiths, *Consistent Quantum Theory* (Cambridge University Press, 2002).

Printed in the United States
by Baker & Taylor Publisher Services